Lepidoptera and Conservation

Lepidoptera and Conservation

T.R. New
Department of Zoology
La Trobe University
Melbourne
Australia

WILEY Blackwell

Library of Congress Cataloging-in-Publication Data
New, T. R.
    Lepidoptera and conservation / T.R. New.
        pages cm
    Includes bibliographical references and index.
    ISBN 978-1-118-40921-3 (cloth)
    1. Lepidoptera–Conservation.   2. Lepidoptera–Habitat.   3. Butterflies–Conservation.   4. Wildlife conservation.   I. Title.
    QL542.N495 2014
    595.78–dc23

                                                            2013022175

A catalogue record for this book is available from the British Library.

Contents

Preface

This book is an overview of the place of Lepidoptera, butterflies and moths, in conservation, and of what we have learned about conservation from concerns over their declines and losses and attempts to redress these. It builds on the burgeoning – and increasingly diverse and scattered – number of scientific papers, symposium proceedings and government agency reports throughout the world that deal with conservation of individual species or assemblages of these insects within most key terrestrial habitats. It emphasises also the enormous ecological variety and diversity amongst the Lepidoptera, and the importance of incorporating whatever is known of that variety into practical conservation planning and management. Advocacy, particularly that flowing from liking for butterflies – a group described as having powerful 'resonance' for many people – is a critical aspect of this conservation. Not least, this support is important because many informed sympathisers may contribute to conservation surveys or practical site or resource management, whilst local or regional interest groups can assemble, coordinate and disseminate relevant information effectively and rapidly to a receptive audience.

In common with many other invertebrates, promoting the conservation of insects is hampered severely by their public image, whereby they are seen widely as harmful and as objects for destruction and suppression. The major exception to this broad generalisation is the butterflies, almost universally seen as popular, harmless, associated with wellbeing of 'nature', and valued as cultural (or, sometimes, collectable) objects, and with their losses and declines accepted as worthy of redress. However, butterflies are only one rather small segregate of the large order Lepidoptera, with the other members (moths) often evoking less sympathetic responses: if they are considered at all, moths have historically suffered from a tendency of people to regard them as pests and inimicable to human interests, another generalisation masking their enormous ecological and taxonomic variety. Consequently, these major components of the order (with the exception of members of a few 'flagship groups', mostly in northern temperate regions) have lagged considerably behind butterflies in their appearance and representation on conservation agendas.

Biases toward butterflies and the northern temperate regions are inevitable in this book – with the butterflies of Western Europe, and especially of the United

Kingdom, providing the foundation of many conservation paradigms in the discipline. It is true also that within the tropics and southern temperate regions, the larger Lepidoptera are perhaps the only representatives of *any* species-rich insect group that are amenable to relatively easy sampling and survey coupled with a high level of accurate taxonomic recognition to species/genus levels. This, linking with historical interest and experience, has rendered them accessible tools in many exercises in evaluating biodiversity. Such studies continue to proliferate and diversify. Simply 'keeping up' with the burgeoning information flow on European butterflies, in particular, would be a full-time (although absorbing!) occupation, and I can make no pretence at having done so. Many recent ecological papers provide further and increasingly penetrating and well focused insights for Lepidoptera conservation, and range from novel approachs and detail, increasing taxonomic and ecological variety, and challenging ideas and perpective. Edited volumes such as those by Settele et al. (2009), Dover et al. (2011), Boggs et al. (2003) and Ehrlich and Hanski (2004) provide much thought-provoking perspective and background. Specialist journals, such as the Journal of Insect Conservation, include relevant contributions in almost every issue. Many others (available up to early 2013) are also cited in this book, and range from single-author studies to the wider 'Proceedings' of major symposia on regional faunas or individual taxa. The latter are exemplified by the massive literature on individual butterfly taxa, such as the Wanderer (or Monarch, *Danaus plexippus*) in North America, the large blues (*Maculinea*) in Europe, and checkerspots (Melitaeinae) in both these regions: these intensively studied species must figure highly in any overview of conservation within the order. Thus, much has been written already about conservation of Lepidoptera – proceedings of some major symposia (such as Pullin 1995, Dover et al. 2011) demonstrate the wide interest present, as do a number of books dealing mostly with butterfly conservation and emphasising different parts of the world. The discipline is developing rapidly, with ever-more sophisticated detail of biological and political subtlety. In this book, I have cited simply a selection of the most relevant literature known to me by very early 2013; inevitably this selection is subjective, and many equally important and informative examples are excluded by space rather than deliberate slight. Another author might write a book on these topics with rather limited overlap with the references cited here!

More widely, the demand for yet more books on Lepidoptera seems insatiable. In his survey of books dealing with British butterflies alone, for example, Dunbar (2010) enumerated some 600 titles up to 2008. Perusal of almost any natural history book dealer's current catalogue reveals proliferating titles aiding identification or recognition of particular genera or families, with many of the books including magnificent colour illustrations to aid this, and some of the more ambitious publishing projects forecasting numerous serial volumes to deal with larger groups or regional faunas. Few of these emphasise conservation, although their perusal can aid evaluation of 'rarity' and enhance biological and distributional knowledge considerably.

However, whilst the discipline of Lepidoptera conservation has expanded enormously over the last few decades, many of the limitations noted in a review by New et al. (1995) persist, not least in the form of strong regional and taxonomic biases in conservation efforts and outcomes. The elegant research flowing from studies on European butterflies, in particular, has 'set the scene' for much conservation practice of far wider application for Lepidoptera in much of the rest of the world, where the necessary foundations of basic documentation and research have yet to be consolidated and the resources to do so are very limited.

Much of the wider current attention to insect conservation has its foundation in concern over fates of butterflies, with studies and sympathies on these insects of paramount conservation interest being critical drivers of establishing insects more widely in conservation policy and consideration.

There are many components to appreciation of Lepidoptera in conservation. In July 2008, Butterfly Conservation Europe issued a document entitled 'Why butterflies and moths are important', in which the major attributes that help to render them serious foci for conservation were encapsulated. The reasons were grouped under a series of headings: aesthetic value, ecosystem value, educational value, health value, economic value, intrinsic value and scientific value. Ranging from idealistic to practical, these broad categories display the variety of values, one or more of which is broadly evident or implicit in almost every case that has attracted attention. Collectively, they cover a highly persuasive suite of justifications for need and action.

The book is global in scope, and the 13 chapters fall into several main groups. The first five chapters provide historical, political and biological background on Lepidoptera and the progressive documentation and understanding of their conservation needs, and their roles as both targets for conservation and signals of wider conservation needs. Chapters 6–10 encompass the ecological scenarios of Lepidoptera conservation and some of the major approaches to their conservation. Chapters 11 and on move more toward the practical pursuit of conservation, emphasising the roles of legislative support, clear definition of conservation needs and actions, and how these may be monitored as a basis for adaptive management. The final chapter assembles the many aspects of progress to suggest how they constitute a template for the future and how priorities for that future may be set. I have tried to include sufficient general background for the book to be useful to non-lepidopterists, as well as for more specialist readers.

Lepidoptera are one of the 'big four' orders of holometabolous insects, each with several hundred thousand species and widely distributed across the world. This book complements the conservation perspectives given for two others of these orders, Coleoptera (New 2010) and Hymenoptera (New 2012b). However, it can draw on a much more complex and informed – at times, controversial – basis of interest, experience and documentation. Fundamental lessons on ecological relationships (particularly between insect herbivores and plants, and mutualisms between Lepidoptera and other insects), population structures and dynamics, impacts of alien taxa and the consequences of climate change are simply examples of the many ways in which studies on butterflies and moths have contributed to general principles of wide, even universal, relevance in insect conservation management. Their biological variety, coupled with high levels of local or regional endemism and ecological specialisation, and the naturally low abundance and high vulnerability of many relatively well documented taxa have all contributed to conservation concern. Energetic debate continues over the absolute and relative importance of many apparent threats. Whether the endemic Hawaiian species of *Eupithecia* moths, with their unique fly-catching caterpillars, are susceptible to alien parasitoids and predatory ladybirds introduced as biological control agents (Sheppard et al. 2004), or selected local butterflies are vulnerable to over-collection by hobbyists both exemplify themes that engender heated controversy. The latter has led to much conflict between collectors and conservationists, with the high prices paid for rare (and, in many cases, officially protected) butterflies – as well as for beetles and some others – a powerful incentive for transgressing any formal protective measures for the sake of financial reward. Historically, the desire to obtain specimens of rare species has extended to practices that have occasionally confused distributional recording in relation to more natural recent range expansions attributed to climate changes – most notably through the desirability of many the rarest British Lepidoptera (some of them sporadic and rare migrants to Britain and seen there only occasionally) inducing importations of livestock from mainland Europe for release for purported capture in southern England, with high rewards for the 'smugglers' from collectors seeking the prestige of exhibiting 'real British specimens' in their cabinet. The 'Old Moth-Hunter', P.B.M. Allan (1943), gave an entertaining account of the subterfuges in a chapter entitled 'The Kentish Buccaneers', which remains a fascinating insight into collector psychology up to the mid-twentieth century.

These topics, and many others are dealt with in this book, but much of it deals with more universal threats associated with loss of habitats and changes within them and how human interference and impacts, both current and anticipated, may be countered or ameliorated. The book is thus a contribution to the wider development of insect conservation, and how this most popular of all insect orders has played central roles in the development and evolution of the discipline.

Two comments on nomenclature are needed. First, the largest moth superfamily (Noctuoidea) has recently been extensively re-arranged in order to reflect more natural phylogenetic relationships revealed by morphological and molecular studies (Zahiri et al. 2011), so that (amongst other changes at the family level) many of the species previously included in Noctuidae are now placed in Erebidae. These changes have not yet penetrated conservation studies widely, and any species previously referred to as included in Noctuidae is referred to as such in this book. Second, the detailed attention to some butterflies has led to use of more than one generic name for the same species, or ambiguity over species/subspecies status, and some possible confusion may ensue. I have used the names given in individual publications, but the following are noted as possible ambiguities: *Euphydryas/Eurodryas aurinea*; *Melitaea/Mellicta athalia*; *Maculinea/Phengaris arion* (and related large blues); *Lysandra/Polyommatus*; *Plebejus samuelis/Lycaeides melissa samuelis*.

T.R. New
Department of Zoology
La Trobe University
Melbourne

References

Allan, P.B.M. (1943) Talking of Moths. Montgomery Press, Newtown.

Boggs, C.L., Watt, W.B. & Ehrlich, P.R. (eds) (2003) Butterflies. Ecology and Evolution Taking Flight. University of Chicago Press, Chicago and London.

Butterfly Conservation Europe (2008) Why Butterflies and Moths are Important. Wageningen.

Dover, J., Warren, M. & Shreeve, T. (eds) (2011) Lepidoptera Conservation in a Changing World. Springer, Dordrecht.

Dunbar, D. (2010) British Butterflies – a History in Books. The British Library, London.

Ehrlich, P.R. & Hanski, I. (eds) (2004) On the Wings of Checkerspots. A Model System for Population Biology. Oxford University Press, Oxford.

New, T.R. (2010) Beetles in Conservation. Wiley-Blackwell, Oxford.

New, T.R. (2012b) Hymenoptera and Conservation. Wiley-Blackwell, Oxford.

New, T.R., Pyle, R.M., Thomas, J.A., Thomas, C.D. & Hammond, P.C. (1995) Butterfly conservation management. *Annual Review of Entomology* **40**, 57–83.

Pullin, A.S. (ed.) (1995) Ecology and Conservation of Butterflies. Chapman & Hall, London.

Settele, J., Shreeve, T., Konvicka, M. & Van Dyck, H. (eds) (2009) Ecology of Butterflies in Europe. Cambridge University Press, Cambridge.

Sheppard, S.K., Hennema, M.L., Memmott, J. & Symondson, W.O. (2004) Infiltration by alien predators into invertebrate food webs in Hawaii: a molecular approach. *Molecular Ecology* **13**, 2077–2088.

Zahiri, R., Kitching, I.J., Lafontaine, J.D. et al. (2011) A new molecular phylogeny offers hope for a stable family-level classification of the Noctuoidea (Lepidoptera). *Zoologica Scripta* **40**, 158–173.

Acknowledgements

The following organisations, individuals and publishers are thanked for advice and granting permission to use or modify material to which they hold copyright: Cornell University Press; the Ecological Society of America Inc.; the Edinburgh Entomological Club (Dr G.E. Rotheray); Elsevier, Oxford; European Journal of Entomology; IUCN Publications; the Lepidoptera Research Foundation Inc.; Minnesota Agricultural Experiment Station, St Paul; the Natural History Museum, London; the Netherlands Entomological Society (Dr R. de Jong); Pensoft Publishers (Dr L. Penev); the Royal Zoological Society of New South Wales, Mosman; Springer Science and Business Media b.v., Dordrecht; Surrey Beatty and Sons, Baulkham Hills; Taylor and Francis; Wiley-Blackwell, Chichester. Every effort has been made to obtain permissions for such use. The publisher apologises for any inadvertent errors or omissions, and would welcome news of any corrections that should be incorporated in future reprints or editions of this book.

I also thank Lucinda Gibson for allowing me to use her excellent pictures of a female Golden sun-moth, a notable endemic species in South Eastern Australia. The book has been supported enthusiastically by Ward Cooper as Commissioning Editor. Also at Wiley, I have very much appreciated the support and patient advice from Kelvin Matthews as the project developed.

Ken Chow has dealt efficiently with production matters. The help of Nancy Arnott (Project Manager for Toppan Best-set Premedia Ltd) is also greatly appreciated, and the careful copy-editing by Joanna Brocklesby has improved the text.

1

Lepidoptera and Invertebrate Conservation

Introduction

Lepidoptera, the butterflies and moths, have for long been familiar both to naturalists and people in many other walks of life. Butterflies, arguably the most popular of all insect groups, have been a major focus for collectors and other hobbyists, as symbols of the wealth and health of the ecosystems that support them – and those interests have also contributed to concerns arising from their declines and, in a few cases, well publicised extinctions. The clearly documented losses of taxa such as the Large copper (*Lycaena dispar dispar*) from the fens of eastern England in the mid nineteenth century (Duffey 1977, Feltwell 1995) and reported decline of the Xerces blue (*Glaucopsyche xerces*) a decade or so later in the western United States (where it became extinct later: Pyle 2012), for example, each mark the beginnings of concern for insect conservation in those regions. More widely, the popularity of butterflies and later extinctions (such as of yet another lycaenid, the Large blue, *Maculinea arion*, in Britain as recently as 1979: Thomas 1991) have led to studies on these insects forming the strongest foundation of the developing science of insect conservation. Several factors contribute to this – simple aesthetics are important in creating a liking and sympathy for conspicuous insects, whether they are tiny lycaenids, as the above cases, or large and spectacular tropical swallowtails or birdwings (Papilionidae) such as those that enthralled explorers of then remote parts of the world during the Victorian era, and continue to do so. That era saw the proliferation of natural history documentation, prompted in part through the 'philatelic approach' to collecting, with progressive accumulation of distribution records, biological and life history details. These interests induced production of increasingly complete and sophisticated illustrated handbooks that enabled hobbyists to identify their study objects with reasonable certainty and summarise biological

Lepidoptera and Conservation, First Edition. T.R. New.
© 2014 John Wiley & Sons, Ltd. Published 2014 by John Wiley & Sons, Ltd.

and distributional information, and so to confidently contribute further to the record of fact and inference that has provided a vital legacy to present-day students. This legacy is geographically biased, of course, but the 200 years and more of accumulated information has rendered the butterflies of Britain, followed by those of some other parts of the northern temperate zones, the best documented of all regional insect faunas. In short, they are informative as examples and models for emulation and understanding to biologists seeking a foothold in the daunting world of invertebrate diversity. Importantly, they are accessible to non-specialists, encouraged by the wealth of well illustrated identification guides and authoritative but non-technical information available.

Butterflies are unusual, also, in their cultural connotations, with artistic roles since pre-Columbian years (Pogue 2009) including representation in the ancient art from many parts of the world, as well as presence in literature, myth and religion – the latter including symbolised connection to the soul in several distinct cultures. That, in general, people 'like butterflies' and do not fear them as harmful renders them popular and powerful ambassadors for the wealth of insect life. However, there is also suggestion that the appeal of such insects may link to 'academic disapproval' and deter young scientists from taking up study of the group. Study of such aesthetically pleasing insects is occasionally associated with second-rate intellect, so that supervisors may lead potential lepidopterist graduate students to turn their focus to 'insect taxa that are judged to be more academically respectable' (Kristensen et al. 2007). Simplistically, butterflies, together with a few families of larger showy moths (notably hawkmoths, Sphingidae, and silk moths, Saturniidae) and the brightly coloured day-active burnets (Zygaenidae), are commonly regarded as 'beginners' bugs', simply because they are attractive and accessible easily by non-specialists. The reality is far different, as much recent literature – some cited in later chapters – demonstrates!

However, the butterflies are only a small component of one of the largest insect orders. They comprise only some 20,000 named species, a total surpassed by each of several individual families of moths which comprise, perhaps, a further 350,000 species. Powell (2003) estimated global Lepidoptera species richness as 'certain to exceed 350,000 species' with considerable uncertainty over what the real total may be, and rather more than 160,000 species having been named. These are distributed amongst about 124 families grouped into 47 superfamilies (Kristensen 1999). More recently, and incorporating the uncertainty implied here, Kristensen et al. (2007) estimated that 'There are considerably more than a quarter-million Lepidoptera species, probably in the order of magnitude of half a million, but there are not a million – let alone several millions'. The theme of taxonomic diversity is revisited in Chapter 2, and is noted here simply to emphasise that we are dealing with an enormous group of insects – confidently amongst the four largest orders of insects as they are at present understood, and probably the smallest of the four – and amongst which biological and taxonomic knowledge is very uneven (Scoble 1992). Estimates of species numbers are difficult, not least because of the great variety of species concepts used in modern biology, and the transition from simple morphospecies to greater appreciation of intrinsic variation may affect the number of entities recognised very considerably.

Our confidence from studies on butterflies dissipates rapidly when confronted with our ignorance over many moths. The 'accessibility' of butterflies contrasts markedly with the confusions that flow from many small moths, and is coupled also with the very different image many people have of moths, as annoying, drab, nocturnal pests: each a sweeping generalisation to which there are many exceptions! The masterly introductory chapter to 'Moths' (Majerus 2002) gives much background to this dichotomy of perceptions. Progressive incorporation of moths to augment the conservation perspective founded in butterflies (Chapter 4) is enriching the themes underpinning much insect conservation and enlarging appreciation of the biological templates against which insect diversity can be appraised. Majerus (2002) ventured that, for Britain, the strong faunistic knowledge has rendered Lepidoptera the most suitable group for studying the impacts of anthropogenic changes on terrestrial fauna. Many others have expressed similar confidence.

Biological background

Lepidoptera are not an ancient order. Unlike the Coleoptera, accepted widely as the most diverse of all insect orders and which occur in the fossil record from the Permian era some 250 million years ago, Lepidoptera proliferated only in the Cretaceous period, developing and radiating in parallel with the flowering plants and so broadly 'only' about 100 million years old. The fossil record is very sparse: Kristensen and Skalski (1999) estimated that only 600–700 specimens of fossil Lepidoptera were then known, a high proportion of them in resins, and including Baltic amber as a major source. Although some fossils believed to be Lepidoptera occur nearly 200 million years ago in the Jurassic, the major lineages of the order seemingly developed much more recently. Details will continue to be refined as further evidence accumulates, as will how angiosperm development really fostered diversification of Lepidoptera (Powell 2009). However, the Lepidoptera constitutes perhaps the largest single evolutionary lineage adapted to depend on living plants (Powell & Opler 2009).

Those early coevolutions with angiosperms apparently founded the two major ecological roles associated with modern Lepidoptera. Many adult Lepidoptera feed on plant nectar, and collectively display a range of features that render many of them effective, and sometimes highly specific, pollinators. Larvae, caterpillars, of most Lepidoptera are chewing herbivores and, whilst most feed on or in foliage, particular taxa may exploit virtually any part of a plant. Lepidoptera are widely considered the most important insect group of defoliators. Many species are very specialised in feeding habit, and strict host plant specificity is common; that specificity may extend from plant taxon to tissue, growth stage, season, degree of exposure (sun or shade environments) and many other restrictions that may influence resource accessibility and suitability, and which must be considered in conservation management. The key realisation is simply that every species of Lepidoptera comprises two very different biological entities, with larva and adult disparate in form and habits; they occupy different habitats (commonly at different times of the year with little or no seasonal overlap) and exploit different resources, so have different ecological pressures and needs for conservation. For many species, the larva, although less often observed, is the dominant stage, far longer lived than the relatively transient adult stage. In essence, the two stages are 'twin' organisms and the needs of both are central to conservation. Those needs include, as examples, how adults track nectar sources for food and find suitable oviposition sites, and how larvae find and exploit plant or (rarely) other foods, withstand depredations of predators and parasitoids and later find suitable pupation sites. Adults may need to disperse actively, sometimes over large distances as seasonal migrants, with most caterpillars in contrast dispersing rather little from where they eclose. Intricate behavioural cues and ecological strategies and specialisations are rife amongst Lepidoptera, and understanding and heeding these is another important component of conservation. Activity of both stages is influenced strongly by temperature and a range of other environmental features, as well as the accessibility of key foodstuffs – often very specific – that have led to highly characteristic seasonal and spatial patterns of development. Dispersion of the key resources (Chapter 7) from local to landscape scales is thus a critical aspect of conservation, with many of the trophic associations long entrenched over evolutionary time.

In common with many other insect herbivores, a particular obligatory food plant species may not always be suitable, but factors such as water or nitrogen content, exposure of the plant and presence of plant defensive chemicals influence local or seasonal exploitation. Figure 1.1 illustrates some of these constraints and, as Slansky (1993) commented, 'For a caterpillar . . . feeding involves

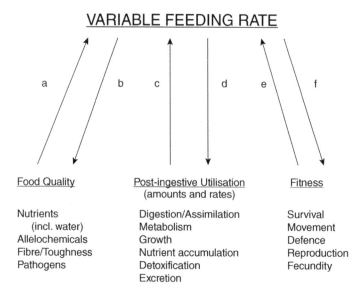

Fig. 1.1 Interactions between feeding rate and food quality, food utilisation and fitness in insect herbivores: (a) food quality can affect feeding rate, and (b) feeding may affect food quality; (c) post-ingestive food utilisation can affect feeding rate, and (d) the converse; (e) fitness components may affect feeding rate, and (f) feeding affects fitness (Source: Slansky 1993. Reproduced with permission of John Wiley & Sons.).

much more than merely filling its gut with the nearest available plant tissue'. However, it is commonplace for a suitable food plant species to be much more widely distributed than a specialist insect herbivore, and reasons for herbivore absence at places well within its range and dispersal capability are often anomalous. 'Host quality' factors affect plant suitability in time and space, and also influence incidence and abundance of the Lepidoptera (Dennis 2010).

Understanding these factors is critical in fine-scale conservation management, and their elucidation may involve very detailed study. The Sandhill rustic moth (*Luperina nickerlii leechi*, Noctuidae, Chapter 5, p. 51) is known only from one site in Cornwall, southern England, where caterpillars are associated with sand couch-grass (*Elytrigia juncea*). There are large areas on the site where the plant occurs but the moth does not (Spalding et al. 2012). *Luperina* is associated with abundant host plant cover and high numbers of stems and rhizomes – but is also restricted to areas with bare ground and levels of disturbance, so that the suitability of the host plants represents 'a fine balance between disturbance and vegetation condition'. Discussed in detail by Spalding et al. 2012, maintenance of suitable conditions involves attention to creating areas of bare ground with coarse shingle and extensive, vigorous patches of *Elytrigia*, and few other competing plants. Occasional disturbance may be beneficial, deterring establishment of competing vegetation.

Most present-day associations with host plant taxa are outcomes of long evolutionary associations (Powell et al. 1999), but many geographical gaps in knowledge persist, and much of the interpretation of host–plant relationships has arisen from northern temperate region studies. A key presumption in conservation is that natural associations involving native plants and native insects are the ideal target for sustainability. In other cases, however, alien plants (such as weeds or ornamentals) may be adopted as resources and add further, sometimes complex, dimensions to conservation management. Native Lepidoptera may 'switch' to utilise such alien hosts, which can become important components of the species' ecology – for example, as substitutes for natural hosts that have declined or been lost to development. Thus, in North America, the endangered Taylor's checkerspot butterfly (*Euphydryas editha taylori*, Nymphalidae) has switched, from unknown original native host(s), to becoming entirely dependent on the alien *Plantago lanceolata* (Severns & Warren 2008). The decline of *E. e. taylori*, to the extent that it was long believed to be extinct in Oregon, where it is now known from two sites (together with a possible one on Vancouver Island and one large and several small populations in Washington State), is attributed to impacts of invasive alien grasses and shrubs overgrowing and reducing abundance of native forb foodplants, with grasses also reducing access of butterflies to basking sites (Severns & Warren 2008). Control of alien grasses was seen as the key need to prevent the Oregon populations becoming extinct. In parallel the encouragement or, at least, tacit tolerance, of the alien *Plantago* is the only known key to sustaining the butterfly. Such adoption of a novel host plant can occur quite rapidly, as possible hosts become available through changed land use patterns. The classic example of *E. editha*, explored by Singer et al. (1993), showed two populations to have independently adopted such hosts: in one case of *Plantago lanceolata* introduced by cattle ranchers and in the other of *Collinsia torreyi* appearing in clearcut patches created by logging. That some individual butterflies tested rejected their normal ancestral host in favour of the recently adventive species created a conservation scenario incorporating rapid genetic adaptation to human-induced changes, and perhaps dependence of rare species on continued availability of these novel hosts. Adoption of the introduced Chilean needle grass (*Nassella neesiana*) in Australia by the endemic Golden sun-moth (*Synemon plana*, discussed later in this chapter) is amongst other such contexts for which parallel experimental investigations to determine any changes in preference are yet to be made.

However, it is also suggested widely that new associations of 'alien plant–native lepidopteran' are not generally as beneficial as the original 'native plant–native lepidopteran' associations that are being replaced or compensated. Alien invasive plants may have varied adverse impacts on native Lepidoptera, falling broadly into three main categories: (1) competitive exclusion and elimination of native host and/or nectar resource plants; (2) negative impacts without direct loss of native hosts, such as the latter being overgrown, not detected so easily, or the alien plants being more attractive to ovipositing females but not as suitable in sustaining offspring; and (3) creating structural changes that modify microclimates and influence interaction, such as those between mutualistic caterpillars and ants. These are not always easy to discover or discern, but some are reasonably unambiguous. The introduced South American vine *Aristolochia littoralis* (more familiarly termed *A. elegans* in the region) in subtropical eastern Australia is super-attractive to females of the Richmond birdwing butterfly (*Ornithoptera richmondia*, Chapter 9, p. 173) and acts as a decoy inducing females to lay on it rather than on their native foodplant vine, *Pararistolochia praevenosa* (Sands et al. 1997). *Aristolochia elegans* foliage, however, is toxic to the hatchling caterpillars, so that they are condemned to death by such oviposition, and spread of this alien vine into native forests is a major threat to the butterfly.

The native grassland habitats of the Golden sun-moth (*Synemon plana*, Castniidae, Chapter 9, p. 174) in south-eastern Australia are invaded by the alien Chilean needle grass, *Nassella neesiana*, a declared noxious weed – a status that legally obliges eradication attempts wherever it occurs. Current strong inference that it may be an important alternative host for subterranean *S. plana* caterpillars on some sites on which native grass hosts have become scarce continues to be debated amidst the conservation controversies between protecting a critically endangered moth by assuring a valuable food source and eliminating this invasive weed (New 2012a). As for *Plantago* and Taylor's checkerspot, discussed earlier, the practical dilemma of tolerating and, even, fostering an invasive alien food plant in order to conserve a notable lepidopteran species, will almost inevitably generate disagreements and strong opinion for or against any such action.

Whilst the full complexities of caterpillar foraging and nutritional ecology are beyond the scope of this account, a conservationist must be aware that many subtleties exist and may need to be considered carefully in optimising food supply in species management. More generally, both caterpillars and adult butterflies or moths have numerous constraints on their activity, imposed by the structure and condition of their environment, the spatial distribution of their resource needs, and complexities of interactions with other species, including

Table 1.1 The variety of biological features amongst caterpillars, as exemplified through a broader survey of herbivorous insects by Damman (1993). Reproduced with permission of John Wiley & Sons.

Characteristic	Categories
Feeding behaviour	Leaf chewers; leaf tiers; leaf miners; gall makers; stem borers; seed predators; fruit borers; root feeders
Feeding position	External (exposed to natural enemies, not bound to any sort of shelter); internal (living inside plant tissue or surrounded by shelter)
Gregariousness	Solitary (larvae feed independently); gregarious (larvae feed in groups, at least during early instars)
Specialisation	Specialists (restricted to one plant family or, in extreme cases, one plant species); generalists (feed on more than one plant family) [terms used at different levels by different authors]
Growth form of food plant	Almost any plant material in terrestrial environments and, more rarely, shallow freshwater environments; every gradation from short-lived highly seasonal annuals to long-lived perennials, and herbaceous to woody

competitors. These needs differ between species and, even, different caterpillar instars – so that, for example, fast growth may enable large size to be attained and be a refuge from natural enemies, or create resistance to some plant defences, or proffer competitive advantage over competing herbivores. Many different lifestyles occur among Lepidoptera (Table 1.1), and each represents a compromise between various ecological pressures (Damman 1993) to dictate a balance that must be sustained for the species to persist. The adaptations of many caterpillars are influenced strongly by the variety, distribution, condition and nutritional state of plant or other foods, and also by the influences of natural enemies, predators and parasitoids. As counters to these, many caterpillars have evolved to become cryptic (reducing detection), aposematic (advertising distastefulness and often exploiting plant chemicals to do so) or evolving physical (toughness, hairiness) barriers to attack or activity patterns that reduce exposure: 'enemy-free space' is itself a critical resource for many herbivorous insects. Any such lifestyle may be influenced or disrupted by competing species within the same milieu – so that, in addition to native species, introductions of alien plants, herbivores or higher-level carnivores (such as classical biological control agents, Chapter 12) may affect long-evolved balances between such naturally occurring trophic groups. Many such introductions are potential threats to native species. Two long-recognised forms of competition may occur, again discussed by Damman (1993): exploitation, in which the species involved compete directly for shared resources, here mainly for the same food plant species or tissues; and interference, in which access to a resource by one species necessitates actively excluding others (such as by aggressive behaviour or territoriality) or in which the species continually impede each others' efforts to forage at the same time. Richmond birdwing caterpillars, discussed earlier in this chapter, are cannibalistic, a habit perhaps of considerable value in reducing competition for food on vines with relatively small amounts of foliage (Sands & New 2013) – each caterpillar of this large butterfly needs to eat about a square metre of leaves during its development! Much of the intricate behaviour of caterpillars may have evolved to reduce competition, facilitate access to food, or to avoid attacks by natural enemies. Nocturnal activity, for example, has an appealingly simplistic function, far more difficult to prove, of avoiding diurnally active predators and parasitoids. Another frequent development has been for caterpillars to become endophytic, occupying leaf mines, galls, stems or roots, in addition to others that make shelters or retreats of various kinds, such as by tying leaves together.

Most Lepidoptera are terrestrial, but caterpillars of a few – notably amongst some pyralid moths, Nymphulinae, feed on aquatic plants in fresh water environments. Herbivory (perhaps ancestrally on lower plants such as bryophytes – a habit that persists in some primitive moth lineages, but now mostly on angiosperms), although the predominant larval feeding habit, is not quite universal, with various forms of aphytophagy involving predation and feeding on exudates of Homoptera or ants having arisen independently in several lineages (see Hinton 1951, Cottrell 1984, Pierce 1995 for background) adding to the spectrum of ecological associations within the order. Whilst the vast majority of Lepidoptera (>99%: Common 1990) are herbivores, a complex and varied array of other larval feeding habits also occur, and these have defied attempts at easy categorisation (such as Hinton's (1951) 'biological groups' based on feeding habits) because of numerous intergrading levels. Pierce (1995) regarded Lepidoptera as 'remarkably unadventurous' as predators, with limited prey range and feeding methods; prey are almost always of arthropods (largely Homoptera or ants) which are sedentary or found near the caterpillars' food plants. Myrmecophily has arisen independently on a number of occasions (in at least eight lineages of lycaenids alone, for example: Cottrell 1984), and now involves many subtle and obligate mutualisms, with those within the Lycaenidae most intensively investigated (Pierce et al. 2002) and of considerable importance in conservation management. Thus, the dependence of large blue butterflies (*Maculinea*) on particular ant species, with caterpillars feeding on ant brood, is a key element in conservation management (Chapter 7, p. 127). At one extreme within the Lycaenidae, the subfamily Miletinae appear to be entirely carnivorous on ants. Both predation and cannibalism have developed repeatedly in different groups of Lepidoptera, with cannibalism perhaps a more widespread opportunistic response to food shortage in some taxa. Predation may become highly focused, with particular species of prey, whether Homoptera or ants, necessary. Some Australian species of *Stathmopoda* (Oecophoridae: Stathmopodinae) are specific predators on *Eriococcus* scale insects on eucalypts (*S. melanochra*) or spider egg-sacs (*S. arachnophora*); and the few species of Epipyropidae and Cyclotornidae (both Zygaenoidea) are ectoparasites of leafhoppers. *Cyclotorna monocentra* (an Australian endemic) lays eggs on trees infested with eurymelid leafhoppers (Dodd 1912), but the later instars are carried into ant nests where they feed on ant larvae. Obligatory predators or parasites occur in eight superfamilies of Lepidoptera, in most of them only very sporadically. However, a few such cases have captured wide interest. The unique 'ambush' by caterpillars of *Eupithecia* species (Geometridae) in Hawaii involves them feeding only on live-caught insects and spiders that venture within range (Montgomery 1982), and depends on tactile stimuli as these possible prey contact caterpillars on vegetation. In common with numerous other endemic Hawaiian Lepidoptera, and as Montgomery foreshadowed, detailed study of this unusual adaptation may be thwarted by loss of the taxa as native ecosystems continue to be lost or invaded by alien taxa.

In some taxa, feeding habit changes markedly as larvae develop; some lycaenid butterfly caterpillars are initially phytophagous, but grow only slightly until they later switch to feeding on ant brood (*Maculinea arion*: Thomas 1989); less specifically, caterpillars of some Swift moths (Hepialidae) are initially detritivores or mycovores, but later feed on or in living plants (Grehan 1989). Such changes add further complexity to resource needs in conservation. As with strictly herbivorous taxa, any such relationships, also, may be very specialised, and in any of these the conservation of mutualisms is a key feature in conservation of the taxa involved. One well known example, a classic obligate mutualism, is the interdependence of yucca plants and yucca moths (*Tegeticula* spp., belonging to the small archaic family Prodoxidae). The moths are the only pollinators of the yuccas, whilst the developing seeds are the sole food of the developing caterpillars; even here, however, recent taxonomic studies have revealed a far greater richness of the moth species than earlier supposed (Pellmyr & Huth 1994).

Very broadly, the three traditional 'functional groups' of herbivorous Lepidoptera adopted widely by ecologists have been based largely on caterpillar feeding habits, as (1) specialists, using plants of a single species or genus; (2) oligophages, using multiple plant species within a single family or other restricted lineage; and (3) generalists, with more cosmopolitan feeding habits. More comprehensively, in contrast, Summerville et al. (2004, also Summerville & Crist 2002) classified forest moths into life forms of the host resource as a guide to interpreting changes in moth assemblage composition. He nominated five such guilds as (1) woody plant feeders; (2) herbaceous feeders; (3) dead or decaying vegetation feeders; (4) encrusting flora feeders; and (5) generalised feeders that transcend two or more of the above categories. As an example of this application, loss of trees in forests of Ohio was associated with lowered numbers of group 1 taxa, but overall richness was compensated by gains of additional herbaceous plant feeders. Knowledge of feeding guilds is clearly advantageous in interpreting changes in diversity, and in part helps to explain unexpectedly high levels of moth richness in some disturbed forest habitats. This theme is revisited in Chapter 7. The pattern detected by Summerville and Crist also reflected species turnover along a gradient of increasing habitat loss, in which different guilds replaced one another, with herbivory shifting from the canopy to the understorey layer, particularly in smaller plots.

The widespread habit of herbivory is also the major feature establishing some Lepidoptera as 'pests' through their depredations on crops, ornamental plants, stored plant products and other organic derivatives, with many species causing widespread and severe economic losses. The converse situation also occurs – that some plant-specific Lepidoptera are valued as feeding on pest plants (weeds) and can be potent biological control agents in pest management. Most species involved as pests are moths, as a further factor influencing their public image, whilst very few butterflies (such as the cabbage whites, Pieridae, on brassica crops) are regarded as damaging pests. However, and in close parallel to Orthoptera (as noted by Samways and Lockwood 1998), serious pest species of moths or butterflies may have very close relatives that are respectively innocuous to human interests and severely threatened, in need of conservation attention. Some such closely related and generally similar-looking taxa may be very difficult to differentiate.

Unusually amongst larger groups of holometabolous insects, for which species-level identification relies very largely on the adult insects alone, the early stages of many Lepidoptera can also be recognised reliably to species level, particularly amongst the northern temperate region fauna. Substantial biological information accrues from studies of both caterpillars and adults, as the two active life history stages with very different biologies, and the resource needs of both are central to conservation management. Well illustrated synopses enabling identification of European butterflies and larger moths flow from the mid nineteenth century, largely the outcomes of hobbyist zeal, and have been followed by comprehensive guides for recognition of many caterpillars, in particular for those of other parts of Europe and North America – with outlines or some descriptions available for many other parts of the world or for particular taxa. The early impetus for this development was also from hobbyists, many of whom rear field-collected or captive-bred caterpillars to obtain cabinet-quality adult specimens, and gain considerable expertise in the discovery, identification and and husbandry of immature stages. In parallel, information on seasonal developmental patterns, food plants and habits of many species has also accrued. Less definitive information, reflecting their lower general profile, is available for many Microlepidoptera (Chapter 2, p. 16), but compendia of biological information (such as that by Ford 1949, for Britain) and recognition guides based on indirect features (such as the form of leaf mines for some endophytic groups) still provide much stronger and more reliable specific information than on any other large insect group. Considerations of species' ecology and of assemblage richness and community interactions can commonly incorporate both caterpillars and adults, with either stage potentially available for evaluation as an intergenerational marker of change (Chapter 13, p. 242). However, as for adults, geographical bias in knowledge of immature stages is strong, and details of most tropical Lepidoptera, in particular, remain undocumented. The large-scale attempt to inventory the Lepidoptera of the Guanacaste World Heritage Site of Costa Rica (Miller et al. 2006) was based on light trap catches of adults and direct searches for caterpillars, and extended over almost 30 years from 1978. Every caterpillar found was isolated and documented individually, and reared to determine its fate in yielding either an adult or a parasitoid. Miller et al. estimated that about 9500 species of larger free-living Lepidoptera occurred at Guanacaste, and this total excluded a possible several thousand leaf-mining species. This long-term and thorough investigation remains unique for tropical biomes.

Sources of information

Publications on Lepidoptera biology and identification are thus far more extensive and varied than for most other insects (rivalled only by those on Coleoptera), and also more fragmented into different 'interest groups'. Continuing documentation on Lepidoptera thereby has several distinct foci, in addition to that dealing directly with conservation matters.

1   Natural history observations, from hobbyists and others, stemming from records of incidence in time and space, as well as focus on life history details and ecology, with notes in a variety of outlets extending to compilation of handbooks and identification manuals for adults and larvae accessible to non-specialists. Much fundamental information has accrued, for example, from hobbyists in Britain through a variety of journals (some extending for more than a century of publication) and newsletters, nowadays in electronic or print formats.
2   More detailed ecological or 'scientific' contributions on systematics, biology and behaviour, often with quantitative/semiquantitative treatment of testable hypotheses. Studies range from those on single species or assemblages primarily addressing the insects themselves, to those employing Lepidoptera as 'tools' to help interpret ecological topics and questions. Relevant scientific contributions also encompass those dealing with studies of Lepidoptera in evolutionary biology and biogeography studies, in helping to illuminate much of the historical record that has determined the incidence and distribution of the recent fauna.
3   'Applied entomology' contributions, extending from the above to aid understanding of the population dynamics, feeding ecology and specificity, dispersal and phenology of individual pest or beneficial species. The aim is to provide information that can be used to predict the species' performance and impacts, and which can be used to suppress or enhance those impacts through practical management in relation to human needs.
4   Use of Lepidoptera more widely as models or surrogates, as signals of environmental wellbeing or change, through comparative studies of diversity or of presence and abundance of particular species across different sites or times and as potential tools to predict impacts of future changes (such as climate change). Any such study may apply to single species, assemblages or communities, particular natural or anthropogenic habitats or wider landscapes, and so focus on any of a wide range of scales.

Practising insect conservationists can gain much from the massive 'applied literature' in ecology, simply because details of knowledge on many key pest or beneficial taxa, including Lepidoptera, can be supported by economically justified funding and the need for that detail to refine management. Such detail is much harder to pursue for many taxa that are innocuous, and difficulty is enhanced for many species of conservation concern because their scarcity renders them intrinsically difficult to study or survey. Thus, much of the basic understanding of insect population dynamics (Chapter 6), has arisen from interpreting long-term painstaking standardised counts of life stages of key pests in forests, orchards or field crops over many consecutive generations.

As one early classic example, comparison of the life systems of codling moth (*Cydia pomonella*) in Canada and Australia (summarised by Clark et al. 1967) showed how local environmental conditions may influence a key pest species. As another case, the introduced European cabbage white butterfly (*Pieris rapae*) has been studied in Australia far more intensively than almost any native Australian butterfly (Jones 1981, Waterhouse & Sands 2001), and its biology can be compared constructively with that in its native Britain (Richards 1940). Management for pests (aiming to reduce numbers) and for conservation (aiming to sustain or increase numbers) draws on the same kinds of ecological information, and the perspectives revealed by intensive studies on pests can at times complement the (usually far less) information available on rare species targeted for conservation. Contexts such as clarifying the roles of natural enemies, food quality and local environmental conditions, and their influences on abundance and fitness are important aspects of threat evaluation in conservation, and of understanding pest dynamics. A universal problem is how to interpret observed changes in numbers, and whether abundance varies naturally and significantly across generations – or, conversely, whether any marked change may be a real trend, either upward or downward, and lead to relief or concern (Chapter 6). The forms of numerical changes in pest moths (exemplified by classic studies on several univoltine pine-defoliating species in Europe, for which numbers in samples were plotted over 60 years, 1880–1940: Fig. 1.2) revealed substantial

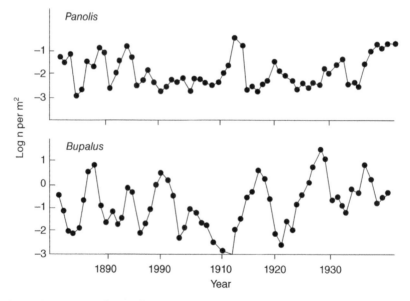

Fig. 1.2 Long-term population fluctuations interpreted from annual census counts of pupae of two univoltine pine forest moths in Germany over 60 years. The two representative species shown are (a) *Panolis flammea* (Noctuidae, the Pine beauty) and (b) *Bupalus piniarius* (Geometridae, the Bordered white) (Source: Varley et al. 1973. Reproduced with permission of John Wiley & Sons.).

natural variations and implied cyclic fluctuations over numbers of generations. Each species occasionally increased to serious defoliation-causing levels but at other times was relatively scarce (discussed by Varley et al. 1973). From a conservationist's viewpoint, normally supported by – at most – a few generations of information, such long-term variations are simply not apparent, and whether the limited data available are from an upward or downward part of the natural cycle may imply very different conservation need.

In general, information on 'how Lepidoptera work' based on such studies of common or easily available species and from hobbyist interests in many more obscure taxa, has provided most of the background available to guide studies for conservation, through demonstrating both general principles and the detailed idiosyncrasies of individual taxa. By their very elusiveness, many of the taxa of greatest conservation concern are difficult to study – and this is by no means restricted to those small of stature. Although the smallest member of its genus, the endemic Australian Ward's atlas moth (*Attacus wardi*, Saturniidae) is one of the country's largest moths, and was not seen for some 60 years after its initial discovery in Darwin (Northern Territory) (Braby & Nielsen 2011). Now known from six localities, it seems to occur very patchily over a small part of north-western Australia. Only very recently has its larval foodplant been determined (as *Croton habrophyllus*, Euphorbiaceae), and this grows along edges of monsoon forests, so giving clues on the critical habitat needed by *A. wardi*.

There remains the literature on Lepidoptera that deals more specifically with conservation, or primarily addresses practical or theoretical themes of conservation interest in the direct context of conservation. As noted in the Preface, books on Lepidoptera abound and well illustrated field guides and related handbooks are a major facilitator of interest in the group. However, until the last decade or so, the term 'conservation' occurred rather rarely in these, or was mentioned only as a brief afterthought. There are notable exceptions, mostly from Britain, western Europe or North America, where the urgency of conservation need was well appreciated and the possible consequences recognised and taken seriously. Thus, the two highly readable volumes on natural history of moths (Young 1997, Majerus 2002) both contain considerable information on moth conservation both as themes running through their text and as more emphatic specific chapters. More awareness, as expected, has been disseminated for butterflies, in many parts of the world, and includes attention to production of directories or lists of threatened taxa (Chapter 11). Many of these, and many research papers and reports, are cited in context here, but it is important to note also the deep roots of Lepidoptera conservation interest in the fates of individual species or, in some cases, local subspecies and, thus, in regional endeavours. Autecological studies on some such taxa have sometimes led to far-reaching appreciations of biological subtlety and of the importance of differences between closely related or co-habiting taxa. The considerable array of peer-reviewed documents is augmented by many consultancy and government agency reports of more limited circulation and which are amongst the 'grey literature' of conservation; these include much valuable information and, in some cases, are the only sources of data on the taxa they treat.

Likewise, the 'snippets' provided in short observational notes from hobbyists and others often include critical and important insights into incidence, behaviour and biology of individual rare species; although it has been fashionable to denigrate 'natural history' in favour of more technical disciplines, good basic observational knowledge and understanding underpins many successful conservation efforts for Lepidoptera. As Young (1997) noted, amateur enthusiasts are 'a mine of information' on the biology of moths, 'often including knowledge that is laboriously "rediscovered" by scientists later'. A major and vital facet of Lepidoptera conservation, setting it apart from much other invertebrate conservation, is the continuing strong cooperation and symbiosis between non-professional interests and scientists. The theme is central to Lepidoptera conservation and recurs in many contexts throughout this book. The continuing study of Lepidoptera and their wellbeing depends to a very large extent on the involvement and commitment of the wide range of people to whom these insects are in some way attractive and important.

References

Braby, M.F. & Nielsen, J. (2011) Review of the conservation status of the Atlas moth, *Attacus wardi* Rothschild, 1910 (Lepidoptera: Saturniidae) from Australia. *Journal of Insect Conservation* **15**, 603–608.

Clark, L.R., Geier, P.W., Hughes, R.D. & Morris. R.F. (1967) The Ecology of Insect Populations in Theory and Practice. Methuen, London.

Common, I.F.B. (1990) Moths of Australia. Melbourne University Press, Melbourne.

Cottrell, C.B. (1984) Aphytophagy in butterflies: its relationship to myrmecophily. *Zoological Journal of the Linnean Society* **79**, 1–57.

Damman, H. (1993) Patterns of interactions among herbivore species. pp. 132–169 in Stamp, N.E. & Casey, T.M. (eds) Caterpillars. Ecological and Evolutionary Constraints on Foraging. Chapman & Hall, London.

Dennis, R.L.H. (2010) A Resource-Based Habitat View for Conservation. Butterflies in the British Landscape. Wiley-Blackwell, Oxford.

Dodd, F.P. (1912) Some remarkable ant-friend Lepidoptera of Queensland. *Transactions of the Entomological Society of London* **1911**, 577–589.

Duffey, E. (1977) The reestablishment of the large copper butterfly *Lycaena dispar* (Haw.) *batavus* (Obth.) at Woodwalton Fen National Nature Reserve, Huntingdonshire, England 1969–73. *Biological Conservation* **12**, 143–158.

Feltwell, J. (1995) Conservation of Butterflies. Wildlife Matters, Battle.

Ford, L.T. (1949) A Guide to the Smaller British Lepidoptera. South London Entomological and Natural History Society, London.

Grehan, J.R. (1989) Larval feeding habits of the Hepialidae. *Journal of Natural History* **33**, 469–485.

Hinton, H.E. (1951) Myrmecophilous Lycaenidae and other Lepidoptera – a summary. *Proceedings of the South London Entomological and Natural History Society* (1961), 111–175.

Jones R.E. (1981) The cabbage butterfly, *Pieris rapae* L.: 'a just sense of how not to fly'. pp. 217–228 in Kitching, R.L. (ed.) The Ecology of Pests. Some Australian Case Histories. CSIRO, Melbourne.

Kristensen, N.P. (ed.) (1999) Lepidoptera: Moths and Butterflies. Vol.1. Evolution, systematics and biogeography. Handbuch der Zoologie, Vol. 4, part 35. W. de Gruyter, Berlin and New York.

Kristensen, N.P. & Skalski, A.W. (1999) Phylogeny and palaeontology. pp. 7–25 in Kristensen, N.P. (ed.) Lepidoptera. Moths and Butterflies, Vol. 1. Handbuch der Zoologie, Vol. 4, part 35. W. de Gruyter, Berlin and New York.

Kristensen, N.P., Scoble, M.J. & Karsholt, O. (2007) Lepidoptera phylogeny and systematics: the state of inventory in moth and butterfly diversity. *Zootaxa* **1668**, 699–747.

Majerus, M.E.N. (2002) Moths. Harper-Collins, London.

Miller, J.C., Janzen, D.H. & Hallwachs, W. (2006) 100 Caterpillars. Portraits from the Rainforests of Costa Rica. Harvard University Press, Boston, Mass.

Montgomery, S.L. (1982) Biogeography of the moth genus *Eupithecia* in Oceania and the evolution of ambush predation in Hawaiian caterpillars (Lepidoptera: Geometridae). *Entomologia Generalis* 8, 27–34.

New, T.R. (2012a) The Golden sun-moth, *Synemon plana* Walker (Castniidae): continuing conservation ambiguity in Victoria. *Victorian Naturalist* 129, 109–113.

Pellmyr, O. & Huth, C.J. (1994) Evolutionary stability of mutualism between yuccas and yucca moths. *Nature* 372, 257–260.

Pierce, N.E. (1995) Predatory and parasitic Lepidoptera: carnivores living on plants. *Journal of the Lepidopterists' Society* 49, 412–453.

Pierce, N.E., Braby, M.F., Heath, A. et al. (2002) The ecology and evolution of ant association in the Lycaenidae (Lepidoptera). *Annual Review of Entomology* 47, 733–771.

Pogue, M.G. (2009) Biodiversity of Lepidoptera. pp. 325–355 in Foottit, R.G. & Adler, P.H. (eds) Insect Biodiversity. Science and Society. Wiley-Blackwell, Oxford.

Powell, J.A. (2003) Lepidoptera (moths and butterflies). pp. 631–664 in Resh, V.H. & Cardé, R.T. (eds) Encyclopedia of Insects. Academic Press, San Diego.

Powell, J.A. (2009) Lepidoptera (moths and butterflies). pp. 557–587 in Resh, V.H. & Cardé, R.T. (eds) Encyclopedia of Insects (2nd edn). Academic Press, San Diego.

Powell, J.A. & Opler, P.A. (2009) Moths of Western North America. University of California Press, Berkeley and Los Angeles.

Powell, J.A., Mitter, C. & Farrell, B. (1999) Evolution of larval food preferences in Lepidoptera. pp. 403–422 in Kristensen, N.P. (ed.) Lepidoptera: Moths and Butterflies. Vol.1. Evolution, systematics and biogeography. Handbuch der Zoologie, Vol. 4, part 35. W. de Gruyter, Berlin and New York.

Pyle, R.M. (2012) The origins and history of insect conservation in the United States. pp. 157–170 in New, T.R. (ed.) Insect Conservation: Past, Present and Prospects. Springer, Dordrecht.

Richards, O.W. (1940) The biology of the small white butterfly (*Pieris rapae*) with special reference to the factors controlling its abundance. *Journal of Animal Ecology* 9, 243–288.

Samways, M.J., Lockwood, J.A. (1998) Orthoptera conservation: pests and paradoxes. *Journal of Insect Conservation* 2, 143–149.

Sands, D.P.A. & New, T.R. (2013) Conservation of the Richmond Birdwing Butterfly in Australia. Springer, Dordrecht (in press).

Scoble, M.J. (1992) The Lepidoptera: Form, Function and Diversity. Oxford University Press, Oxford.

Severns, P.M. & Warren, A.D. (2008) Selectively eliminating and conserving exotic plants to save an endangered butterfly from local extinctions. *Animal Conservation* 11, 476–483.

Singer, M.C., Thomas, C.D. & Parmesan, C. (1993) Rapid human-induced evolution of insect-host associations. *Nature* 366, 681–683.

Slansky, F. (1993) Nutritional ecology. The fundamental quest for nutrients. pp. 29–91 in Stamp, N.E. & Casey, T.M. (eds) Caterpillars. Ecological and Evolutionary Constraints on Foraging. Chapman & Hall, London.

Spalding, A., Young, M. & Dennis, R.L.H. (2012) The significance of host plant-habitat substrate in the maintenance of a unique isolate of the Sandhill Rustic: disturbance, shingle matrix and bare ground indicators. *Journal of Insect Conservation* 16, 839–846.

Summerville, K.S. (2004) Functional groups and species replacement testing for the effects of habitat loss on moth communities. *Journal of the Lepidopterists' Society* 58, 129–132.

Summerville, K.S. & Crist, T.O. (2002) Effects of timber harvest on forest moths (Lepidoptera): community, guild and species responses. *Ecological Applications* 12, 820–835.

Thomas, J.A. (1989) The return of the Large blue butterfly. *British Wildlife* 1, 2–13.

Thomas, J.A. (1991) Rare species conservation: case studies of European butterflies. pp. 149–197 in Spellerberg, I.F., Goldsmith, F.B. & Morris, M.G. (eds) Scientific Management of Temperate Communities. Blackwell Publishing, Oxford.

Varley, G.C., Gradwell, G.R. & Hassell, M.P. (1973) Insect Population Ecology. An Analytical Approach. Blackwell Scientific Publications, Oxford.

Waterhouse, D.F., Sands, D.P.A. (2001) Classical biological control of arthropods in Australia. ACIAR Monograph no 77, Canberra.

Young, M. (1997) The Natural History of Moths. T. & A.D. Poyser, London.

2

The Diversity of Lepidoptera

Introduction

The major categories of Lepidoptera most familiar to hobbyists transcend strict taxonomic groupings, but are important to acknowledge as having rather different traditions of study and markedly different involvement in conservation. In essence, they constitute a 'gradient' of relatively well known to poorly known, which parallels their abundance from lower to higher, and appeal from higher to lower, so differing in their amenability to sound conservation, and the ways in which they are perceived. These groupings are the butterflies, macromoths and Microlepidoptera (Table 2.1). Butterflies are indeed taxonomically defined, as members of three well defined superfamilies. Two of these (Hesperioidea, Papilionoidea) are the most traditionally accepted butterflies, and are now accompanied formally by a small and more anomalous South American group, Hedyloidea, as their closest relatives. 'Macromoths' include all the groups most familiar to hobbyists and are also reasonably well documented. They comprise the majority of named Lepidoptera species, mostly belonging to the more advanced lineages of moths but with some members of the remaining more primitive evolutionary groups tacitly or traditionally included by hobbyists through being colourful, diurnal or large. The 'Microlepidoptera', the wealth of mostly tiny moths, are by far the most diverse section of the order, and also by far the least documented and least popular. They include more than three quarters of the 47 recognised superfamilies of Lepidoptera (Kristensen et al. 2007) and, other than for parts of Europe, knowledge is highly incomplete. However, at higher taxonomic levels the phylogenetic relationships amongst, and between, many of the major groups are also still debated: the predominant group, Ditrysia, including perhaps 98% of all lepidopteran species, is particularly complex but

Lepidoptera and Conservation, First Edition. T.R. New.
© 2014 John Wiley & Sons, Ltd. Published 2014 by John Wiley & Sons, Ltd.

Table 2.1 Major 'informal groupings' of Lepidoptera used commonly by hobbyists and naturalists. Note that these are not formal taxonomic categories and that the two moth categories, in particular, transcend phylogenetic relationships: more formal synopses and discussions of relationships are outlined in specialist accounts such as those by Scoble (1992) and Kristensen et al. (2007), in which up to 47 superfamilies are recognised.

---

1. 'Butterflies'. A relatively small group, including the most popular and best-studied of all insects. About 20,000 species described, and comprises the superfamilies Hesperioidea and Papilionoidea (the 'conventional' butterflies) augmented recently by Hedyloidea. A presumed monophyletic suite within the 'Macrolepidoptera'
2. 'Macromoths'. The larger and more conspicuous moths, including almost all of those targeted commonly by hobbyists. The majority are Ditrysia (the more advanced moths) but some more primitive groups (such as swift moths, Hepialoidea) are traditionally included in this category. The best documented groups of moths, and containing the majority of currently described moth species
3. 'Micromoths'(or 'Microlepidoptera'). The mostly small and relatively inconspicuous moths including most of the more primitive lineages, many of which only rarely attract attention from non-professional lepidopterists. In most parts of the world, these are by far the least documented sections of the Lepidoptera, and numerous species are still undescribed. More than three-quarters of the superfamilies fall into this enormous group

---

with substantial understanding coming from recent molecular approaches to their study (Regier et al. 2009).

Kristensen et al. suggested that the geographical bias in knowledge and interest for conservation reflects rather different local influences and capability, contrasting scenarios in Europe and North America to illustrate this idea. In Europe, the 'average local collector' is prone to extending interest from butterflies initially to macromoths and then to micromoths of his/her country of residence, rather than to the butterflies of adjacent or other countries, because of cultural differences and political and language barriers. This contrasts with North America, where Microlepidoptera are relatively poorly known and where many butterfly enthusiasts continue to work with butterflies, simply extending their compass to wider areas of the Nearctic rather than diversifying their interests to include moths. Parallel incentives to the European impediments to abandon butterflies and extend interest to moths of a limited region do not occur. The outcome of this traditional difference in focal interest has been that many European Microlepidoptera have been the subjects of illustrated identification guides and biological accounts paralleling those on butterflies in completeness and detail, and tiny moths have become increasingly 'accessible' for study through intensive appraisals of local faunas. Such comprehensive accounts are scarce for other regional faunas, and most contain significant gaps in coverage. A pioneering guide to smaller moths of south-east Asia, for example, noted that probably fewer than a third of the more than 6000 species known to the authors had been described (Robinson et al. 1994). The number of as yet undiscovered species there is also likely to be high: Robinson et al. suggested that this could be at least the same number as those already known, and that the omission of Microlepidoptera from many popular books on Lepidoptera reflects perceptions that they are ' "too difficult", or "too dull" or "too small" '. An appraisal (Powell & Opler 2009) of a major field guide to North American Lepidoptera considered that Microlepidoptera had been 'slighted', with only token representatives included.

Elsewhere in the world, correlation between hobbyist interest and accumulated knowledge is very evident, with butterfly interests predominating strongly. In addition, the number of hobbyists living in many parts of the tropics, in particular, is low and much of the background information has derived from activities of expatriate visiting collectors rather than experienced residents. In common with many aspects of documenting biodiversity and accepting conservation beyond the more affluent temperate regions, the 'luxury' of such basic activity, unless it is directly and economically important to local human welfare, is simply not affordable. The traditional rise of 'butterfly collecting' reflected the rise of an affluent leisure class with time and resources to pursue such hobbies; for much of the world, pressing environmental problems of food production and health care for burgeoning human populations clearly take precedence and may leave little for such more tangential or hobbyist 'luxury' pursuits.

The wider legacy is that butterflies predominate on conservation schedules on which Lepidoptera are noticed, and are followed by some macromoths. It is indeed rare for any micromoth to be singled out for conservation attention, except in the very best documented faunas: several have been listed as of conservation concern in the United Kingdom, for example (Shirt 1987). One notable case is for a tiny Scottish nepticulid, the Sorrel pygmy moth (*Enteucha acetosae*), which was a key influence in conservation of its sole raised bog habitat (Auchennines Moss) in that country (Stubbs & Shardlow 2012). The novelty value of this moth, within a culture of ready acceptance of values of Lepidoptera conservation, gave it 'respectability'. This case raised an important issue of effective communication: initially known only by its scientific name, the public showed little sympathy for *Enteucha*, but once a common name was given, 'character' and appeal rapidly ensued. Other microlepidopteran novelties have been noted. For North America, Pyle (2012) noted the remarkable sand dune-frequenting *Areniscythrus brachypteris* (Scythrididae), with its recently diagnosed close relatives the only continental moths in the world with both sexes flightless, adapted strongly to leaping and with its scuttling movements on sand likened to those of miniature silverfish. The various undescribed sibling species each occur on restricted and isolated sand dune systems along the west coast and – following Powell (1976) – may be vulnerable to shifting sand caused by human disturbance.

Despite their low public profile, Microlepidoptera include all the most primitive and isolated lineages of Lepidoptera. Many of these contain rather few species, and their phylogenetic positions may imply some priority for conservation attention, as contrasted with focusing on still more representatives of diverse and well established lineages. The debate over whether it is 'better', faced with inevitable shortage of resources for conservation activity, to (1) select further butterflies and so build on the mass of useful information already available for the group or (2) select other less understood taxa and so expand the foundation knowledge, will continue.

Some Lepidoptera, representing many different groups, are ecologically isolated in their dietary habits, with any such specialisation potentially leading to vulnerabilty. Thus, caterpillars of the tineid genus *Ceratophaga* feed on horny material, keratin, rather than on vegetation. In Florida and neighbouring southern United States, the only New World species, *C. vicinella*, feeds only on the shells of dead gopher tortoises (*Gopherus polyphemus*), apparently the only caterpillars that do this – although the feeding habits of some of the African species of *Ceratophaga* are still unknown. *C. vicinella* does not feed on shells of living tortoises, and the caterpillars feed from within silken tunnels that extend into the sandy soil on which the scutes lie (Deyrup et al. 2005). *G. polyphemus* is declining throughout its range, due primarily to habitat loss, as well as predators, poaching for trade and disease, and is federally listed in the western areas in which it occurs (Tuberville et al. 2005), so also rendering the moth threatened. In this example, as in many others, the critical need for conservation is assuring the wellbeing of the specific food species, whether plant or animal.

Distinguishing taxa

Capability to assess incidence of particular taxa, and to assess national or local richness depends (as for any group of organisms) on accurate and consistent identification, with the criteria understood clearly and applied in a similar manner by all participating observers. From this viewpoint, many Microlepidoptera remain inaccessible, with contributions to documenting their diversity and distributions almost wholly from specialists in individual families or investigating fauna associated with individual plant taxa, and most species identifiable only by the few specialists working on that family or genus. Macromoths have far more devotees and attention to their conservation is increasing rapidly, especially in the northern temperate regions, and through some selected 'flagship groups' (Chapter 4, p. 44) in other places. However, only the butterflies, so far, can be used widely and relatively consistently and confidently to provide basic locality or wider inventories of sufficient quality (accuracy, reliability, completeness) to enable comparisons of richness, representation and changes in time and space across most places where they occur. Standard survey methods (Chapter 5) that are easily understood and carried out, with coordinated recording of observations, are the core of reliable comparative sampling – and in many of the places where such procedures are most advanced, recognition of most species likely to be encountered is straightforward, and individual problems of separations well informed, so that extensive collecting may not be needed. Capture of anomalous individuals for confirmation, or to check small differences between closely similar taxa can often then be followed by their release unharmed after inspection. The goodwill of recorders ethically opposed to collecting can thereby be maintained. Through 'team efforts' and coordination, less experienced participants can receive constructive advice, so that both sampling effort and sample interpretation can be standardised for sensible use.

In many cases involving other Lepidoptera, capture and retention of voucher specimens is needed to avoid errors. That need varies markedly across different groups, and understandably becomes more widespread amongst poorly documented taxa or in regions where baseline information is weak. Extensive collecting may then be needed, including substantial series to interpret known complex groups – and, increasingly, visits to remote areas may be both expensive and logistically complex, so that repeated visits to confirm identities by collecting further specimens may be impracticable. For basic documentation of 'biodiversity', specimen-based study remains the most potent and universal key to identification, with the advent of molecular studies (see later in this section) an increasingly important aspect of diagnoses. Confusions that result from failure to identify species correctly amongst the members of closely similar taxa may have important and costly ramifications. One such example, amongst pest species, is of the two major species of *Helicoverpa* moths (Noctuidae) in eastern Australia (Matthews 1999). Both are polyphagous pests that cause severe depredations on a range of crops, and are amongst the world's most intensively studied pest insects. *Helicoverpa armigera* and *H. punctigera* have rather different diapause regimes and developmental patterns, and correct identification is critical to sound pest management, including management of insecticide resistance. In this example, aided by the investment available to manage key economic pests, a specific field identification kit enabled separation of all life history stages – which are morphologically indistinguishable up to the later caterpillar instars in the two moths. The kit, based on antibody reactions, is simple to apply and interpret (Trowell et al. 1993).

Such individual attention is of course unusual, and rarely possible, but illustrates the kind of interpretative problem that is far more widespread and difficult to address. In this example, also, there is no doubt that the two taxa are distinct species. This is not always the case. Many taxonomic problems with enumerating Lepidoptera species devolve on ambiguities over (1) definition and delimitation of a species; (2) whether infraspecific categories (such as many butterfly 'subspecies') are valid, represent full species or are simply trivial variants, perhaps changing in frequency along a continuous cline (for example along elevational or latitudinal gradients); and (3) whether 'diversity' is defined adequately in any such taxonomically based format or is better characterised by some other form of variety that encompasses evolutionary categories and isolated populations more comprehensively. There is no single or universally accepted opinion on these complex topics, but each viewpoint may markedly influence evaluations of 'richness' and be influential in inventory assessments. And, at a level fundamental in much practical conservation, 'It is not easy to assess the conservation status of a [butterfly] species when you don't know what that species is.' (Lewis & Senior 2011).

Interpreting variation in adult Lepidoptera has always been problematical – and part of the 'poor reputation' of Lepidoptera alluded to by Kristensen et al. (2007, Chapter 1, p. 2) results from differences in colour pattern from the vestiture of small scales covering the wings leading to proliferation of names based solely on appearance. Kristensen et al. noted that vestiture 'has invited literal superficiality in Lepidoptera taxonomy', with the philatelic tradition of butterfly collectors, in particular, generating enormous numbers of varietal names, some applying only to individual freak insects and never intended to designate any formal taxonomic entity. Some such names are now associated firmly with features of local populations, and for which formal subspecies status may be valid – but any such trinomial name, whether based in individual features or consistent population characteristics, has equivalent formality under the International Code of Zoological Nomenclature.

The precise status of many formal subspecies in Lepidoptera is by no means settled, but much discussion has acknowledged that many names have been erected somewhat subjectively and without full formal systematic analysis. In practice, ambiguity and continuing debate over the relevance of names as designating biological entities or more trivial variants has considerable relevance in evaluating taxon richness, and how those taxa may be treated in legislation (Chapter 11). The Australian butterflies exemplify the differences that may occur: Sands and New (2002) noted 427 full species in this relatively well known fauna, but this expanded to a total array of 654 recognisable taxa once named subspecies were included. More recent accounts have changed these totals, and the balance between them, with continuing debate over status of some taxa, predominantly amongst complex endemic radiations across fragmented landscapes suggesting that conventional 'typological' characters are inadequate to reach any sound consensus. As Braby et al. (2012) put it, 'the utility of subspecies in taxonomy and conservation biology is hampered by inconsistencies by which they are defined conceptually' as well as by 'a lack of standardised criteria or properties that serve to delimit their boundaries', together with 'their frequent failure to reflect distinct evolutionary units according to population genetic structure'. They suggested that the subspecies concept may no longer be useful in conservation biology. However, the common use of subspecies or broader infraspecies entities, and many similar examples, some extending to moths, imply that comparisons between surveys, or lists of taxa across sites or countries may vary considerably in extent and reliability according to individual predilections, which may be evident to lepidopterists but not to many of the managers who rely on such reports for important decisions on land use.

Recent developments and refinement of DNA analysis approaches to assessing diversity, most frequently based on sequences of mitochondrial cytochrome c oxidase (CO1) ('barcoding') have been heralded widely as vital tools in adding objectivity to some otherwise subjective taxonomic decisions. The patterns of diversity revealed are commonly hard to interpret: transitions between species, possible patterns of divergence, possible sibling or otherwise cryptic species, individuality of populations and many other evolutionary themes are indeed revealed – but a common trend is simply to increase markedly the number of clades (as consistently distinct evolutionary lineages) that demonstrate that many species are indeed far more intrinsically diverse than implied from typological assessment alone. A frequently discussed example involves investigation of 'one' conventionally accepted but variable species of neotropical skipper (*Astraptes fulgerator*), revealed by CO1 analyses to comprise 10 species (Hebert et al. 2004). As with the difference between including or omitting subspecies in butterfly enumerations, a conservationist aware of the political power of higher numbers, or an evolutionary biologist may elect to emphasise the variety present. A pragmatic manager may be tempted to disregard it, and the challenge remains to interpret it realistically in relation to its significance in any individual conservation context.

The case of *Astraptes*, in the preceding paragraph, has stimulated awareness of such cryptic variety and how it may be detected and interpreted in both formal taxonomy and expressing biological variety. These themes have commonly been under-regarded, so that specimens that in appearance superficially seem to belong to a named species may be attributed to it, albeit uncritically. Burns et al. (2008) noted that the repeated application of a specific name has, in this way, tended to broaden the concept of that species – with a common outcome that some widespread, somewhat variable but easily recognised taxa have become uncritically accepted as 'too familiar to generate further taxonomic interest'. This had apparently been so for *A. fulgerator* and only detailed study of this complex, from specimens from Guanacaste (Chapter 1, p. 10) revealed the real diversity present, detected initially by clues from examination of adults reared from different larval food plants, and sexual differences. However, the actual number of species present may be fewer than 10, and the thoughtful discussion of problems of evaluation by Brower (2010) merits critical reading by any worker contemplating similar interpretations. In naming the 10 entities formally, Brower remarked 'I remain sceptical about the biological reality of these taxa but, whether or not they exist in nature, they "live" in the literature and the web'. These were the first species to be described solely on the basis of DNA barcode features – and Brower's concern over the practice is serious: '. . . simply because species can be identified and publicised on the basis of a few nucleotide differences does not mean that they should be' (Brower 2010: 488).

In short, interpreting barcoding results from Lepidoptera is a complex exercise, with its roles and values debated extensively in recent years as techniques and approaches to analysis continue to diversify, as reviewed by Silva-Brandao et al. (2009). They are progressively revealing numerous previously undetected entities loosely thought of as 'cryptic species' or 'sibling species' earlier treated as identical, and with strong support for the CO1 sequence to diagnose novel taxa (Hajibabaei et al. 2006). However, there have also been some notable failures from this method to separate forms that are abundantly distinct on more conventional characters. Many Lepidoptera groups clearly need a combination of study approaches to differentiate them unambiguously, but there remains little doubt that the order contains numerous taxa that have not yet been diagnosed. Some are likely to be highly labile as representing rapid incipient speciation.

With spatially or seasonally phenotypically variable insects – amongst which butterflies are paramount, with numerous slightly different forms accorded formal 'varietal' or subspecific names – any such additional tools to help elucidate biological reality are invaluable. As for morphological characters, problems may arise in establishing taxonomic boundaries from DNA data, and can be confounded further by undetected hybridisation, as discussed for European butterflies by Descimon and Mallet (2009). A 'working scheme' used for several animal groups, including Lepidoptera, has been that if DNA sequences display divergences of at least 2%, such divergences may designate a lineage equivalent to a cryptic species. An unusually complete regional survey involved a DNA barcode investigation of almost the entire butterfly fauna of Romania, namely 180 species (Dinca et al. 2011). Ninety per cent of species could be identified reliably, with the remainder involving nine pairs of closely related taxa that either can hybridise regularly or whose precise taxonomic status remains controversial. Intriguingly, outcomes of this approach were somewhat higher for identification than use of either wing morphology or male genitalic details, as more common primary character sets used in butterfly recognition. Dinca et al. noted the nine pairs as being 'worst case scenarios', for which the same dilemmas arise from morphological differentiation. They include, as elsewhere, taxa that are currently differentiating rapidly; examples elsewhere include neotropical Ithomiinae (Nymphalidae) (Elias et al. 2007) and the lycaenid genus *Agriodiaetus* (Wiemers & Fiedler 2007).

In principle, DNA barcoding approaches enable rapid and reliable identification, and discovery of undescribed taxa. They can also help to associate early stages with corresponding adults, otherwise possible only from careful individual rearing. DNA can be extracted satisfactorily even from caterpillar frass and cast exuviae (Feinstein 2004), with those sequences corresponding closely with analyses of parental adults of *Pieris rapae* and *Vanessa cardui*. Because of increasing interest in using DNA analyses to clarify the status of putative threatened species, in particular, increasing attention has also been paid to use of non-lethal sampling, with the need to clarify that any such technique is indeed not harmful. The spectacular European silk moth *Graellsia isabellae* has been studied to compare the suitability and possible adverse impacts of obtaining DNA from severed mid-legs or the 'tails' of the long hind wings (Vila et al. 2009). Both were useful, but DNA yields were higher from the wing samples and, whilst leg clippings did not affect survival, it had some impact on female mating success. The other 'advantage' of using wing clippings was that it was believed to be very similar to normal 'wear-and-tear' likely to be experienced by moths in the wild; it was recommended as the technique to be adopted.

In conservation, DNA analyses have particular importance in helping to reveal cryptic species, and indicating lineages that could represent distinct species or 'evolutionarily significant units' (ESUs). Amongst many Lepidoptera undocumented cryptic species, commonly undetected or undetectable amidst overlapping biologies and patterns of physical and temporal appearance, are undoubtedly numerous. Small differences may indicate the presence of these, with subsequent focused mtDNA investigation then aiding their recognition and diagnosis. But extending the number of individuals analysed – for example by including those from additional localities – will tend to increase the number of differences found, so differentiating local populations more evidently through additional mtDNA haplotypes. DNA barcodes can thus be valuable both in identifying cryptic species (sometimes indicated by earlier morphological or biological study) and initiating awareness of other variation. Interpreting transition series, for example, and inferring places of origin and from which dispersal and differentiation have occurred may have conservation importance in designating 'significant' populations of a taxon. As Clarke (2000) discussed for the Golden sun-moth (*Synemon plana*, Castniidae) in southern Australia, mtDNA sequence data (and allozyme analyses) render it possible to infer historical population processes such as bottlenecks and fragmentation (Chapter 6), so have applications well beyond the concern for taxonomic integrity emphasised here.

Continually increasing numbers of accounts reveal genetically distinct populations within apparently well defined species, and progressively help to indicate patterns of gene flow, past dispersal and relatedness. The variations and unexpected novelties detected emphasise the complex genetic structures of many endangered 'species' and can help to explain differences across a species' range, where different populations may be considered of very different conservation significance. Thus, the two populations of the Mormon metalmark butterfly (*Apodemia mormo*) in Canada have been given different status as 'endangered' (British Columbia population) and 'threatened' (Saskatchewan population). Proshek et al. (2013) showed that these populations are not closely related, with the Rocky Mountains apparently an important barrier to gene flow. Although the populations do not merit specific separation, differences in habitat and host plants indicate some clear and consistent ecological distinctions. The Saskatchewan population is the more genetically diverse, so that the depauperate genetic structure of the British Columbia population indeed signals need for a higher conservation status. Proshek et al. therefore endorsed the above difference in ranking for the populations, whilst noting that both merit continued protection because of their regional distinctiveness.

Gaining consensus on taxonomic limits in order to define 'species' generally precedes delimiting ESUs in conservation, but the two levels of investigation are often intricately intertwined. The nature of practical application is illustrated through a DNA study of the endangered Karner blue butterfly (Lycaenidae) in North America (Gompert et al. 2006). Although now formally designated a full species (*Plebejus samuelis*), Karner blue is more commonly known as a subspecies of the Melissa blue (*Lycaeides melissa melissa*), as *L. m. samuelis* (Packer et al. 1998). The Karner blue has undergone about 99% range-wide decline, mostly over the last decades of the twentieth century, and remnant populations are now restricted to a few localities in the eastern half of the continent (Fig. 2.1). *L. m. melissa* remains widespread and is not threatened. Ambiguity has occurred over the precise taxonomic status of the western populations of Karner blue, to the west of Lake Michigan, with the possibility that the lake might have been a barrier preventing mixing of these populations with the more easterly ones. If these groups are indeed different they may need to be treated separately for conservation. Gompert et al. (2006) found little evidence for separate origins of the two regional groups, and suggested that they do not need separate management – although they recommended further study before undertaking any interpopulation translocations across the lake.

Nevertheless, cases paralleling *A. 'fulgerator'* are probably not exceptional in the complexity they reveal. More general suggestions over several decades that in many insect groups cryptic species may substantially outnumber 'recognisable species' emphasise needs to consider these as consistently as possible in conservation evaluation, not least in defining 'threatened taxa' in some uniformly acceptable manner. Evaluating diversity is far more than simply counting names, but whatever names exist must be applied as correctly as possible. For many groups

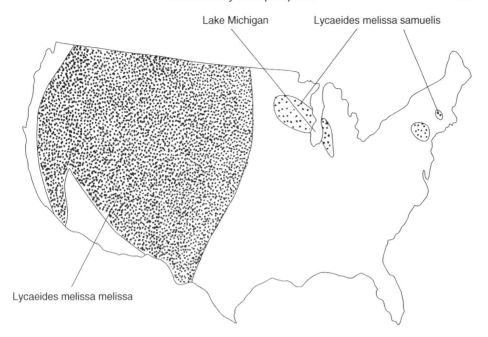

Fig. 2.1 Population distributions of the Karner blue, *Lycaeides melissa samuelis* (paler shading, east), and the more widespread *L. m. melissa* (darker shading, west) (Source: Gompert et al. 2006. Reproduced with permission of John Wiley & Sons.).

of Lepidoptera, and many others, this can be achieved only after considerable revisionary study. Revisions commonly reveal synonyms (so reducing the number of real species) as well as new taxa. However, and important because the biases may not be appreciated widely, the number of species presented from an analysis can vary enormously depending on the 'species concept' adopted and the tendency of the enumerating specialist to amalgamate or separate similar forms. Over-separation, for example, by elevating local populations to distinct subspecies or species can create 'political distinctiveness' for conservation, but may in some cases have little biological significance. The species concept employed is often not clear from reading accounts, and Frankham et al. (2012) recommended that this should be remedied in the interests of clarity in cases where (1) new threatened species are described; (2) where taxonomic disputes are resolved; and (3) where boundaries for threatened species are described. Their essay, emphasising that an ideal species concept should consider carefully the genetic consequences of translocations (Chapter 10), renders the degree of reproductive isolation a critical consideration. Many such ambiguities arise within Lepidoptera, for which species and infraspecies limits are most commonly not considered in this way, other then by inference.

Historically, very many Lepidoptera have been described and named from very few specimens (many from a single individual); they cannot reflect the possible variations present. Isolated descriptions are also outside the context of revisionary work on the families or genera involved, in which all available material and, wherever possible, newly collected specimens are investigated: new material is often critical to augment and redress historical inadequacies. Thus, for Borneo, in an 18 volume account of immense value in interpreting the larger moths of south-east Asia, Holloway (1986–2011) incorporated historical information with large new collections to produce new perspectives on the fauna. For a representative group within one of the largest families, Geometridae: Ennominae, that study led to description of 88 new species or subspecies, but also involved 56 new synonyms of previously described taxa. At a higher taxonomic level, 15 new genera of Ennominae were accompanied by 35 new generic synonyms – so that increased species richness was accompanied by reduced genus-level diversity, but in addition the concept of numerous species recognised by previous authors has changed considerably.

Simply because of relative levels of attention, numbers of future synonyms are likely to be fewer for many Microlepidoptera, especially in relation to greater numbers of new taxa. However, a few groups have demonstrated the complexities that may occur, mostly either or both of pest taxa or apparently polyphagous taxa with supposed 'host races' feeding on different plant species, and for which (as for *Helicoverpa*, p. 20) correct identification is needed for focused management. The Small ermine moths (*Yponomeuta*, Yponomeutidae) are one classic example, and were studied intensively in Europe from the 1970s to resolve the ambiguity of whether the complex comprised one polytypic species or a number of different entities (Menken 1989, for summary). Five distinct biological species were confirmed, with the different host–plant 'races' indeed distinct.

Enumerating valid species is a key facet of documenting diversity and, for Lepidoptera as for many other taxa, the actual species concept used may differ between authors or studies, hampering or confusing comparative interpretations. The above are simply examples of problems that may arise, and demonstrate the need for consensus or, at least, awareness that differences arise from reliance on different species concepts. The greatest need in practical conservation is simply awareness of such interpretative problems and ambiguities. As each individual case arises, it is wise to seek the most informed specialist advice available. This itself may be very limited and tentative for some groups or regions but, conversely, may be sufficient to lend confidence and authority to the endeavour. In particular, many designated subspecies of Lepidoptera may represent 'evolutionarily significant units' but clearly provide lower numbers than, for example, use of the phylogenetic species concept that recognises all/any diagnosable populations as species – so that isolated subspecies such as those on many islands (and parallel 'island habitats' such as mountains) are then full species. The distinction has much practical relevance in planning to reduce losses, and in selection of areas for protection toward that end. Whatever boundaries or criteria are elected for distinguishing taxa, these must be made very clear to users of that information and, for the foreseeable future, conventional morphology-based ('typological') species boundaries accompanied by acceptance of some subjectivity in infraspecific categories supported by molecular and other data is likely to prevail, as the major current practice. A 'working consensus' on identity and integrity of a taxon by experts is accepted for recognition by some legal instruments in designating and listing threatened taxa. However, there are historical cases of putative subspecies and other categories nominated for listing being rejected on grounds of imprecision. One case is for the Altona skipper butterfly (*Hesperilla flavescens flavescens*) in Victoria, Australia, for which a nomination for listing under the state Act of the distinctive and highly restricted populations representing 'the phenotype equivalent to the type' was rejected as representing an entity below subspecies. Adoption of the full subspecies epithet in a later nomination overcame this setback. More generally, the relevance of subspecies in conservation depends on being able to distinguish them and, as Braby et al. (2012) have suggested, each should have significance in contributing to evolutionary understanding.

As emphasised earlier, information on butterflies is the most reliable, but with new taxa continuing to be described from both reorganisation of 'existing' species and novel discoveries. However, in the opinion of Kristensen et al. (2007), butterflies are presumably 'the largest invertebrate clade for which the species inventory is nearing completion', a status that renders them invaluable in conservation and biodiversity studies. In addition they illustrate well some further, sometimes confusing, aspects of diversity that have caused historical confusions in species recognition, and in some instances continue to do so. Many butterfly species occur in several very different-looking forms, as polytypic, not simply as subspecies but reflecting within-population biological traits such as seasonal polyphenism (so that the adults from different seasonal generations look dramatically different, and have sometimes been designated separate species), or polymorphism (such as in some mimetic complexes, particularly in the tropics). Hybridisation can also cause interpretative problems and, even amongst the well known European butterflies, confusion and disagreement over status of some taxa persists and seems unlikely to be resolved firmly in the near future (Descimon & Mallet 2009).

Such lessons from the best known regional group of Lepidoptera indicate the likely magnitude of parallel complexity within the lesser known moths. However, the predominant variations that have drawn attention to butterflies, of colour and wing pattern – whilst paralleled and documented in many diurnal moths – may be particularly evident in those, because diurnal taxa rely heavily on visual signals and communication for their wellbeing (Kristensen et al. 2007). In that case, variation might be generally less amongst crepuscular and nocturnal moths, notwithstanding wide dependence on crypsis whilst they are at rest.

Information on diversity and phylogeny, as well as distribution patterns, constitutes much of the templates for characterising local to regional faunas and distinguishing the unique or special features of each. Many families are cosmopolitan, but very few species extend as widely. Local endemism is high, with some taxa (1) very isolated taxonomically, (2) constituting diverse local radiations as characteristic elements of the region they inhabit, (3) or being unique denizens with complex associations with local biota. Many are also scarce, or rarely encountered, and narrow range (even, point) endemism is inferred frequently, with numerous species being known from single individuals or single populations and – in fewer cases – targeted searches failing to reveal additional specimens or incidences. Local radiations can be enormous. The moth subfamily

Oecophorinae (Gelechioidea: Oecophoridae) in Australia includes at least 20% of the continent's moth fauna (Common 1994), with many of the more than 5000 species so far recognised of these 'mallee moths' known from few localities or specimens. For these, as for so many other insects, lack of intensive collecting renders it impossible to confirm whether these moths are as restricted or as rare as they appear to be, or simply under-investigated. Such inferences extend for various moth groups to virtually all of the world beyond parts of the northern temperate zone. Assessing conservation status on the primary grounds of small distribution and low abundance is often unreliable; many such taxa are necessarily 'data deficient' and in need of further focused attention.

Drivers of diversity

In very general terms, high diversity of insect life is most closely associated with resource-rich environments within climate regimes that support development and activity. For Lepidoptera, this broadly parallels decreases from highly productive tropical environments along latitudinal and elevational gradients, as reflecting vegetation richness and progressively cooler and increasingly seasonally variable physical regimes. Such patterns can be investigated by studies of widely distributed groups, and clear definition of the predominant resource needs: Lepidoptera and plants are a very suitable template, so that correlations of butterfly and moth species richness and distributions with those of flowering plants are a key general theme in understanding their ecology, and the wellbeing of the plants often critical in conservation. The various forms of aphytophagy in Lepidoptera all represent divergences from phytophagous lineages.

Within this broad, and overly simplistic, framework, several main scenarios relevant to conservation emerge.

1   Vegetation systems that are rich in species and structurally diverse may support many more Lepidoptera than simpler or more structurally uniform systems.
2   Local or endemic plant radiations or taxa may form co-evolutionary systems with insect herbivores, so that Lepidoptera may diversify or radiate in parallel with the plants.
3   Any plant species or other taxon may support species of Lepidoptera that are monophagous, and depend wholly or very largely on the host(s). Other Lepidoptera are polyphagous, but in many species that polyphagy is taxonomically restricted within angiosperm groups.
4   Any restricted vegetation type, and any other defined or recognisable biome may support characteristic Lepidoptera restricted to it – so that systems such as peat bogs, mangrove swamps or alpine grasslands (and many others) each have characteristic communities of Lepidoptera that occur only in those biomes.

These themes clearly overlap in many ways. 'Extreme environments' support relatively few species, which may have highly unusual adaptations to persist and thrive there. High latitudes, for example, constrain insect development by cool temperatures, short growing seasons for few species of potential food plants, and seasonal photoperiods. The Arctic woolly bear moth (*Gynaephora groenlandica*, Lymantriidae) is the most northerly of all Lepidoptera and has been described as the largest terrestrial invertebrate reaching the northern limits of vegetation (Kukal 1993). Caterpillars feed for only about 3 weeks in June, following snowmelt, on the longlived prostrate dwarf arctic willow, *Salix arctica*. At that time food quality is high, parasitoid attack is low and the larvae can bask openly to thermoregulate. Early suggestions that development could take 14 years were later modified (Morewood & Ring 1998) to an estimated 7 years with many of the seven larval instars taking a year to grow, and each being inactive for most of the year.

Elsewhere, lower diversity climax or subclimax vegetation types may support local radiations of Lepidoptera, many of them likely to be local endemics. For example, the tussock grasslands of New Zealand's South Island harbour a rich moth fauna: White (2002) reported 446 light-trapped species from his surveys there and, whilst many of these also occur elsewhere, a substantial proportion of species are restricted to this vegetation type. Many families are represented.

The second theme above is one that receives considerable attention in considering patterns of evolution and endemism, with it implicit that any lepidopteran specialised to exploit localised plants has its maximum distribution defined by the greatest range of that/those plant(s), and often with climatic boundaries influencing whether occupancy is possible. Again from the southern hemisphere, the modern vegetation of Australia is highly diverse, but dominated by proliferation of two rather different and widespread genera, *Eucalyptus* (and its close allies, Myrtaceae) with around 600 species and *Acacia* (Mimosaceae) with, perhaps, more than 1000 species, these together comprising more than 10% of the vascular plant species present. Both are linked with complex diversification of Lepidoptera, with considerable radiations of species associated with these predominant flora, and parallel cases throughout the world are major components of locally endemic Lepidoptera. Australia's abundant Oecophoridae, discussed earlier in this chapter, include many species restricted to Myrtaceae and, unusually, adapted to feeding on dead foliage (including fallen leaves) as an important suite of decomposers through this specialised form of herbivory. Dead eucalypt foliage can otherwise persist for many months, and Oecophoridae are an ecologically significant contributor to facilitating material recycling in these environments. Other such keystone roles, often undocumented, are a powerful incentive for conservation under guises such as 'ecosystem services', many of direct benefit to humanity.

In general, Lepidoptera species richness tends to increase with increased habitat heterogeneity, supporting a general paradigm of resource/structural diversity supporting greater biotic diversity – because most species do not occur in all biotopes, addition of new resources is associated with increased regional Lepidoptera richness. However, Kerr et al. (2001) showed that within individual biotopes, Lepidoptera species richness increases with potential evapotranspiration independently of habitat heterogeneity, although the latter remains the primary determinant of species richness.

References

Braby, M.F., Eastwood, R. & Murray, N. (2012) The subspecies concept in butterflies: has its application in taxonomy and conservation biology outlived its usefulness? *Biological Journal of the Linnean Society* **106**, 699–716.

Brower, A.V.Z. (2010) Alleviating the taxonomic impediment of DNA barcoding and setting a bad precedent: names for ten species of 'Astraptes fulgerator' (Lepidoptera: Hesperiidae: Eudaminae) with DNA-based diagnoses. *Systematics and Biodiversity* **8**, 485–491.

Burns, J.M., Janzen, D.H., Hajibabaei, M., Hallwachs, W. & Heberet, P.D.N. (2008) DNA barcodes and cryptic species of skipper butterfly in the genus *Perichares* in Area de Conservacion Guanacaste, Costa Rica. *Proceedings of the National Academy of Sciences* **105**, 6350–6355.

Clarke, G.M. (2000) Inferring demography from genetics: a case study of the endangered golden sun moth, *Synemon plana*. pp. 213–225 in Young, A.G. & Clarke, G.M. (eds) Genetics, Demography and Viability of Fragmented Populations. Cambridge University Press, Cambridge.

Common, I.F.B. (1994) Oecophorine Genera of Australia. I. The *Wingia* group (Lepidoptera: Oecophoridae). CSIRO, Melbourne.

Descimon, H. & Mallet, J. (2009) Bad species. pp. 219–249 in Settele, J., Shreeve, T., Konvicka, M. & Van Dyck, H. (eds) Ecology of Butterflies in Europe. Cambridge University Press, Cambridge.

Deyrup, M., Deyrup, N.D., Eisner, M. & Eisner, T. (2005) A caterpillar that eats tortoise shells. *American Entomologist* **51**, 245–248.

Dinca, V., Zakharov, E.V., Hebert, P.D.N. & Vila, R. (2011) Complete DNA barcode reference library for a country's butterfly fauna reveals high performance for temperate Europe. *Proceedings of the Royal Society, series B* **278**, 347–355.

Elias, M. Hill, R.I., Wilmott, K.R. et al. (2007) Limited performance of DNA barcoding in a diverse community of tropical butterflies. *Proceedings of the Royal Society, series B* **274**, 2881–2889.

Feinstein, J. (2004) DNA sequence from butterfly frass and exuviae. *Conservation Genetics* **5**, 103–104.

Frankham, R., Ballou, J.D., Dudash, M.R. et al. (2012) Implications of different species concepts for conserving biodiversity. *Biological Conservation* **153**, 25–31.

Gompert, Z., Nice, C.C., Fordyce, J.A., Forister, M.L. & Shapiro, A.M. (2006) Identifying units for conservation using molecular systematics: the cautionary tale of the Karner blue butterfly. *Molecular Ecology* **15**, 1759–1768.

Hajibabaei, M., Janzen, D.H., Burns, J.M., Hallwachs, W. & Hebert, P.D.N. (2006) DNA barcodes differentiate species of tropical Lepidoptera. *Proceedings of the National Academy of Sciences* **103**, 968–971.

Hebert, P.D.N., Penton, E.H., Burns, J.M., Janzen, D.H. & Hallwachs, W. (2004) Ten species in one: DNA barcoding reveals cryptic species in the neotropical skipper butterfly *Astraptes fulgerator*. *Proceedings of the National Academy of Sciences* **101**, 14812–14817.

Holloway, J.D. (1986–2011) The Moths of Borneo (complete in18 volumes published over this period; variously Malayan Natural History Society, and Southdene Sdn Bhd, Kuala Lumpur).

Kerr, J.T., Southwood, T.R.E. & Cihlar, J. (2001) Remotely sensed habitat diversity predicts butterfly species richness and community similarity in Canada. *Proceedings of the National Academy of Sciences* **98**, 11365–11370.

Kristensen, N.P., Scoble, M.J. & Karsholt, O. (2007) Lepidoptera phylogeny and systematics: the state of inventory in moth and butterfly diversity. *Zootaxa* **1668**, 699–747.

Kukal, O. (1993) biotic and abiotic constraints on foraging of arctic caterpillars. pp. 509–522 in Stamp, N.E. & Casey, T.M. (eds) Caterpillars. Ecological and Evolutionary Constraints on Foraging. Chapman & Hall, New York.

Lewis, O.T. & Senior, M.J.M. (2011) Assessing conservation status and trends for the world's butterflies: the Sampled Red List Index approach. *Journal of Insect Conservation* **15**, 67–76.

Matthews, M. (1999) Heliothine Moths of Australia. CSIRO, Melbourne.

Menken, S.B.J. (1989) Electrophoretic studies on geographic populations, host races and sibling species of insect pests. pp. 181–202 in Loxdale, H.D. & Den Hollander, J. (eds). Electrophoretic Studies on Agricultural Pests. Oxford University Press, Oxford.

Morewood, W.D. & Ring, R.A. (1998) Revision of the life history of the High Arctic moth *Gynophora groenlandica* (Wocke) (Lepidoptera; Lymantriidae). *Canadian Journal of Zoology* **76**, 1371–1381.

Packer, L., Taylor, J.S., Savignano, D.A., Bleser, C.A., Lane, C.P. & Sommers, L.A. (1998) Population biology of an endangered butterfly, *Lycaeides melissa samuelis* (Lepidoptera: Lycaenidae): genetic variation, gene flow, and taxonomic status. *Canadian Journal of Zoology* 76, 320–329.

Powell, J.A. (1976) A remarkable new genus of brachypterous moth from coastal sand dunes in California (Gelechioidea, Scythrididae). *Annals of the Entomological Society of America* 69, 325–339.

Powell, J.A. & Opler, P.A. (2009) Moths of Western North America. University of California Press, Berkeley and Los Angeles.

Proshek, B., Crawford, L.A., Davis, C.S., Desjardins, S., Henderson, A.E. & Sperling, F.A.H. (2013) *Apodemia mormo* in Canada: population genetic data support prior conservation ranking. *Journal of Insect Conservation* 17, 155–170.

Pyle, R.M. (2012) The origins and history of insect conservation in the United States. pp. 157–170 in New, T.R. (ed.) Insect Conservation: Past, Present and Prospects. Springer, Dordrecht.

Regier, J.C., Zwick, A., Cummings, M.P. et al. (2009) Toward reconstructing the evolution of advanced moths and butterflies (Lepidoptera: Ditrysia): an initial molecular study. *BMC Evolutionary Biology* 9: 280. doi: 10.1186/47 1-21 48-9-280.

Robinson, G.S., Tuck, K.R. & Shaffer, M. (1994) A Field Guide to the Smaller Moths of South-East Asia. Malayan Nature Society, Kuala Lumpur.

Sands, D.P.A. & New T.R. (2002) The Action Plan for Australian Butterflies. Environment Australia, Canberra.

Scoble, M.J. (1992) The Lepidoptera: Form, Function and Diversity. Oxford University Press, Oxford.

Shirt, D.B. (ed.) (1987) British Red Data Books 2. Insects. Nature Conservancy Council, Peterborough.

Silva-Brandao, K.L., Lyra, M.L. & Freitas, A.V.L. (2009) Barcoding Lepidoptera: current situation and perspectives on the usefulness of a contentious technique. *Neotropical Entomology* 38, 441–451.

Stubbs, A., Shardlow, M. (2012) The development of Buglife – the Invertebrate Conservation Trust. pp. 75–105 in New, T.R. (ed.) Insect Conservation: Past, Present and Prospects. Springer, Dordrecht.

Trowell, S.C., Lang, J.C. & Garsia, K.A. (1993) A *Heliothis* identification kit. pp. 176–179 in Corey, S.A., Dall, D.J. & Milne, W.M. (eds) Pest Control and Sustainable Agriculture. CSIRO Publishing, Melbourne.

Tuberville, T.D., Clark, E.E., Buhlmann, K.A. & Gibbons, J.W. (2005) translocation as a conservation tool: site fidelity and movement of repatriated gopher tortoises (*Gopherus polyphemus*). *Animal Conservation* 8, 349–358.

Vila, M., Auger-Rozenberg, M.A., Goussard, F. & Lopez-Vaamonde, C. (2009) Effect of non-lethal sampling on life-history traits of the protected moth *Graellsia isabellae* (Lepidoptera: Saturniidae). *Ecological Entomology* 34, 356–362.

White, E.G. (2002) New Zealand Tussock Grassland Moths. A taxonomic and ecological handbook. Manaaki Whenua Press, Lincoln, New Zealand.

Wiemers, M. & Fiedler, K. (2007) Does the DNA barcoding gap exist? A case study in blue butterflies (Lepidoptera: Lycaenidae). *Frontiers in Zoology* 4: 8, doi:10.1186/1742-9994-4-8.

3

Causes for Concern

Introduction: Historical background

Concerns for the wellbeing of Lepidoptera are not new, but have accelerated markedly since the middle of the twentieth century, as a component of generally heightened awareness of the problems faced by so many organisms in the face of human advance. They have become the major and most influential component of conservation concern for insects. Evaluating or listing the numerous factors that either have led to declines, or have potentials to do so, incorporates a litany of human cupidity and other environmental changes. Many of the main 'threats' are outlined in Chapter 12; this short chapter is simply to introduce the wider perspective that has generated major interests in Lepidoptera conservation and driven actions to recover or preserve taxa and the habitats they frequent.

During the nineteenth century, groups of collectors – including the forerunners of the earliest entomological societies – provided initial focus for exchange of information and ideas, and comments on declines of particular, mainly desirable butterflies commenced from that era. Some became extinct. The most notorious, and most intensively discussed of these is the British subspecies (*Lycaena dispar dispar*, Chapter 9, p. 170) of the Large copper butterfly (Lycaenidae), as one of the earliest documented extinctions. Energetic debate persisted over whether this loss, from the fenlands of eastern England, was due primarily to habitat drainage or to overcollecting (Duffey 1968, Pullin 1995). Consensus generally supports the first of these as the main cause of decline and caused fragmentation of the copper's range into small isolated populations, whence secondary causes (including overcollecting and stochastic influences) contributed to extinction. This general trend of species vulnerability to habitat loss pervades much of insect conservation concern (Chapters 7 and 8). However, whilst it is

Lepidoptera and Conservation, First Edition. T.R. New.
© 2014 John Wiley & Sons, Ltd. Published 2014 by John Wiley & Sons, Ltd.

suspected that many Lepidoptera may have disappeared globally, rather few cases have been as clearcut as the Large copper. The United Kingdom is one of very few places where such extinctions can be documented reliably, simply because of interest in all then known moths and butterflies flowing from at least the middle of the nineteenth century. Fifty-eight species of moths have become extinct, or possibly extinct, in the twentieth century (Parsons 2003, 2004), and interpreting the causes of such declines is a central conservation theme. They are paralleled elsewhere, occasionally dramatically, as previously superabundant taxa are lost. The Cotton leafworm moth (*Alabama argillacea*, Noctuidae) was at one time among the most prolific migratory insects in eastern North America, but has now all but disappeared from the United States. The decline was likened (Wagner 2009) to losses of the passenger pigeon and Rocky Mountain locust. The moth is not known to have any diapausing state and breeds throughout the year: with heavy pesticide applications to cotton crops and high levels of Bt-cotton varieties, accompanied by moves away from cotton growing, conditions throughout much of its former US range are no longer supportive. Wagner commented that 'the species' days may be numbered' there following its catastrophic collapse from the mid-twentieth century.

The first reported loss of a butterfly in North America was the satyrine *Cercyonis sthenele*, from Marin County, California, in the late 1800s (Arnold 1983a).

Extinctions and declines

Large populations of Lepidoptera can decline rapidly, usually because of habitat changes, but a combination of environmental and genetic factors is commonly involved, so that the precise or predominant cause of loss is often not wholly clear. Species' life-history traits are a central consideration, and may relate closely to features of favourable habitats through long honing of adaptations. The loss of the Large blue (*Maculinea arion*) in Britain has been documented fully (Thomas 1991) as reflecting unexpectedly high ecological specialisation of caterpillars and need for finely regulated grazing regimes to maintain the short sward height sufficient to support the host ant, *Myrmica sabuleti*, knowledge that came too late to prevent the butterfly's demise. The decline of *Chazara briseis* (Nymphalidae) in steppeland grass areas of central and eastern Europe, has been clarified by a multifaceted study investigating life-history pattern, demographics and population structure, using mark–release–recapture, direct behavioural observations and allozyme electrophoresis (Kadlec et al. 2010). Many female butterflies (around half the total present) die during a prolonged period between mating and oviposition, decreasing the effective population size markedly. Viable populations must be sufficiently large to withstand that level of loss, so may need correspondingly large habitat patches, with these succumbing progressively to land use changes. However, Kadlec et al. implied that patch fragmentation had not yet led to reduced genetic diversity, suggesting that management to increase and restore habitat may still be able to 'capitalise' on this favourable condition.

Most such declines, however, have been noticed most clearly for the insects that people actively seek or notice, so that collector 'lore' has for long been a major source of concern through information and inference and, correspondingly, that much of the information has come from observations on butterflies in regions with many resident enthusiasts and well established collector tradition of annual (or other regular) visits to traditional localities where rarer or more desirable species occur. Many of the species (and, for butterflies, putative subspecies) most effectively documented are those traditionally deemed 'rare' (and, so, desirable) and exhibiting one or more of low abundance, very limited distribution (often as narrow range endemics) and ecological specialisation. They are thus also generally amongst the most vulnerable and easily threatened taxa, with changes likely to be observed by hobbyists, and become of conservation concern.

The closest North American parallel to *L. dispar* is another lycaenid, the Xerces blue (*Glaucopsyche xerces*), that declined around San Francisco due to habitat changes from around 1875 (Pyle 1995, 2012) and became nationally extinct in the early 1940s.

In parallel with *L. dispar*, and around the same time, the fenland Rosy marsh moth (*Coenophila subrosea*, Noctuidae) also became extinct in England – and, seemingly, from similar causes. However, the subsequent conservation trajectories of the two species differ considerably. *L. dispar* became the focus of long-term reintroduction attempts using European subspecies and habitat enhancement and protection (Chapter 9). In contrast, *C. subrosea* was rediscovered in the mid-1960s, in an estuarine mire in west Wales (chronicled by de Worms 1968), leading to considerable interest and exploration amongst lepidopterists keen to obtain specimens. Further discoveries increased its known distribution to include five raised bog sites in Wales, so that subsequent conservation has focused on these isolated but resident populations (Fowles et al. 2004).

Declines amongst the British Lepidoptera have been well documented for the last century and more, with a recent perspective of larger moths (Fox et al. 2013) revealing that a further four species might have become extinct there over the first decade of this century. Although countered by 27 species newly recorded from Britain over the same period, the declines of 61 species by more than 75% (including 20 by 90% or more) implied from light trap surveys (Chapter 5, p. 55) over the last 40 years is dramatic. The most extensive example cited by Fox et al. (2013) is of the V-moth (*Macaria wauaria*, Geometridae), with an estimated 99% decline, possibly linked to decreased cultivation of its major food plants (currants and gooseberries) and insecticide use. Despite rather less complete information, it seems that many Microlepidoptera in Britain may also be of conservation concern, with the overview by Davis (2012) listing tentative status comments on some 1660 species.

In a few cases beyond Britain, long-term interval inspections have also revealed declines. Thus, for southern Sweden, repetition of a survey made from 1904–1913 was undertaken over 2001–2005 (Nilsson et al. 2008), to evaluate the butterflies and burnets (Zygaenidae) present, and any changes that had occurred in relation to changed land uses. Of 48 resident species found early in the twentieth century, 21 had become locally extinct, and five additional species were found in the recent surveys – some of these probably representing northward expansions of range. Species with narrow habitat needs (such as several formerly most abundant in flower-rich forest glades, which no longer existed in the area), short flight periods and narrow European distribution declined most. Losses of hay meadows and other open grasslands, with changes in traditional grazing regimes, were also linked to losses. As another example, assessment of grassland butterflies in Europe from national trends across 17 species collectively showed about 50% decline since 1990 (van Swaay & Warren 2012). However, whilst the reality of species declines are incontrovertible, as are losses of wider diversity of Lepidoptera, the precise causes of these are often very difficult to pinpoint, and clarifying these is a major and recurrent theme throughout this book. Thus, in considering the well documented declines of many larger British moths (Chapter 5, p. 69), Fox (2013) noted that even direct evidence of impacts of historical habitat loss, decreased habitat quality or increased habitat fragmentation on moth diversity or abundance is largely absent – but leaves strong suggestion that changes in land use and management, with consequent losses of breeding sites, are the major cause of moth declines. Likewise, agricultural chemicals have a strongly advocated probability of contributing to declines but, again, in the arena in which the many aspects of chemical pollution have not yet been proven to contribute. These, and other ambiguities, led Fox (2013) to strong inference that multiple factors have driven the declines and, using this best known of all moth faunas, to suggest six major issues to help understand and reverse those declines (Table 3.1): he noted that the requirements are 'substantial and challenging'. Assessing 'risk of extinction' is complex, and hindered by lack of detailed population data. In an effort to help overcome this, Mattila et al. (2006, 2008) explored whether selected ecological features or traits might in some way 'predispose' species to decline and, if so, whether the sharing of those traits with other (currently non-threatened) relatives might indicate priorities for future need. Examination of two large moth families in Finland (Noctuidae: 306 species; Geometridae: 284 species) emphasised that single traits may

Table 3.1 The six major issues and questions posed by Fox (2013) to help understand and reverse the decline of moths in Britain, recognising that the main drivers of change all occur and interrelate.

1. What is the complete picture of change for Britain's moths? (need investigation of contrast in declines between northern and southern regions; determination of primary causes; greater emphasis on micromoths, and expansion of coverage of trends so far investigated for fewer than half the macromoths)
2. Improved understanding of impacts of different elements of agricultural management
3. More research into land management techniques that attempt to mitigate against biodiversity loss (many outcomes at present unclear and mechanisms poorly understood)
4. Targeting of agri-environment schemes for maximum benefit and cost-effectiveness
5. Investigation of the impacts of outdoor, artificial and background light pollution (requires urgent ecological research)
6. Impacts of climate change generally very poorly understood

be inadequate for reliable predictions – but also indicated that such themes merit further investigation.

In a different fauna, the butterflies of south-east Asia, a broad survey of the factors associated with 'extinction proneness' indicated that several factors independently increased risk (Koh et al. 2004). The major positive correlations were with larval host plant specificity (so that the fate of the consumer is linked intricately with that of its key resource species), and adult habitat specialisation (with specialisation or functional dependence on restricted biotopes creating vulnerability). These themes are discussed later (Chapters 7–9). The presence of suitable larval food plants may be a more widespread constraint than adult dispersal or colonisation ability in many Lepidoptera. Ecological correlates with habitat change are clearly highly influential, but the wider spectrum of additional threats, again mostly related to human activities and noted in Chapter 12, must also be considered in determining causes of declines and loss.

Whilst many declines have been noted only relatively recently, the underlying loss of habitat has often been a far longer process – but specific long-term parallel studies of Lepidoptera and key habitat losses are rather few. In an area of Belgium, calcareous grassland declined from 171,306 ha (1905) to 8159 ha (2004), and average patch size from 7.79 ha to 0.82 ha (Polus et al. 2007). At intervals over this period, a pool of 83 butterfly species was inspected, with about 35% of these being specialists on this grassland biotope. Nine of these specialists disappeared over that period, but no generalists became extinct. The pattern of habitat loss, mirrored widely elsewhere as a prime cause of occupant losses, was (1) initial fragmentation of large patches, so that fragment (patch) number increased; and (2) as fragments became smaller, continued habitat loss eliminated fragments, leading to decrease in total number. The perhaps more general outcome of this process, the replacement of specialists by generalists in assemblages is a recurring theme in this book, and was referred to by Polus et al. as 'ecological drift'.

Small distributions so frequent amongst Lepidoptera (and other insects) are likely to increase vulnerability, not least from stochastic impacts such as fires or storms, several examples of which are cited later. Woodhall (2005) used the term 'rarity trap' to describe such potential losses resulting from episodic calamities, and suggested that a number of South African butterflies with extremely restricted distributions may have been lost unnoticed – a supposition becoming more widespread in other places as the extent of narrow range endemism amongst Lepidoptera becomes clearer. Local or more widely encompassing studies of biogeography of Lepidoptera reveal many examples of such narrow range endemism as that noted for *C. subrosea* or, at least, of such being the only distribution known. For many, those distributions are associated clearly with particular biotopes or food resources. Almost all regional faunas contain unique elements (both single taxa and radiations or centres of diversity or evolution), some of which have been explored systematically for patterns of value in setting conservation priorities. Centres of endemicity for several butterfly groups in the Neotropics, for example, can sometimes be seen to coincide with regions in which tropical forests are believed to have persisted during glaciations (Brown 1991), so are local refugia within which suitable conditions persisted. Montane areas in Europe and parts of the tropics have similar origins and significance, as 'archi-pelagic' habitats (Holloway and Nielsen 1999: 439) that foster evolution of distinctive taxa on each isolated site through intrinsic, mainly allopatric, speciation.

As in many aspects of interpreting evolution and biogeography of Lepidop-tera, studies on the European butterflies have suggested generalities that may be much more widespread, and merit investigation elsewhere. Thus, endemic taxa are restricted to latitudes below 52°N, coinciding with southern limits of ice cover during the last major glaciations (Gutierrez 2009). Gutierrez drew on interpretations by Dennis et al. (1995) to comment on three possible lines of evidence for changes during glacial/interglacial cycles: (1) that butterfly habitats in southern Europe have been present for longer than those at higher latitudes, so that northern regions were recolonised during interglacial phases; (2) popula-tions currently isolated in southern Europe were perhaps more isolated during glacial phases than they are now; and (3) endemicity is more frequent in families with substantial proportions of poorly mobile habitat specialists, limiting the extent of evolutionary change resulting from (and depending on) dispersal ability. Climate features then (as now, Chapter 12) were important influences on distribution, with latitudinal and elevational gradients demonstrating some par-allel trends – but with many examples of greatest richness at intermediate eleva-tions, rather than at the lowest levels (Gutierrez 2009). Distribution patterns of European butterflies might reflect both the 'evolutionary time hypothesis' (that species within certain sites have not yet had sufficient time to evolve) and the 'ecological time hypothesis' (that species potentially able to live in certain sites occur, but have not yet had sufficient time to disperse to them) (Dennis et al. 1995, for discussion).

Many authors have commented on the unevenness of insect distributions, diversity and documentation in different regions, but the regional contrasts in knowledge and ability to pursue conservation between (1) northern and southern temperate zones and (2) temperate and tropical regions are both significant (Stewart and New 2007, Lewis and Basset 2007), and are recurrent themes in assessing the origins and spread of conservation interest and practicality in, even, the best known insect taxa. Despite many gaps in knowledge, Lepidoptera are far better documented than any other insect group over much of the tropics, so that distribution of some of the best known families can be analysed in consider-able confidence for patterns of local and regional diversity and endemism, and their peculiarities ranked for conservation merit and need.

Assessing such needs may become far more complex than initially anticipated. Whilst it is widely evident that extinctions or declines are strongly related to habitat fragmentation and loss, leading to impacts on individual taxa and changes in richness and assemblage composition, the observed rapid extinctions may represent only part of the real impacts. Declining populations may persist for some time after the habitat (site, patch) has diminished. Their continued pres-ence under these circumstances can lead to underestimation of the level of endangerment so, in turn, underestimation of the impacts of changes made to their environment. This has been referred to as 'extinction debt' (Bulman et al. 2007, Kuussaari et al. 2009) and has been suggested to be quite common – but with the proviso that, as long as the species is/are permitted to persist, there is still time to institute conservation measures to redress their fate, for example, through habitat restoration. Kuussaari et al. (2009) formally described extinction debt as follows (p. 564), but noted also that its evaluation is 'fraught with uncertainty': 'In ecological communities, the number or proportion of extant specialist species of the focal habitat expected to become extinct as the community reaches a new equilibrium after environmental disturbance. . . . In single species, the number or proportion of populations expected to eventually become extinct after habitat change'. One of the most detailed appraisals of the consequences involved modeling of the metapopulation processes (Chapter 6) for the Marsh fritillary, *Euphydryas aurinia*, across six extant metapopulation networks in Britain (Bulman et al. 2007), for which simulations based on patch occupancy and colonisation/extinction rates allowed calculation of the minimum habitat needed to prevent extinction. Alarmingly, they inferred that there was a greater than 99% chance that one or more of those metapopulations would become extinct within a century – even without any further habitat loss. Four of the six networks carried an extinction debt, with simulations giving median times to extinction of 15–126 years, and with colonisation rates insufficient to sustain them. The major conservation lessons are that extinction debt may be far more widespread than presumed, even if it is considered at all in conservation planning, and that metapopulation modelling may help to distinguish the landscapes that should be targeted for priority. The long 'lag time' of extinctions following habitat loss and fragmentation provides opportunity for remedial actions to reduce extinction rates and increase colonisation rates across patches.

References

Arnold, R.A. (1983a) Ecological studies on six endangered butterflies (Lepidoptera: Lycaenidae): island biogeography, patch dynamics, and the design of habitat preserves. *University of California Publications in Entomology*, 99. Berkeley.

Bulman, C.R., Wilson, R.J., Holt, A.R. et al. (2007) Minimum viable population size, extinction debt, and the conservation of a declining species. *Ecological Applications* 17, 1460–1473.

Davis, A.K. (2012) Are migratory monarchs really declining in eastern North America? Examining evidence from two fall census programs. *Insect Conservation and Diversity* 5, 101–105.

Dennis, R.L.H., Shreeve, T.G. & Williams, W.R. (1995) Taxonomic differentiation in species diversity gradients among European butterflies: contribution of macroevolutionary dynamics. *Ecography* 18, 27–40.

de Worms, C.G.M. (1968) The recent discovery in Wales of the rosy marshy moth *Coenophila subrosea* (Stephens) (Lepidoptera, Noctuidae). *Entomologists' Gazette* 19, 83–89.

Duffey, E. (1968) Ecological studies on the large copper butterfly, *Lycaena dispar batavus*, at Woodwalton Fen National Nature Reserve. *Journal of Applied Ecology* 5, 69–96.

Fowles, A.P., Bailey, M.P. & Hale, A.D. (2004) Trends in the recovery of a rosy marsh moth *Coenophila subrosea* (Lepidoptera: Noctuidae) population in response to fire and conservation management on a lowland raised mire. *Journal of Insect Conservation* 8, 149–158.

Fox, R. (2013) The decline of moths in Great Britain: a review of possible causes. *Insect Conservation and Diversity* 6, 5–19.

Fox, R., Parsons, M.S., Chapman, J.W., Woiwod, I.P., Warren, M.S. & Brooks, D.R. (2013) The State of Britain's Larger Moths 2013. Butterfly Conservation and Rothamsted Research, Wareham, Dorset.

Gutierrez, D. (2009) Butterfly richness patterns and grasslands. pp. 281–295 in Settele, J., Shreeve, T., Konvicka, M. & Van Dyck, H. (eds) Ecology of Butterflies in Europe. Cambridge University Press, Cambridge.

Holloway, J.D. & Nielsen, E.S. (1999) Biogeography of the Lepidoptera. pp. 423–462 in Kristensen, N.P. (ed.) Lepidoptera: Moths and Butterflies. Vol.1. Evolution, systematics and biogeography. Handbuch der Zoologie, Vol. 4, part 35. W. de Gruyter, Berlin and New York.

Kadlec, T., Vrba, P., Kepka, P., Schmitt, T. & Konvicka, M. (2010) Tracking the decline of the once-common butterfly; delayed oviposition, demography and population genetics in the Hermit, *Chazara briseis*. *Animal Conservation* **13**, 172–183.

Koh, L.P., Sodhi, N.J. & Brook, B.W. (2004) Ecological correlates of extinction proneness in tropical butterflies. *Conservation Biology* **18**, 1571–1578.

Kuussaari, M., Bommarco, R., Heikkinien, R.K. et al. (2009) Extinction debt: a challenge for biodiversity conservation. *Trends in Ecology and Evolution* **24**, 564–571.

Lewis, O.T. & Basset, Y. (2007) Insect conservation in tropical forests. pp. 34–56 in Stewart, A.J.A., New, T.R. & Lewis, O.T. (eds) Insect Conservation Biology. CAB International, Wallingford.

Mattila, N., Kaitala, V., Komonen, A., Kotiaho, J.S. & Paivinen, J. (2006) Ecological determinants of distribution decline and risk of extinction in moths. *Conservation Biology* **20**, 1161–1168.

Mattila, N., Kotiaho, J.S., Kaitala, V. & Komonen, A. (2008) The use of ecological traits in extinction risk assessments: a case study on geometrid moths. *Biological Conservation* **141**, 2322–2328.

Nilsson, S.G., Franzen, M. & Jonsson, E. (2008) Long-term land-use changes and extinction of specialised butterflies. *Insect Conservation and Diversity* **1**, 197–207.

Parsons, M.S. (2003) The changing moth and butterfly fauna of Britain during the twentieth century. *Entomologist's Record and Journal of Variation* **115**, 49–66.

Parsons, M.S. (2004) The United Kingdom Biodiversity Action Plan moths – selection, status and progress in conservation. *Journal of Insect Conservation* **8**, 95–107.

Polus, E., Vandewoestijne, S., Choutt, J. & Baguette, M. (2007) Tracking the effects of one century of habitat loss and fragmentation on calcareous grassland butterfly communities. *Biodiversity and Conservation* **16**, 3423–3436.

Pullin, A.S. (ed.) (1995) Ecology and Conservation of Butterflies. Chapman & Hall, London.

Pyle, R.M. (1995) A history of Lepidoptera conservation, with special reference to its Remingtonian debt. *Journal of the Lepidopterists' Society* **49**, 397–411.

Pyle, R.M. (2012) The origins and history of insect conservation in the United States. pp. 157–170 in New, T.R. (ed.) Insect Conservation: Past, Present and Prospects. Springer, Dordrecht.

Stewart, A.J.A. & New, T.R. (2007) Insect conservation in temperate biomes: issues, progress and prospects. pp. 1–33 in Stewart, A.J.A., New, T.R. & Lewis, O.T. (eds) Insect Conservation Biology. CAB International, Wallingford.

Thomas, J.A. (1991) Rare species conservation: case studies of European butterflies. pp. 149–197 in Spellerberg, I.F., Goldsmith, F.B. & Morris, M.G. (eds) Scientific Management of Temperate Communities. Blackwell Publishing, Oxford.

van Swaay, C.A.M. & Warren, M.S. (2012) Developing butterflies as indicators in Europe: current situation and future options. De Vlinderstichting/Dutch Butterfly Conservation, Butterfly Conservation UK, Butterfly Conservation Europe, Wageningen. Report no. VS2012.012.

Wagner, D.L. (2009) Ode to *Alabama*: the meteoric fall of a once extraordinarily abundant moth. *American Entomologist* **55**, 170–173.

Woodhall, S. (2005) Field Guide to the Butterflies of South Africa. Struik Publishers, Cape Town.

4

Support for Flagship Taxa

Introduction

That 'people like butterflies' is not merely a trite observation but one that continues to have immense importance in fostering public and political goodwill and concern over their declines. However, 'liking' and 'wider values' tend to go hand-in-hand, so that the aesthetic appeal of butterflies flows to other aspects of appreciation, including values as tools in science and environmental investigations. These wider values of butterflies in scientific studies, and their critical relevance in clarifying aspects of evolution contributing to human welfare were summarised by Ehrlich (2003) in emphasising the urgency of using these insects in monitoring ecosystem health and assessing the conservation significance of various areas. Some of the examples discussed in this book have been pivotal catalysts for insect conservation interest, with butterflies the most potent 'flagship' group of insects and providing foci for its wider development and acceptance. Two representative gatherings of interest, both now established for more than 40 years, demonstrate the positive outcomes based in initial concerns for butterflies amongst better known faunas. Recent historical accounts of organisations such as 'Butterfly Conservation' in Britain (Warren 2012) and the 'Xerces Society' in North America (Pyle 2012) trace the expansions of interest, the increasingly diverse and expanding memberships, and the influences of these leading invertebrate conservation organisations and the variety of activities they now pursue. 'Butterfly Conservation' has recently expanded to become 'Butterfly Conservation Europe', enabling this important regional fauna to be considered holistically, and not as limited by the strictures of political boundaries and individual legislations. Within the United Kingdom, its activities have expanded to incorporate macromoths. The Xerces Society has a broader remit for invertebrates, but many of its members have interests in Lepidoptera and its foundation was largely driven through concerns over butterflies and moths.

Lepidoptera and Conservation, First Edition. T.R. New.
© 2014 John Wiley & Sons, Ltd. Published 2014 by John Wiley & Sons, Ltd.

Many entomological societies now have embedded 'conservation groups' or committees, amongst which butterflies are amongst the most frequent concerns. An intriguing exception is New Zealand, where butterflies have gained only a low conservation profile, reflecting both the low number of taxa present (although with very considerable variety amongst some complexes: Patrick & Patrick 2012), and the existence of an alternative and accessible superlative flagship insect group, the spectacular weta (Orthoptera) (Watts et al. 2012). However, more generally, butterflies tend to dominate lists of insects designated or scheduled for legal protection as 'threatened' (Chapter 5), as a direct consequence of historically accumulated knowledge founding confidence in assessment of conservation needs, and the reasons for this – including willing community interest in, and support for, their conservation.

In short, the awareness and knowledge of butterfly biology and distributions, especially in the northern temperate regions, renders them both acceptable and amenable to conservation, with public and political sympathy likely to exist and be garnered readily. This chapter examines how that interest has been incorporated into conservation practice.

Community endeavour

Only with effective community networks and willing volunteers is it possible to undertake the intensive or long-term surveys needed to plot distributions of Lepidoptera and assess trends in abundance and conservation need. Fostering that interest is a core component of their conservation, and many local groups dedicated to conservation of particular sites or taxa, as advocates and 'citizen scientists', are essential to augment the very limited professional capability available. Whilst garnering community interests may be far easier to support butterflies than for many other invertebrates, the immediacy of need for some conservation programmes may lead to impulsive inputs. Whether their activities are primarily political activism, survey or active management in site maintenance, two rather different contexts arise. In many places, suggestions to conserve a moth or butterfly are still relatively novel, and the first context is the initial 'excitement' generated by a notable discovery, often accompanied by purported urgency of action if the inhabited site is scheduled for loss, with attendant high profile and media attention and publicity. The second context is the sustained endeavour needed, perhaps over decades, to undertake a continuing practical conservation management programme, likely to comprise routine tasks and largely lacking media interest or sensationalism. The general issues involved (New 2009, 2011) are demonstrated well through butterfly cases, not least because such attractive local icons commonly engender local pride, and this may be enhanced by a patronymic common name that can be adopted readily by a local support group. A 'sense of ownership' by a community group enables local people to identify with the issues involved, and their importance, and may be fostered by representation of all community constituency groups on the species' or site's management committee. Examples such as the longest-running butterfly conservation programmes in Australia (for the Eltham copper, *Paralucia pyrodiscus lucida*, Lycaenidae, in Victoria and the Richmond birdwing, *Ornithoptera richmondia*, Papilionidae, further north) have depended on sustaining community interest and support throughout their existence, now each for more than 20 years. Both the above common names reflect geographical occurrence, and a further Australian case takes this principle further. *Paralucia spinifera*, highly localised in New South Wales, is known both as the Bathurst copper and the Lithgow copper, from the two major 'rival' towns within its range.

Almost every species of butterfly or moth brought to conservation attention in conservation-aware countries garners a support group or defence group in some way. However, for many parts of the world where conservation need is greatest, this support is rare, even impossible: such commitment of leisure time and resources is essentially a product of societies with sufficient leisure and with spare resources that can be devoted to activities that do not immediately and tangibly support individual wellbeing.

However, community support need extends well beyond looking after individual taxa or sites to wider aspects of field recording through participation in multispecies surveys (Chapter 5, p. 56), and annual inspections or counts of selected and easily recognised taxa. The North American 'Fourth of July Butterfly Counts' have become an invaluable annual estimator of butterfly populations across many sites in the continent. Commenced through the Xerces Society (p. 40) in 1975, and now organised through the North American Butterfly Association (NABA), they are based on local initiatives by volunteers who select a 15-mile (25-km) diameter circle within which a group of people conducts a one-day census of all butterflies seen. In 2011, 427 United States counts included 47 states plus the District of Columbia, and 25 Canadian sites across two large provinces. Number of counts varies from year to year and the survey period, although centred on the major holiday date, spans several weeks each side of this to allow for vagaries of weather and of observer availability. The annual summary reports are a dynamic record of butterfly incidences, with details for many natural and anthropogenic biotopes. The initial purposes of the surveys were 'a pleasant and educational experience' and 'an informal tool for studying and monitoring butterfly populations' (Swengel 1990). Clearly, these counts are not fully quantitative, and are subject to varying levels of sampling effort – but they nevertheless provide useful observations on trends, and sustain public involvement. Data from such surveys have many applications in conservation, flowing directly from the quality of information accumulated by observers. Thus, overlaying butterfly data collected through NABA, standardised through rarefaction analyses, with landscape composition (cropland, grassland, woodland, wetland, urban land) showed some clear associations (Meehan et al. 2012). The rarest butterflies, perhaps unsurprisingly, were associated with woodland-, grassland- and wetland-dominated landscapes, those likely to support the specific food plants they need. Parallel trends were found by Kocher and Williams (2000) in butterfly richness being lower in highly altered landscapes, those more than half composed of agricultural, residential, commercial or urban land. Both these North American studies, and others, support the general theme of species losses with intensive land development.

Unlike the Fourth of July counts, the more recently initiated European Moth Night now coordinated annually since 2004 is held at different times each year, to yield some wider information on macrolepidoptera abundance. It has provided information of considerable conservation value, and operates through lepidopterists throughout Europe recording nocturnal moths over one or more of several designated consecutive nights within a 3–5-day period. The data (with requirement that any records of species only doubtfully identified be omitted, and also not to transgress legislation by collecting protected species) are submitted in a clear form, such as an Excel spreadsheet, for ease of handling. This project was initiated jointly by the Jozsef Szalkay Lepidopterological Society of Hungary and the Entomological Society of Luzern and has several stated purposes (Rezbanyai-Reser et al. 2004), namely (1) to bring together a large number of European lepidopterists and others to 'strengthen unity and mutual understanding, while also offering an insight into the lepidoptera-fauna of the different countries and the local methods of collecting; (2) to present a wide-ranging snapshot on the macromoths (used in the traditional sense) with particular attention to species possibly or actually needing protection and the ones traditionally considered as migratory species; with (3) the results, data and their evaluation to be made available to the general public'. The first European Moth Night attracted 154 participants in 21 countries, involving 159 localities in 23 countries and yielding 850 species over 3 nights. By the fourth annual count, 549 participants from 29 countries dealt with 621 localities in 33 countries with 546 moth species recorded across 5 days. More than half the European species of macromoths had been recorded in the first 5 years, and the exercise has continued to expand and contribute to inventory knowledge, with annual evaluations and overviews containing much information relevant to conservation. As with other such surveys even amongst well known faunas, surprises have emerged – perhaps most notable being records of a New Zealand geometrid (*Pseudocoremia suavis*) in southern England, where it is likely to have been imported in plants, perhaps through a nursery near the recorder's site (James, in Rezbanyai-Reser et al. 2009).

The history of the United Kingdom Butterfly Monitoring Scheme was summarised by Asher et al. (2001) and from 1976 this has operated at more than 100 sites throughout the UK, as an invaluable foundation for systematic distribution mapping in the modern era. The information was fundamental to the later 'Butterflies for the New Millennium' project launched in 1995 and which had five major purposes: (1) inform and support statutory measures in nature conservation; (2) inform practical action for the conservation of species and habitats; (3) obtain data for comparison with similar data sets for other fauna and flora and other environmental data; (4) develop better methods and encourage more people to record butterflies effectively; and (5) make appropriate information about butterflies, their habitats and their conservation readily available. The information was thereby accrued within the contexts of education and conservation evaluation, as well as promoting ways in which butterfly information could be integrated more widely rather than being simply 'stand alone'. The resulting Millennium Atlas (Asher et al. 2001) is one of the most significant contributions to butterfly status evaluation, with the extensive historical records unambiguously demonstrating the declines and range changes of many species. The later Butterfly Conservation initiative, the 'Action for Threatened Moths Project', is doing much to raise the profile of UK moths and their conservation, and involves a range of different organisations in implementation of action plans and related documents: Parsons (2004) noted that many thousands of people have become involved in recording and observing moths. The UK's National Moth Recording Scheme aims to 'encourage interest in moths throughout the United Kingdom' and 'improve knowledge and conservation' and is organised through Atropos and Butterfly Conservation. It is paralleled by the United States-initiated 'National Moth Week' to 'promote moths . . . organising events at the local park, environmental education centre, university or homes'. In 2012, participants from 49 United States and 28 other countries took part in a wide variety of events, many of them surveys. Such festivals are doing much to spread awareness of moths and to encourage interest in 'citizen science' through this focus.

The extensive community participation in such programmes can be incorporated into wider appraisals, using butterflies in monitoring environmental changes (Chapter 13, p. 242) and combining recent observations with data from archival collections to gain a longer-term perspective. Many major museums hold large collections of butterflies, whose individual data labels span a century or more of capture date and locality information. As one example, Polgar et al. (2013) used museum records (1893–1985) and recent field survey data (1986–2009) to explore impacts of climate change on date of flight season start for selected Lycaenidae in eastern North America. Various biases may, of course, be hard to detect retrospectively, but Polgar et al. suggested that date of first appearance (a measure of change from climate warming) might be represented adequately in museum accumulations because collectors are at their most diligent in seeking and reporting at that time.

Flagships

The essential feature of 'flagship taxa' is simply their appeal – that, in this context, they help to overcome the generally poor image problems of insects and enlist sympathy and support for conservation (New 2011). Promotion of species as flagships, and extending public goodwill to political willingness to conserve them, demands advocacy and acceptance by local communities. The species must, in some way, be both notable and noticeable. 'Notable' equates largely to conservation need, but can incorporate any unusual or idiosyncratic feature of biology, appearance, distribution or habit likely to promote interest. 'Noticeable' draws on any of these through publicity and widening of awareness. However, the impacts of flagships clearly reflect the local societal context and needs, and have proved of greatest values amongst 'westernised societies'. They have only rarely been appraised elsewhere, and such appeal may be highly culture-specific (Barua et al. 2012), or simply aesthetic. In a wider study, based on questionnaire surveys, that compared responses of tourists and rural area residents in India, Barua et al. found that butterflies were clearly the most popular invertebrates from the selection of different groups proffered for reaction. Large size and bright colours contributed to this, with preference by both residents and tourists based largely on aesthetic appeal amongst the 15 butterfly species presented. As examples, in selecting the 'top three' species by appeal, people listed the birdwing *Troides helena* (Papilionidae) because it was 'dazzling', 'good looking' and 'large in size'; in contrast dull coloration and small size contributed to low rankings. The Rice swift (*Borbo cinnara*, Hesperiidae) and a Bush brown (*Mycalesis* sp., Nymphalidae) were ranked low as 'moth-like', 'dull', 'ugly', and with the image of hesperiids apparently tainted by their angular wings and bulbous eyes (Barua et al. 2012). In contrast to adult butterflies, caterpillars were ranked very low, and Barua et al. noted that negative attitudes towards caterpillars might counter positive ones towards the more familiar and appealing adults.

The widespread goodwill for many butterflies can also have important 'flow-on' effects with the aid of sound and accessible information. Perhaps the most notable group of flagship moths is the hawkmoths (Sphingidae), which Beck et al. (2006) regarded as second only to butterflies in the amount of information available for conservation assessment. Some colourful diurnal moths, for example, have gained the status of 'honorary butterflies' – in some cases even advertised under this enticing but misleading epithet, which is nevertheless welcome in encouraging interest! It is intriguing to ponder implications of reversing this perspective to reflect greater evolutionary reality, so that butterflies become 'honorary moths'; the contrasting public image would almost certainly be detrimental to conservation interests.

A major tourist attraction on the island of Rhodes is the 'Valley of the Butterflies', with large numbers of visitors viewing the spectacular aestivating aggregations of the Jersey tiger moth (*Euplagia quadripunctaria*, Arctiidae) present from June to September. The phenomenon figures prominently in advertising of hotels and tourist bodies for the island and the moth's appeal is clearly a substantial contributor to local revenue. However, the massive number of moths present necessitates some regulation of tourist numbers and behaviour, with tourism activities themselves contributing to decline of *Euplagia* through direct disturbance and trampling of vegetation, as well as unauthorised collection (Petanidou et al. 1991). These authors estimated that, conservatively, 100,000 individual moths were removed by tourists each year, so that the abundant opportunities for providing information on conservation and moth biology, together with rehabilitation of the *Liquidambar* forest habitat, also incorporate features such as fencing (to exclude people and grazing goats), increased vigilance and controls over noises that induce restlessness amongst moths.

Such 'superabundance' of particular species is highly attractive to tourists as a spectacle, and to local proprietors as a source of economic reward, but such concentrations also create vulnerability (Chapter 6, p. 108). Aggregations constitute critical phases of the species' ecology. By far the best known are the overwintering populations of the monarch or wanderer butterfly, *Danaus plexippus* (Nymphalidae, Danainae) in California and Mexico, where different geographical segregates of this strongly migratory North American insect gather in enormous numbers. Both those overwintering areas are vulnerable to human pressures, and each is vital because 'entire populations coalesce from almost continent-wide distributions to a few tiny overwintering areas. . . . It is these overwintering areas that are threatened by our sociological, political and economic inability to solve ecologically simple conservation problems' (Malcolm 1993). More has been written on this designated 'threatened phenomenon' (terminology of Wells et al., 1983, in which 'aggregations or populations of organisms that together constitute major biological phenomena, endangered as phenomena but not as taxa') than on any other butterfly species of conservation concern (for example, see Malcolm & Zalucki 1993; Oberhauser & Solensky, 2004). And, as the latter put it 'No other nonpest insect has attracted as much attention as the monarch butterfly'.

The two taxa noted in the preceding paragraphs are both brightly coloured diurnal species. In contrast, Australia's Bogong moth (*Agrotis infusa*, Noctuidae) is nocturnal and rather drab, but has also gained notoriety through its massive aestivating aggregations in the subalpine country of the south-east mainland, traditionally a source of food for local aboriginal people. Its recent fame stems, rather, from large numbers being attracted to light during the annual migrations, so that it becomes regarded as a nuisance through invading venues as diverse as the Commonwealth Parliament building in Canberra (with a Parliamentary Report addressing problems of controlling its entry: McCormick, 2005) and major sporting arenas hosting evening events. Any such notoriety involving public notice, even if not wholly sympathetic, accords flagship values; as a phenomenon paralleling those noted above, the Bogong moth has become one of the very few Australian moths notionally familiar to many politicians! It is unfortunate that such familiarity is associated primarily with its nuisance value rather than admiration of its migratory accomplishments.

References

Asher, J., Warren, M., Fox, R., Harding, P., Jeffcoate, G. & Jeffcoate, S. (2001) The Millennium Atlas of Butterflies in Britain and Ireland. Oxford University Press, Oxford.

Barua, M., Gurdak, D.J., Ahmed, R.A. & Tamuly, J. (2012) Selecting flagships for invertebrate conservation. *Biodiversity and Conservation* 21, 1457–1476.

Beck, J., Kitching. I.J. & Linsenmair, K.E. (2006) Effects of habitat disturbances can be subtle yet significant: biodiversity of hawkmoth assemblages (Lepidoptera: Sphingidae) in south-east Asia. *Biodiversity and Conservation* 15, 451–472.

Ehrlich, P.R. (2003). Introduction: butterflies, test systems and biodiversity. pp. 1–6 in Boggs, C.L., Watt, W.B. & Ehrlich, P.R. (eds) Butterflies. Ecology and Evolution Taking Flight. University of Chicago Press, Chicago and London.

Kocher, S.D. & Williams, E.H. (2000) The diversity and abundance of North American butterflies vary with habitat disturbance and geography. *Journal of Biogeography* 27, 785–794.

Malcolm, S.B. (1993) Conservation of monarch butterfly migration in North America: an endangered phenomenon. pp. 357–361 in Malcolm, S.B. & Zalucki, M.P. (eds) Biology and Conservation of the Monarch Butterfly. Natural History Museum of Los Angeles County, Los Angeles.

Malcolm, S.B. & Zalucki, M.P. (eds) (1993) Biology and Conservation of the Monarch Butterfly. Natural History Museum of Los Angeles County, Los Angeles.

McCormick, B. (2005) Bogong moths and Parliament House. Research Brief No. 5. Department of Parliamentary Services, Canberra.

Meehan, T.D., Glassberg, J. & Gratton, C. (2012) Butterfly community structure and landscape composition in agricultural landscapes of the central United States. *Journal of Insect Conservation* DOI 10.10007/s 10841-012-9523-4.

New, T.R. (2009) Insect Species Conservation. Cambridge University Press, Cambridge.

New, T.R. (2011) Butterfly Conservation in South-eastern Australia: Progress and Prospects. Springer, Dordrecht.

Oberhauser, K.S. & Solensky, M.S. (eds) (2004) The Monarch Butterfly. Biology and Conservation. Cornell University Press, Ithaca and London.

Parsons, M.S. (2004) The United Kingdom Biodiversity Action Plan moths – selection, status and progress in conservation. *Journal of Insect Conservation* 8, 95–107.

Patrick, B. & Patrick. H. (2012) Butterflies of the South Pacific. Otago University Press, Dunedin.

Petanidou, T., Vokou, D. & Margaris, N.S. (1991) *Panaxia quadripunctaria* in the highly touristic Valley of Butterflies (Rhodes, Greece): conservation problems and remedies. *Ambio* 20, 124–128.

Polgar, C.A., Primack, R.B., Williams, E.H., Stichter, S. & Hitchcock, C. (2013) Climate effects on the flight period of Lycaenid butterflies in Massachusetts. *Biological Conservation* 160, 25–31.

Pyle, R.M. (2012) The origins and history of insect conservation in the United States. pp. 157–170 in New, T.R. (ed.) Insect Conservation: Past, Present and Prospects. Springer, Dordrecht.

Rezbanyai-Reser, L., Kadar, M., Petranyi, G. & Kocsy, G. (2004) 1st European Moth Nights, August 13th–15th, 2004, a scientific overview (Lepidoptera: Macrolepidoptera). Lepidopterological Society of Hungary/ Entomoloogical Society of Luzern.

Rezbanyai-Reser, L., Kadar, M. & Schreiber, H. (2009) 4th European Moth Nights, 11th–15th October, 2007, a scientific evaluation (Lepidoptera: Macrolepidoptera). Lepidopterological Society of Hungary/ Entomoloogical Society of Luzern.

Swengel, A. (1990) Monitoring butterfly populations using the Fourth of July Butterfly Count. *American Midland Naturalist* 124, 395–406.

Warren, M. (2012) Butterfly Conservation; the development of a pioneering charity. pp. 133–154 in New, T.R. (ed.) Insect Conservation: Past, Present and Prospects. Springer, Dordrecht.

Watts, C., Stringer, I. & Gibbs, G. (2012) Insect conservation in New Zealand: an historical perspective. pp. 213–243 in New, T.R. (ed.) Insect Conservation: Past, Present and Prospects. Springer, Dordrecht.

Wells, S.M., Pyle, R.M. & Collins, N.M. (1983) The IUCN Invertebrate Red Data Book. IUCN, Gland and Cambridge.

5

Studying and Sampling Lepidoptera for Conservation

Introduction

Determining the conservation status and needs of any species is ideally founded in sound knowledge of the biology of that species, and information on changes in its abundance and distribution. Similarly, the vulnerability of communities or assemblages is often reflected in changes in their richness or wider diversity, so that the number of species present is commonly used as a measure of fitness or need, together with their distribution and abundance. The roles for field work in Lepidoptera conservation are very varied, and range from autecological studies of individual taxa or populations, most commonly rare and elusive species and small and area-restricted populations, and so of primary conservation interest, to inventory studies to evaluate and compare local taxon representation and diversity (most commonly as species richness) in time and space. Very broadly, they encompass four main themes, as (1) determining presence or absence of particular focal taxa; (2) detailing distribution of species or populations, in some cases also determining trends in distribution or abundance; (3) seeking causes of such variations and clarifying basic biology from which to formulate management; and (4) elucidating diversity (most commonly as the number of species present) rather than studying single species, by emphasising and enumerating the richness of taxa present in an assemblage or community, or associated with a specific site or biotope. Any of these themes, and the many intermediate needs, occur at a variety of scales from local (single site) to global (equating to 'range-wide or 'region-wide') and from 'one-off' surveys to repeated surveys using identical methods, as monitoring to determine numerical and distributional trends. In most cases, the work must be undertaken with very limited resources, with considerations of costs and availability of the requisite skills dictating the

Lepidoptera and Conservation, First Edition. T.R. New.
© 2014 John Wiley & Sons, Ltd. Published 2014 by John Wiley & Sons, Ltd.

limits of the study. A practical matter, however, is that many conservation surveys are time-bound, and initiated by some form of crisis-management need, commonly stimulated by imminent danger of a habitat or site being changed and concerns for notable species present (or possibly present) there. As in many such contexts, needs and pressures for valid 'short-cut' approaches to aid rapid assessments may drive and restrict investigations. Most such work on Lepidoptera has focused on the conspicuous adult stage alone – with, again, the values of butterflies and diurnal moths demonstrably paramount.

Thorough consideration of methods available, such as whether standardised sampling or inspection approaches are suitable or more innovative measures may be needed, and for the proposed approach to be informed by review of any previous studies on the same or similar taxa or habitats are clear planning needs, together with very clear definition of the objectives of the study. Those objectives dictate the sampling or survey approach needed. Focused questions to formulate or monitor conservation needs and management are far preferable to simple data-gathering exercises, however important and attractive those may be: indeed for poorly known taxa, elucidating the basic biology by such fundamental research may be the highest priority in leading to informed conservation, and to formally establishing some 'conservation status' as a prelude to legal recognition of need. Such targeted research on which to found management is far different from simply accumulating information.

Two rather different objectives of surveys are particularly relevant in conservation. The first is simply 'detectability', confirming the presence of notable focal species, whilst the more complex 'monitoring' necessitates obtaining more stringent numerical data sufficient to detect trends, as described earlier. More detailed or interventionist approaches, mostly involving some form of mark–release–recapture, may be needed to explore demographics of a population (Chapter 6), and provide more reliable quantitative data on, for example, population sizes, and longevity and movements of individuals. The former may need specialised search by operators knowing the methods most likely to detect the species (Hudgins et al. 2011), while the latter may extend over a considerable period with samples needed at regular, often intergenerational or annual, intervals. For demographic studies, multiple surveys during a single generation flight period may be necessary. Multispecies studies, particularly to clarify changes in composition of assemblages over time, may require sampling over many generations or years or at different seasons to establish phenological patterns. Techniques overlap considerably between single species studies and multispecies surveys. As Haddad et al. (2008) commented, 'Determining which approach to use involves weighing complex trade-offs among bias, precision, and cost'. It follows that the primary purpose of any survey must indeed be defined carefully.

Because the great majority of studies have focused wholly on adult insects, well documented survey methods have been developed for these relatively accessible and often conspicuous targets. Relatively few studies have emphasised surveys of immature stages, despite their abundance and that they are often available for much longer periods than adults, and quantitative/semiquantitative sampling approaches lag behind those for adults. One practical problem is that, whereas adults are generally recognisable and identifiable to species, many caterpillars must be held in captivity and reared to adults before confident identification is possible. In some cases, however, distinctive and easily detected caterpillars can be used. The communal behaviour of some Nymphalidae, such as some checkerspots in which caterpillars form groups and construct a conspicuous silken web, has led to web counts being used to indicate relative population sizes, as easier to evaluate than the mobile adults. Examples are the Marsh fritillary (*Eurodryas aurinia*) in Wales (Lewis & Hurford 1997) and the Glanville fritillary (*Melitaea cinxia*) in England (Thomas & Simcox 1982), and later extensively in Europe (Kuussaari et al. 2004). Empty pupa cases protruding from the ground may be a useful monitoring tool for sun-moths in Australia (Richter et al. 2012), and persist for several weeks.

Occasionally it is feasible to use more indirect 'signs' as evidence of presence. Many Microlepidoptera are leaf-miners, and the domains constructed by caterpillars within foliage are often very characteristic in form and also persistent, lasting well after the larvae have developed. With confidence of accurate recognition within relatively well documented faunas (such as Mediterranean forests: Tribert & Braggio 2011), their diversity can contribute to assemblage assessments, but considerable additional study is needed to enable their more general use.

Sampling methods

The sampling methods available for early stages and adult insects are described in considerable detail elsewhere (Samways et al. 2010, for example), but some brief discussion here aids perspective. The most common approaches for adults of diurnal taxa rely on standardised visual inspections based either on transect walks or timed 'spot surveys'. Transect walk procedures (widely termed 'Pollard walks') pioneered for the British butterflies (Pollard & Yates 1993) have been emulated widely elsewhere and have many applications based on standard monitoring and recording of the target species. Not least, the very existence of a simple, understandable procedure that can be used easily by recorders with little formal training itself led to increased interests in butterflies and their fates as people recognised that they could participate constructively in conservation status assessments. In transect sampling, the observer walks along a predetermined route, which may be compartmentalised by distance into various habitat categories (commonly based on vegetation or exposure) and, with suitable precautions, records by species all the butterflies seen within a fixed corridor (depending on visibility, up to 5 m from the pathway). Substantial areas can be inspected, with outcomes linked to habitat category. Correlative records of changes in abundance and distribution of British butterflies, largely from transect walk data (Asher et al. 2001), constitute the most influential compilation of data on any insect fauna. The same procedure is suitable for diurnal moths (Groenendijk & van der Meulen 2004). As Pollard and Yates emphasised, the basic needs for any monitoring or wider survey method are that it is rapid and easy to use in ways that are well defined, standardised and minimise errors, and the approach provides sound information on representation of species and population sizes, so that repetition may reveal trends of change.

Transect walk approaches usually confine information to those insects recorded within a (defined) few metres of the transect line. However, if detection at greater distances is straightforward and unambiguous, 'distance sampling' may also be relevant to help in assessing densities and population sizes. The method (discussed by Haddad et al. 2008) assumes that the transects are placed randomly with respect to the target Lepidoptera, in turn assuming a uniform density distribution about the transect. Detection at the centre of the transect is then certain, and detection elsewhere is a decreasing function of distance from the central transect line. Conditions for use in estimating population numbers include (1) uniform habitats; (2) high visibility, so that targets are not obscured by dense vegetation or uneven terrain; (3) random transects do not damage habitats, such as by trampling or other disturbance, in contrast to the more usual employment of predetermined regular paths for conventional Pollard walks; and (4) the insects occur in sufficiently large numbers to estimate detection functions. Use of this sampling approach for the conspicuous Regal fritillary butterfly (*Speyeria idalia*) (Powell et al. 2007, Chapter 13, p. 248) evaluated data on individuals from up to 30 m from the central transect line, so that substantial amounts of terrain could be surveyed. As the above authors emphasised, density estimates from distance sampling can be unbiased only if four critical conditions are satisfied: (1) transects are random, as above; (2) all individuals at the transect centre line are detected; (3) perpendicular distances of others are accurate; and (4) individuals are detected at their initial location or their movement is random and slow with respect to the observer. In practice, one or more of these conditions may be very difficult to satisfy, so that using distance sampling for population size estimations needs careful attention.

Conventional transect counts are not as rewarding or suitable for species present only at low densities as for more abundant taxa. The Jersey tiger moth (*Euplagia quadripunctaria*, Arctiidae) occurs at less than an average of 1/km length of transect during its peak flight season in the Netherlands, so that frequent counts are needed and population estimates can be highly inaccurate unless augmented by other methods (Groenendijk & van der Meulen 2004). The general worth of transect counts has also promoted trials of 'torchlight transects' for moths at night. Studies on the very localised Sandhill rustic (*Luperina nickerlii leechi*, Noctuidae) on its only known British site that supports a small closed population (Spalding 1997) also revealed the problems of dealing with only very low numbers and providing any accurate estimate of abundance. Most transects yielded numbers well below the threshold of 40 individuals suggested as valid, for butterflies, by Thomas (1993). Spalding's counts were used to construct an annual 'Index of Abundance' based on means of interval counts (every third day) over the entire limited flight season, of the numbers of resting moths. Unusually, this study did not employ light traps (see later in this chapter), as the most widely used tool for nocturnal moth surveys. Easily recognisable individual species are clearly the most suitable targets of such nocturnal transects, with difficulties of identifying numerous species limiting their use for inventory assessments. These necessitate moths being netted to confirm identity (Birkinshaw & Thomas 1999) and released immediately other than for ambiguous cases for which closer examination was needed. In that study, considered effective in giving useful habitat and behavioural information, a quartz halogen spot lamp powered by a 12 V motorcycle battery was used.

In common with any survey method, it is important to acknowledge limitations of transect walks. A recent critical survey of butterfly transect walks as described in the preceding paragraphs (Isaac et al. 2011) noted that a sizeable proportion of butterflies are missed, and that detectability varies substantially among species and sites. Abundance of several British butterflies of considerable conservation concern was perhaps systematically underestimated, and some were amongst the least detectable of the 17 species scored in that survey. Variation in detectability was much greater among sites than among species.

The major observational alternative (or complement) to transects, of 'spot surveys' or 'timed surveys', can allow for more intensive investigations of small areas and, as for transects can be replicated as required or feasible. The observer stands at a predetermined point (selected randomly, more systematically by GPS or grid reference, or to represent habitat features) and during a standard period, commonly of 10 minutes, records all target insects seen within a known radius. The process can be repeated for spatial or temporal comparisons and, if undertaken at stations along the routes used for Pollard walks, may provide additional data for site evaluations. However, cautions may be needed in interpreting single counts for some taxa, because vagaries of individual species may produce considerable bias. The Golden sun-moth (*Synemon plana*, Castniidae) (Chapter 9, p. 174) demonstrates intricate, currently unpredictable and highly variable spatial patterns over open grassland areas, all apparently suitable for habitation, in which apparent distribution can shift substantially across a site during a flight season. Although the reasons for this remain to be confirmed experimentally, this change may reflect topography and exposure, with soil temperatures differing across the site and influencing rate of development of the subterranean caterpillars. Interpreting such 'microdistribution' in any such species is a formidable problem but, more generally, site aspect and insolation may strongly influence local development and activity. *S. plana* is one of the notable flagship species on native grasslands in south-eastern Australia, with its habitat under continuing threat from urban and industrial expansion around major cities. It is designated nationally as 'Critically endangered' and listed also under the legislations of all three range jurisdictions (Australian Capital Territory, New South Wales, Victoria) that apply. Much of the high conservation profile for the moth arose from it being difficult to discover, so that few populations were known until recently, when the factors circumscribing its activity and detectability became clearer (Gibson & New 2007). Females fly little, whilst males patrol about a metre above the ground and alight to mate once a female is detected. However, males are active for only a few hours in the middle of the day, and in calm sunny weather at temperatures above about 20 °C; should cloud cover or wind speed increase, temperatures fall, or rainfall occur, they settle rapidly. Thus, for much of the day they are naturally inactive and largely undetectable, so that under many conditions and at non-suitable times, observers fail to find the moth. Visits to several sites in sequence during a day almost invariably lead to inspections outside the limited window of opportunity when the moth is active. Once the intricacies of this activity pattern became clear, discoveries of new populations

Table 5.1 Limitations and considerations in sampling the Golden sun-moth, *Synemon plana*, on grasslands in south-eastern Australia (Source: Gibson & New 2007. Reproduced with permission from Springer Science+Business Media.).

1. Gender bias. Only male moths are sufficiently conspicuous and active to include in counts or to detect easily. Females are relatively inactive and fly very little
2. Longevity. Adults live for only 1–5 days, and do not feed. On any site, moths may emerge continually over a flight season of up to 6–8 weeks, so that turnover is very rapid. This limits values of mark–release–recapture exercises, and estimating population size may require multiple visits and become costly
3. Activity. Males fly rapidly but over short distances, and site separation of 200 m is considered isolation for defining populations. Activity is very limited: normally, flight occurs only within the period of about 1100–1500 h (Eastern Daylight Savings Time), and surveys at other times of day may not reveal presence
4. Conditions for flight. Flight occurs only in warm conditions (above about 20 °C) in bright sunlight and in fine calm conditions: rain, cloud, and strong wind all lead to cessation of flight, so that activity (and detection) is highly influenced by weather
5. Comparison between sites. Inter-site comparisons may not be possible on the same day, or under the same conditions: sound comparative information is difficult to obtain, and most evaluations depend on 'presence/absence' data alone
6. Distribution on a site. Adult emergence from the underground early stages can be very patchy on a site, perhaps reflecting local topography and insolation influencing soil temperatures. Concentrations of moths may change in distribution over a flight season, necessitating multiple visits to map distribution on any site
7. Life cycle. The length of the life cycle is unknown, but is thought to be 2, or even 3, years. The moths present in any one season may represent only a single 'population cohort', rather than the entire resident population
8. Objectives. Surveys (by transect walks or spot inspections) mostly aim to detect the moth, rather than estimate population size accurately. For guidance, detection of five moths has been adopted by some operators as inferring a significant population
9. Improvements? Recent attempts to survey *S. plana* by counts of pupal cases protruding from the soil surface need further investigation

increased markedly, and sampling protocols for environmental consultants have been prepared to aid surveys (Table 5.1). The universal principle that the behaviour of a species will influence its detectability, and the suitability of any particular sampling method in turn depend on behaviour, has wide ramifications in surveys of this kind. Thus, Kremen (1994) noted that the strongly flying *Charaxes* butterflies in Madagascar spend much of their time high in the forest canopy, so are rarely recorded in visual samples such as near-ground transects. Use of bait traps high in the canopy revealed them as the dominant butterfly group in that environment.

A third approach, intensive direct searches, may be feasible for some highly restricted and local species on small sites. This involves the observer slowly but carefully exploring the entire site systematically, and has potential to give very useful information on population sizes by directly and reliably counting the individuals present. It was, for example, a rewarding approach for the New Forest burnet moth (*Zygaena viciae*, Zygaenidae) (Young & Barbour 2004).

However, the intensity of interference needed for this method must be balanced by considering possible harm by trampling and other disturbance to the site.

In short, a variety of approaches to survey are usually available for any given case, and objective comparisons between the results from each are relatively rare. In comparing transect surveys and timed surveys, Kadlec et al. (2012) found the latter more useful for inventory studies in which the most comprehensive species lists for sites are needed and the numbers of surveyors limited. More species and more individuals were found in this way, but no differences occurred in detecting poorly visible or less active butterfly taxa.

Sampling intensity and interval are thus important, with (1) repeated samples across a single year or flight season revealing seasonal patterns of species incidence, richness and activity, and the records summing to constitute an inventory of richness of the site(s) investigated; and (2) longer-term survey series transcending generations and providing an intergenerational marker to indicate change. The major alternative survey approach, using more complex mark–release–recapture studies for a species, has many applications in individual species studies, but is far less amenable for broad comparative or monitoring studies. The extent of refinement needed for any method is dictated by the objectives of the study. Considerable effort may be needed simply to detect or confirm the presence of a rare species, and quantitative data may be far secondary to simple presence/absence information. However, whilst 'presence' necessitates only a single sighting or other record to confirm, 'absence' is far more difficult to prove. The need then is to sample as effectively as possible, incorporating guides from all information on the target species that may be relevant to reduce the incidence of spurious zeros in the outcome. *S. plana* illustrates this process well, and it is likely that some other poorly documented species may also prove to be more widespread than currently supposed, once their biology is better understood and sampling approaches refined accordingly. Individual and idiosyncratic behaviour of species commonly affects amenability to sampling by standard means, and creates wrong impressions over presence or abundance. The North American noctuid *Lithophane joannis* had long been regarded as very scarce, reflecting that the adults are not attracted to light, so were not retrieved by this most popular collecting tool. In consequence, it escaped detection in at least 5 years of survey in a major national park, despite it being one of the park's most common Lepidoptera (Wagner 2006)! Caterpillars escape detection because they live in the leaf shelters constructed mainly by Microlepidoptera on buckeye (*Aesculus flava*), and adults can be attracted to baits. Indeed, baiting techniques, such as use of 'sugaring' or 'wine ropes' (Table 5.2), may attract many moths not amenable to light trapping and which are thus commonly overlooked in inventory surveys.

The major survey method for nocturnal moths has been use of light traps, exploiting the well known behaviour of many (but by no means all, as in the previous paragraph) moths being attracted to light, and refining a method long used by collectors. The many factors influencing moth flight activity are thus relevant to interpreting the catches, as for those regulating diurnal flight activity as discussed above. Factors such as temperature, humidity, precipitation, cloud cover, wind speed, atmospheric pressure and moon phase may each be

Table 5.2 Summary of general techniques for collecting/sampling/detecting Lepidoptera for conservation.

Adults

Capture
  Direct capture of individual free-flying insects (the 'butterfly net')
  Sweeping or beating vegetation (some small moths)
  Light traps, based on attraction of many moths to a light source. Use of traps that
    do NOT kill the catch is recommended
  Bait traps: use of fruit, carrion, dung, sugar or alcohol ('sugaring' or 'wine ropes' for
    some moths) as attractants. Many such baits are highly selective
  Visual surveys of flowers to detect nectar-seeking individuals
  Pheromone traps: a highly specific form of 'bait' suitable for some species.
    'Assembling' draws on this principle for some moths, in which virgin females are
    exposed and attract males
  Malaise traps, and other flight traps. Occasionally useful for wider survey, but
    material collected often in alcohol and in poor condition that renders
    identification difficult. This and other methods that kill insects indiscriminately are
    NOT recommended for conservation work

Observation
  Transect walks
  Point surveys
  Binoculars, sometimes useful for easily identified diurnal species

---

Larvae

Direct searches of food plants or other substrates to discover larvae for rearing in
  captivity
Beating or sweeping vegetation, for similar outcome
Trap plants. Expose potted/planted food plants for field oviposition by focal species,
  and inspect at intervals for incidence
Collect samples of leaf litter, galls or leaf mines for retention in captivity and rearing of
  taxa present
Rearing from eggs laid by captured females useful for clarifying identities and
  providing specimens for collector or release

---

influential. In addition, many species have characteristic flight times – such as being crepuscular (flying at dusk, dawn or both) or more widely nocturnal (Muirhead-Thomson 1991). Long-term surveys using light traps, such as their being a major component of the Rothamsted Insect Survey in Britain (described later in this section), demonstrate the formidable amount of information that can be accumulated by such apparently simple means. One major concern with light trap surveys is that they have traditionally been operated to accumulate and kill moths without discriminating between taxa captured, so that the method itself could become a threat in capturing rarer taxa, in addition to the ethical problems that arise over mass slaughter of taxa that include arthropod 'bycatch' of no relevance to the focal project. Live storage and subsequent release of moths has become much more widespread, and live capture is a deliberate conservation strategy in many places where known significant or threatened species are likely to occur, and in surveying for these. Rather than leaving a trap on overnight, collectors and monitoring exercises may opt for shorter collecting periods, in which the light is suspended before a vertical white sheet or within a light tower (see later in this section) and selected moths removed directly from this for retention.

Light traps are used both for short 'spot surveys' and for long-term monitoring, and three basic patterns of trap, each with numerous modifications have been popular (Fry & Waring 2001):

1   the Rothamsted trap uses a 200 W tungsten bulb powered by mains electricity or a generator;
2   the Robinson trap, the pattern most familiar to hobbyists, with a 125 W mercury vapour bulb powered as above;
3   the Heath trap, smaller and more portable, with a 6 W actinic black light supported by a 12 V battery.

Because of needs for a power source, the first two of these are often impracticable for use in remote areas, and for which portability may be needed. Light traps have provided the bulk of data on macromoths from which estimations of diversity and changes in richness and local assemblage composition over time have been gauged. The most intensive surveys have provided numerous records of species incidence and long-term surveys given evidence of declines and changes. The Rothamsted light trap survey (Conrad et al. 2004) has operated since 1968, recovering more than 730 species of macromoths across more than 430 sites throughout Great Britain (c.f. Fig. 5.4). The accumulated database 'provides one of the longest-running and spatially extensive datasets of a species-rich insect group anywhere in the world' (Conrad et al. 2004: 120), and provided a basis from which to estimate 35 year trends for 338 moth species in the region (p. 69). The value of the surveys reflects Holloway's (1977) sentiment that the macromoths (whose taxonomy is reasonably well known) are amongst the world's most easily sampled monitors, for which standardised light trap surveys over extensive periods can provide near-complete inventories of species present in habitats where the traps are deployed. Thus, his extensive surveys on Norfolk Island, undertaken from 1971–1976 and using light traps over more than 120 sites and all major habitats on the island (Fig. 5.1) provided 'perhaps the first instance where a whole community of organisms has been sampled down to the rarest species' (Holloway 1977: 233), and with year-round data informing seasonal appearances. The total sample (of >70,000 moths) contained all resident species, with no additional species being recorded in the final years. This level of definition is rare, but enumerating island faunas has relied extensively on this approach, as have numerous exercises in comparing moth faunas across sites or times, or associated with restricted biotopes or sites.

Surveys of Sant Barbara Island, at 2.6 km^2 the smallest of the California Channel Islands, illustrate some of the difficulties of establishing inventories of Lepidoptera on even small isolated areas from sporadic visits. Powell (2004) reported a cumulative total of about 153 species (8 butterflies and about 145 moths), but no accumulation asymptote was evident, and many species were in

Fig. 5.1 Outline map of Norfolk Island, indicating sites comprising the network of light trapping points yielding data for an inventory of moths of the island (area of island ca 34 km²). (Source: Holloway 1977. Reproduced with permission of John Wiley & Sons).

very low abundance: 47 singletons and 13 as doubletons. Assessing status of such taxa is very difficult – Powell noted that rare records of a species might result from sampling in marginal habitats, sampling too early or too late in the season, at a trough of abundance (either a 'poor year' or shorter-term fluctuation), or as vagrant individuals rather than resident species. 'Calibration' of any such inventory is thus difficult, but Powell's approach was far more critical than used by many other workers. He used three approaches to estimating the true number of species probably present, as (1) comparing the major taxonomic groups (butterflies, macromoths, micromoths, Pyraloidea) to their relative representation on the relatively well documented mainland; (2) comparing the number of species to richness of flowering plants, assuming the island fauna is as rich as a low architecture mainland plant community; and (3) estimating the proportional rarity among recorded species using the Chao 1 index. Each of these makes major assumptions, but the different approaches may be complementary, and projected richness from each was (1) 206, (2) 186, (3) 237: if reliable, the inventory is 65–82% completed. Even on this small island, three moth species appeared to be endemic, and other island endemics occur elsewhere on the Channel Islands.

A central problem in most such surveys is that long-term or year-round sampling is impracticable, and that both biases and inadequacies resulting from this are largely unknown. Although Hebert (1980) asserted that moth communities he sampled in Papua New Guinea were aseasonal (of 166 species represented by 10 or more specimens in December samples, 162 occurred also in August), those samples were relatively small (3 nights/site/occasion). Hebert considered the data sufficient to indicate that most of those species flew throughout the year. Studies in forests in tropical north Queensland, Australia, suggested that patterns of seasonal incidence are rather more complex, with 'season' having a significant effect on representation and abundance of many families (Kitching et al. 2000). In the Queensland samples, nearly three-quarters of the total moth individuals were captured in the wet season. The influences of wet and dry seasons on tropical moth phenology have been discussed extensively: Janzen's (1993) thoughtful appraisal gives much useful background to the complexities of interpreting these. More broadly, series of light trap catches taken at intervals along transects can reveal (1) variations in diversity; (2) zonation patterns or ecological continua; and (3) associations of species that may be related to different biotopes or vegetation types (Holloway & Nielsen 1999). Collectively, these aspects of pattern analysis provide data of enormous value in conservation planning or environmental monitoring (Chapter 12).

With the exception of highly standardised comparative surveys undertaken by the same workers using the same regimes, some biases in sampling outcome from light traps are almost inevitable, and render fully quantitative comparisons difficult. Comparison of different light trap models, for example, may yield rather different outcomes: some of the caveats and concerns were discussed in a comparative survey of moths in Malaysian lowland rainforest (Intachat & Woiwod 1999) and in many remote environments the traps used may be determined by portability and site access, in addition to the actual sampling needs. Several authors have suggested the need to examine and compare results from various patterns of light trap to 'calibrate' comparisons of moth catches from different sources, but this is extraordinarily difficult to achieve reliably (Leinonen et al. 1998). Differences between samples from light traps (automatic capture as moths enter the container) and 'light towers' (with moths collected manually as they settle on a gauze surface) in Costa Rica were much less pronounced than those from Tanzania (see later in this chapter), with one possible reason for this difference noted by Brehm and Axmacher (2006), being that moths were collected easily from the latter, whereas many of the moths attracted to the trap might not actually enter it, but 'escape'. Further confusions arise from the vertical stratification of moth activity in forests (Brehm & Axmacher 2006, Schulze et al. 2001). In addition to encompassing seasonal variations, trapping must be sufficient to consider shorter-term fluctuations in moth activity, for example from lunar periodicity, in order to reflect both short-term and long-term variations. Thus, a study of hawkmoths (Sphingidae) in Tanzania (Robertson 1977a,b) showed that many taxa display short-term variability in 'trappability' as moonlight was associated with lowered attraction to light traps. As for any monitoring exercise, values of short-term surveys, often the only option possible, in relation to more extensive assessments are important to determine, and must be linked to careful definition of the study's objectives. One such informative comparison involved two series of catches of moths in forests of Louisiana (USA), with the aim of estimating the validity of short-term 'intensive' sampling in terms of the proportion of species found in more extensive surveys (Landau et al. 1999). The regimes were: (1) the intensive survey of four traps operated for 2 consecutive nights/week over 4 weeks, and (2) two traps operated about 1 km apart (and at a site about 50 km from that of the first survey) twice a month over an 8-month period. Each thus comprised 32 nightly collections, so had identical sampling effort on that measure. The outcomes (intensive: 4198 individuals, 261 species; long-term: 3155 individuals, 314 species) included about 70% species overlap (species pool: 362), so that the rapid survey approach – other than considering seasonal variations – may provide useful data for richness or assemblage appraisal.

Despite the logistic attractiveness of using light traps that accumulate the attracted moths for inspection the following day, these may not always furnish the most representative samples. In Tanzania, comparison of such traps with manual catches of moths attracted to 'light towers' revealed markedly different outcomes in abundance and composition of the geometrid moths captured (Axmacher & Fiedler 2004). The light tower catches averaged ten-fold more moths than trap catches, with the species distributed as 42 species (both light trap types), 67 (light tower only) and 8 (light trap only), and with relatively consistent differences across three major habitats. The three predominant species at light towers were also within the top four at light traps. The differences appear even more marked when sampling effort is considered; light traps were run all night, and light towers inspected for 3 hours (1900–2200 h), a period that included the peak moth flight activity, as moth numbers declined considerably after 2100 h.

Data from light trap catches of moths are often taken as valid for quantitative studies of diversity and for comparison across sites, but several major reservations over this should be noted. Discussed by Beck et al. (2002), these include (1) that light traps sample selectively rather than randomly from natural assemblages; (2) because they can attract moths from a distance, samples may be 'diluted' to an unknown extent by species not truly native to the focal area; and (3) the effective radius of attraction may depend on habitat structure, such as density of vegetation.

The species pool likely to occur in surveys at sites in Britain or other parts of the northern hemisphere can be predicted quite realistically, so that 'surprises' (such as novel species) from survey exercises are detected easily, together with unexpected absences or changes in composition. This is not the case for much of the world, for which records of species distributions and site species richness are fragmentary or, most commonly, non-existent. Many such exercises must be viewed in isolation, but become valuable when undertaken on some more standardised basis to compare outcomes in different habitats or in relation to land use or disturbance by giving clues to (1) which taxa are patchy in distribution and (2) which taxa may be influenced by disturbance and may have some 'indicator value' (Chapter 13, p. 246). Many such examples exist, and some are noted in context in later chapters. Two of the surveys mentioned earlier indicate contexts for general comparative studies in clarifying ecological patterns in the tropics:

1   Hebert's (1980) Papua New Guinea survey focused on assessing species richness along an elevational gradient in montane tropical rainforest, in which comparisons were made between samples at 200 m intervals from 2200–2800 m. Samples were not taken from lower levels because the lower habitats below this minimum were disturbed extensively by agricultural activity. Five

light traps were deployed along a line of approximately 200 m (so forming a 'light trap transect', as used widely elsewhere) at each sample level. Over three occasions, each of 3 nights, a total of 21,830 moths were collected, representing more than 1250 morphospecies. Many taxa were represented by few individuals but a progressive reduction of richness with increasing elevation included an approximate halving of species number from 774 at 2200 m through 613, 591, to 379 at 2800 m. The decline was attributed to increasingly rigorous climate and reduction in available habitat area at higher elevations.

2   Kitching et al. (2000) compared three vegetation regimes across wet and dry seasons on the Atherton Tablelands of tropical Queensland, gathering a pool of 835 larger moth species. The differences in abundance across the three categories of sites (three never cleared, three with substantial regrowth after clearing and three newly cleared 'scramblerland' sites) and season were substantial (Fig. 5.2). A major outcome was implication that selected series of moth families and subfamilies might constitute a 'predictor set' as an indicator of disturbance (Chapter 12) in these forest remnants. These taxa (Table 5.3) included 10 groups that increased significantly with disturbance and eight others that declined. Any such refined focus is logistically valuable in reducing sample analysis costs to avoid taxa redundant to the purpose of an individual survey.

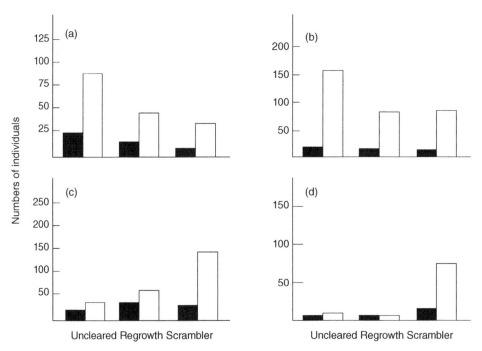

Fig. 5.2 Numbers of individual moths of (a) Geometridae – Ennominae; (b) Lymantriidae; (c) Noctuidae – Amphipyrinae; (d) Arctiidae – Arctiinae in the dry season (open bars) and wet season (black bars) in forest remnants on the Atherton Tablelands, Queensland, Australia. (Source: Kitching et al. 2000. Reproduced with permission of John Wiley & Sons).

Table 5.3 The moth components of a 'predictor set' of taxa to indicate forest disturbance in tropical Queensland, Australia, focusing on remnants of notophyll vine forest on basalt. (Source: Kitching et al. 2000. Reproduced with permission of John Wiley & Sons).

Groups that increased in overall abundance with disturbance:
Arctiidae: Arctiinae
Noctuidae: Amphipyrinae, Catacalinae, Hadeniinae, Heliothinae, Hypeninae, Noctuinae, Plusiinae
Herminiidae
Pyralidae: Phycitinae

Groups that declined in overall abundance with disturbance:
Anthelidae
Lymantriidae
Geometridae: Ennominae, Geometrinae, Larentiinae, Oenochrominae
Pyralidae: Appaschiinae

Such scenarios help to demonstrate another, sometimes not sufficiently emphasised, problem of inventory studies of insects, that of the influence of context on what constitutes 'all species present' in defining a species pool for comparisons. If, as commonly used, asymptotic levelling of richness occurs (or is approached) with increased sampling effort, that asymptote may indeed be realistic for the individual site(s) studied. It may not represent fully the expectation from a wider regional landscape and the outcome will also reflect limitations of the sampling methods used. Each sampled site is itself a sample of greater historical diversity that may not have been measured directly (Bucheli et al. 2006), so that estimates of total expected diversity must be made from surveys of different areas. The surveys of gelechioid moths in Ohio by Bucheli et al. are an instructive approach to values of focal groups sampled over limited regions.

A major contrast between temperate region and tropical region surveys, and, to a lesser extent, paralleled between butterflies and macromoths, is knowledge of seasonal development patterns, voltinism and defined flight seasons, all important components of species detection and monitoring as aids in focusing sampling season and effort for adults. Thus, seasonal patterns amongst tropical Lepidoptera are often not clearly understood, and relatively few systematic surveys have extended over sufficient periods to imply their reality and causes. One such study, on a pool of 137 species of fruit-feeding Nymphalidae in Ecuador, involved monthly samples of these butterflies over 10 years (Grotan et al. 2012), to reveal that seasonal rainfall cycles influence annual cycles in butterfly species diversity and community similarity: greatest species diversity occurred during the dry season.

Caterpillars or other pre-adult stages are sometimes the primary available focus in single species studies, either alone or as a complement to adult surveys as twin intergenerational or other monitoring markers. Exposed caterpillars on foliage may be sufficiently conspicuous for easy counting on vegetation quadrats or along transects. Gregarious caterpillars of the Glanville fritillary (*Melitaea cinxia*), which form exposed webs that are easily counted on their low-growing *Plantago* food plants, were used in a valuable early survey in Britain (Thomas & Simcox 1982). Whilst caterpillars of many Lepidoptera are cryptic, those of a number of other Nymphalidae bask openly on vegetation. As Murphy and Weiss (1988) documented for the Bay checkerspot, *Euphydryas editha bayensis* (also found largely on *Plantago* in short grassland), quadrat counts in that relatively uniform habitat provided a non-intrusive accurate and replicable way to estimate population sizes. The two nymphalids noted are amongst the most intensively studied of all butterflies, and such counts have contributed to many aspects of understanding population structure (Chapter 6) and dynamics (Ehrlich & Hanski 2004). Thus, the annual monitoring of larvae soon after they construct their winter nests is conducted on about 4000 habitat patches in the Aland Islands of Finland by time-standardised direct searches, and compared with a post-winter spring survey (Nieminen et al. 2004; Kuussaari et al. 2004).

The caterpillars of the myrmecophilous lycaenid *Paralucia pyrodiscus* in south-eastern Australia are wholly nocturnal and shelter by day in the underground nests of their obligatory host ant genus, *Notoncus*. They can be counted (using torchlight inspection) when they ascend the hostplant (*Bursaria spinosa*) to feed, attended by ants. However, not all larvae emerge every night and their activity is influenced strongly by weather conditions, so that the surveys, valuable for indicating presence, have only limited quantitative value in assessing population size.

These examples, representing diurnally and nocturnally conspicuous larvae, are complemented by others in which larvae can be detected and counted by indirect methods. Leaf-miners, for example, include large numbers of host-plant restricted Microlepidoptera, many of which construct mines that are both diagnostic in appearance and that persist on the plant for considerable periods (so obviating the need for sampling that is regulated strictly by season, or by weather), and which can be diagnosed *in situ* by direct inspection. A related example is the shelters constructed by 'foliage-tiers' including many moths but also some butterflies. The retreats of some hesperiid caterpillars formed by joining leaves together on sedges are the first visual cues to their incidence sought by collectors, for example, but provide a useful census tool for those taxa.

As for adults, attempts to enumerate caterpillars may involve complementary methods or approaches. Whilst beating or sweeping foliage is the most frequently used sampling approach, direct searches are also common, although often laborious to perform. Thus, beating vegetation and direct examination can be combined in a single two-stage sampling (Bodner et al. 2010). In their study of caterpillars on two species of *Piper* in Ecuador, the sampling comprised (1) a timed visual inspection of each shrub for immature stages, followed by (2) beating the same individual plant, with the outcomes evaluated by (3) a complete leaf-by-leaf defoliation of the entire shrub, with each leaf checked carefully for presence of eggs or caterpillars. Almost half the caterpillars on the 160 shrubs were retrieved through the first two steps.

More generally, Lepidoptera can be sought or collected by a great variety of general methods, many of them devised by hobbyists and largely qualitative. Some are noted in Table 5.2, to indicate a range of possibilities for consideration as additional options for particular contexts. The precise methods employed will reflect whether single species or multispecies samples are sought, whether one-off or repeated surveys are contemplated, and the general logistic strictures over requirements. Thus, surveys for individual species of conservation interest commonly depend on presence/absence information, often augmented by some imprecise estimate of population size (such as 'few' to 'many'). More precise estimates of population size (Chapter 6) are intrinsically difficult for rare species, and most are based on a single life stage on either a single occasion or across sequential generations. Protocols thereby vary with both focal taxon idiosyncrasies and survey objectives. Attempts to increase sampling returns may lead to inclusion of two or more life stages. Thus, protocols for Golden sun-moth in Victoria include those for larvae and for empty pupa cases protruding from the ground, as well as for the adult moths. In this example, particular care is needed because larvae and pupae of *S. plana* have not been formally described, and some possibilities of misidentification persist.

Whatever method or methods are used, sampling effort is critical, with a very broad tendency for the recorded richness of Lepidoptera at any site to increase with extended or intensified sampling. Thus, the records of 324 species of butterflies (about a third of the total reported from Borneo) from an area of about 1 km² in Brunei (Orr & Haeuser 1996) was attributed by the authors largely to sampling over an extended period of nearly 2 years. Details of the sampling regime should be specified clearly in any report or publication, with description of method, timing, procedures and replication needed for any comparative interpretation in which method or effort may be a strong influence on the outcomes. Many cases exist in which the same species has been studied differently by different workers and the results are difficult to harmonise, or in which inventory figures are based on very different levels of sampling effort, sometimes with this unstated. Whilst the standards for recording and documentation are very well defined for the United Kingdom, flowing from the success and experience of butterfly documentation (Harding et al. 1995) and allowing people to contribute by following very clear directions, this is not always the case elsewhere.

Much practical conservation survey focuses directly on rare species, implicitly difficult to sample quantitatively and, in some instances, even to discover. 'Presence' may be all that can reasonably or economically be confirmed, and data on population size only inferred. The Small ant-blue butterfly (*Acrodipsas myrmecophila*, Lycaenidae), then known at only a single site in Victoria (Australia) (but more widespread in some other parts of the country) could be surveyed realistically only by sightings of hilltopping individuals at Mount Piper (New & Britton 1997), but the first three seasons of field work yielded sightings of only five individual butterflies. Many other species are not as elusive, but are 'rare' through occurring in very localised populations or biotopes where individuals may be reasonably common. The endangered St Francis' satyr (*Neonympha mitchellii francisci*, Nymphalidae) occurs in small areas of grassy wetlands in several of the United States. Four such subpopulations on sites of less than 5 ha were studied to compare population size estimations by transect counts and mark–release–recapture (MRR) methods (Haddad et al. 2008). That study strongly emphasised the need to focus clearly on 'what information is needed', with MRR providing data useful for population viability analyses, but necessarily balanced against dangers of possibly harming individuals through handling (Murphy 1987). Haddad and his colleagues suggested that an optimal sampling strategy might incorporate a balance between these methods by using them jointly – so including limited MRR to estimate survival and detection probability in conjunction with frequent, less expensive and less harmful, transect counts. MRR could be used successfully for this satyr, because the populations were very restricted in distribution – the contexts of using MRR for estimating dispersal (Chapter 6) differ considerably, with this approach to measuring population size and changes largely relying on no (or very limited) dispersal.

For rare species that occupy numerous small patches in a landscape, a simple measure of patch occupancy (as 'presence/absence') based on standard appraisal may be sufficient to reveal trends.

Interpretation for conservation

Multispecies samples for inventory or comparison almost always show a pattern of few species being abundant and numerous others being scarce, with a 'tail' of taxa represented by one or two individuals. And, as noted above, the traditional approach has been one, widespread in entomological surveys, of 'catch 'em, kill 'em and count 'em'. Indeed, such patterns, from light trap catches of moths in Britain, were important in formulating the 'log series' distribution of taxa in ecological samples (Fisher et al. 1943). This general trend persists with increasing sample size, accumulating more individuals of the more abundant species, and adding to the number of species represented by singletons or few individuals: only very long-term surveys can approach a full inventory and, for such dynamically mobile organisms as adult Lepidoptera it is perhaps doubtful whether any local inventory can ever become complete. The asymptote of moth species accumulation noted in the previous section for Norfolk Island, and emulated for the Rothamsted samples in Britain, is achievable only from very extensive sampling regimes in well circumscribed faunas. Elsewhere, most surveys, even over extended periods, have not levelled out their species accumulation in that way but, rather, continue to accumulate species as sampling continues. Analysis of such samples, based on richness and relative abundance, has numerous applications in conservation – incorporating interpretation of pattern and distribution, with comparisons across sites, localities or treatments as either natural, or as possible signals of impacts of environmental change or management. In that context, richness and responses of the more abundant taxa are usually most amenable to interpretation, with the scarcest (least represented) species sometimes disregarded as incompatible with statistical appraisal. High proportions of singletons or otherwise very poorly represented species in samples are frequent, with 40% or more of the total species captured as singletons reported on repeated occasions amongst tropical surveys (Miller et al. 2011). It is usually unknown whether this outcome represents sampling artefacts, or presence of genuinely rare species, or transients within the habitat(s) investigated (Novotny & Basset 2000). However, those apparently scarce species may be those of greatest individual significance as conservation targets, and their presence – even in very low numbers, whether genuine or inferred from inadequate sampling – might signal population existence and site importance for the species' survival. A related practical matter is the need for sample appraisal as samples accumulate, so that any unexpected capture of listed or known threatened species can be detected and the sampling, perhaps, modified to avoid killing further individuals or, in extreme cases of potential endangerment from captures, stopped.

In whatever manner the samples are obtained, counts of species (as 'species richness') are one of the most common parameters used in assessing and comparing insect communities and in characterising and ranking sites or biotopes. As Summerville and Crist (2008) noted, for example, research on Lepidoptera in temperate forest systems has concentrated on two main goals: (1) characterising and comparing the diversity of local faunas through inventory studies; and (2) monitoring populations of species of economic importance. North American deciduous forest systems are rich in Lepidoptera (with Noctuidae and Geometridae predominant) but individual forest communities may be dominated by a small subset of species, and the majority of taxa apparently occur only in low abundance. Many such comparative figures are quoted in examples throughout this book. In assessing species richness of floodplain forest moth assemblages in Austria, however, such counts were considered to be 'totally inappropriate' for detecting patterns of diversity (Fiedler & Truxa 2012), largely because of undersampling, whereby numerous species represented by singletons cannot be confirmed as typical of any of the major biotopes surveyed and might be casual strays from the wider regional species pool. The major interpretative problems itemised by Fiedler and Truxa included (1) incomplete sampling of species-rich communities; (2) large differences in sampling success across sites, even with standardised sampling effort; and (3) high mobility of the moths in fragmented landscapes, leading to confusions over vagrants, as above. Rather than relying on simple counts of species, as most commonly used, more sophisticated diversity measures (such as Fisher's α and Shannon diversity) were considered more useful. The concerns raised may have wide relevance. Amongst diverse tropical faunas, comparisons across sites bcome more difficult through lack of biological knowledge and considerable uncertainties over the real size of the local 'species pool' as previously unsampled taxa continue to accumulate. Thus, Barlow and Woiwod (1989) compared the results from a year of light trapping of forest macromoths in Pahang (Malaysia) with those from a more intermittent previous 10-year survey in the same site. As in many such surveys, particular families dominate richness, with the longer survey yielding 918 species of Noctuidae, 546 of Geometridae and close allies, and 294 of Arctiidae. Comparable data from the single year of Rothamsted light trap use were 414 species (Noctuidae), 331 (Geometridae) and 165 (Arctiidae) with the additional inclusion of Pyralidae furnishing 385 species. Such species-rich families may be particularly informative of the assemblages. However, each of these families revealed species in the 1-year survey that had not been found previously, suggesting that even further taxa remained to be detected, so that the real extent of these is unknown. Many of the 77 species of Noctuidae found only in the more recent survey, for example, were Hypeninae primarily associated with low-growing vegetation, and Barlow and Woiwod suggested that some of these may not have been discriminated earlier because of the subtleties of differentiating very similar-looking species. Some may simply have been overlooked.

Whilst much is made in conservation and ecological surveys of 'the number of species present', determining this in any but the very best documented faunas is difficult – for Microlepidoptera in any tropical region, description and comparison across sites 'poses enormous problems' (Robinson & Tuck 1996). They remarked, for illustration, that 'There is no such thing as an absolute number of species present in a particular habitat except at a particular instant' and 'The basis for a total species estimate can be anything we want it to be', based on sampling effort and season, trapping methods, biotope features and the discrimination of taxa. Even when these factors are, as far as possible, standardised, achieving near-complete inventory counts can be a formidable task, far more difficult than commonly assumed. For the (relatively well known) butterflies of a single 250 ha forest fragment in south-eastern Brazil, with standardised sampling in April each year, so that this covered the season of greatest butterfly richness and abundance, even eight consecutive years was insufficient to yield stable species totals for all familes (Iserhard et al. 2013). The 518 species detected represented only 74% of those in accumulated records from this patch.

In general, the assessment of species characteristic of a specified biotope can come only from very large samples and associations with the heterogeneity of the local environment. Comparisons across seasons or sites are influenced strongly by vegetational richness and heterogeneity, itself often very difficult to assess, and for Lepidoptera, their influences complicated by the mobility of many of the adults on which counts are focused.

Priorities amongst species

In selecting individual species for conservation priority, many possible candidates arise. Many of these are 'genuinely rare' but may be naturally and persistently so and not currently threatened with declines or loss by any detectable anthropogenic influence. 'Rarity' and 'vulnerability' are distinct and have very different implications, but are not always easy to differentiate. The latter constitutes conservation need. Exceptions are possible – a rare species may be susceptible to stochastic impacts, for example, so that rarity may be a precursor to vulnerability (Schtickzelle and Baguette (2009) noted Charles Darwin's comment that 'rarity precedes extinction'): but many rare species appear to have changed little in distribution or abundance when left to their own devices. However, impacts of climate change may render many more of these vulnerable as their environments change. The major practical problem is to set priority for attention within the long lists of taxa that may be nominated or scheduled for conservation interest, either on legal schedules or advisory lists (New 2009, Chapter 9). Many different criteria for priority have been discussed, with butterflies perhaps the only terrestrial insect group sufficiently well known for realistic and relatively comprehensive comparative debate across a global fauna. Thus, an 'Action Plan' for Australia's butterflies (Sands & New 2002) included discussion of all taxa nominated as of conservation concern, and remains the only relatively comprehensive such overview for any group of Australian insects. Incomplete knowledge and the stimulus to future interest and discovery necessitates any such appraisals or grades of conservation status being open to revision. Indeed, periodic review and revision of conservation status of 'listed species' is formally required in some legislations, although less regularly undertaken. In Britain, parts of continental Europe, and North America the more advanced knowledge of butterflies and many moths allows for better definition of conservation status and needs than in most other places. That accumulated information has been critical in (1) recognising the extent of problems faced by those taxa, and likely to reflect the plight of many other insect (and other invertebrate) taxa and (2) confirming that selection amongst deserving taxa must indeed occur to distribute the limited support as effectively as possible. The major alternative approach is to focus primarily on levels above single species (Chapter 8), in efforts to increase benefits from the very limited support available.

Although most conservation concern devolves on threatened species, most of them specialist taxa depending on restricted or threatened biotopes, that concern necessarily spreads to encompass the many strongly declining taxa formerly regarded as secure, as 'common generalists'. The sobering outcome of a 16-year survey by van Dyck et al. (2009) in the Netherlands, in which 11 of 20 such butterfly species declined substantially in distribution and abundance, may simply exemplify much more widespread trends. Two such species (*Lasiommata megera*, decline by 55%, *Gonepteryx rhamni*, 52%) achieved 'Endangered' status, and two others, 'Vulnerable'; three further species became 'Near threatened'. Conversely some species increased over that period (Table 5.4). Declines were most severe amongst species frequenting woodland, farmland and urban areas, reflecting changed land use patterns (Fig. 5.3). Similar cases of declines have occurred in Flanders (Maes et al. 2012), and may indeed be far more numerous than actually documented.

With one of the major aims of practical conservation being to prevent currently common species from becoming threatened, such trends are a major concern, and have been echoed for moths in Britain (Conrad et al. 2004). In all these examples, interaction of habitat factors (Chapter 7) and climate change (Chapter 12) render interpretation difficult, with abundance and distribution changing in different ways amongst co-occurring or closely related taxa. Declines are often not detected because surveys are not undertaken at a sufficiently fine

Table 5.4 Changes in the distribution and relative abundance (1992–2007) of 20 butterfly species considered common in The Netherlands (Source: Van Dyck et al. 2009).

	Change in distribution		
Change in abundance	*Increase (5)*	*No trend (6)*	*Decline (9)*
Increase (7)	5	1	1
No trend (5)	–	3	2
Decline (8)	–	2	6

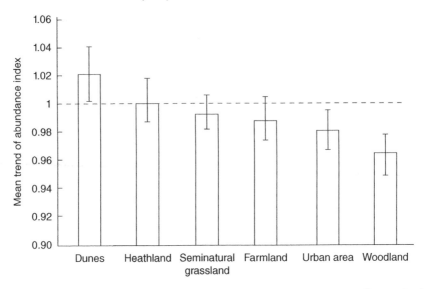

Fig. 5.3 Mean trend of the abundance index for the 20 common butterfly species in The Netherlands (1992–2007) in six vegetation associations. Horizontal line, no change; increased abundance (above line) and decreased abundance (below line) indicated (Source: van Dyck et al. 2009. Reproduced with permission of John Wiley & Sons).

scale. For British butterflies, Thomas and Abery (1995) demonstrated that examination of recording units as a 2 km grid revealed considerably greater declines than using more standard 10 km grid squares – and suspected that even the smaller grid units might mask steep declines on even more restricted scales. The balance needed is one of realistic outcome versus logistics: opting to use larger monitoring areas (with 'too large' leading to non-detection of declines) or smaller areas (in which declines might be detected more readily but with much more intensive monitoring needed – so that spurious declines might result from inadequate monitoring on occasions after presence was confirmed). Whilst noting the need for complete mapping, Thomas and Abery remarked also on the impracticability of mapping even a single British county sufficiently well to accurately assess status changes amongst the most common butterfly species. Some compromise is almost inevitable.

Need for standardised monitoring procedures was emphasised by Roy et al. (2007), but achieving this may be constrained by a recommended procedure becoming labour intensive and, so, expensive. Thus, the major contributions to the United Kingdom Butterfly Monitoring Scheme (Chapter 13, p. 244) have been derived from weekly transect counts on sites, extending over 26 weeks from April to September, to encompass the full flight period of many taxa. These counts are then summed to give an annual total for each species and site. Various forms of reduced intensity monitoring (fewer counts, counts restricted to only part of this extended period) were compared and analysed to reveal that a minimum scheme of three site visits in July–August is a valid approach. Roy et al. noted that this covers the peak period of butterfly abundance, and that three visits – with requirement to double site numbers to achieve precision equivalent to that of the 26 visit scheme – are sufficient to detect long-term trends, whilst the increased site numbers facilitate reduction in site bias towards protected and seminatural areas by enabling farmland and other anthropogenic habitats to be incorporated readily.

One of the best-appraised declines of a formerly common British moth involves the Garden tiger (*Arctia caja*, Arctiidae), a large, conspicuous polyphagous species, for which a 31-year record (1968–1998) of abundance/distribution was obtained through the Rothamsted light trap survey (p. 55) (Conrad et al. 2002). Indices of abundance were derived from records over a total 406 sites (Fig. 5.4, one trap per site), with a variable trapping intensity but above 70 traps

Fig. 5.4 The network of Rothamsted light trap survey sites contributing to the survey of status of the Garden tiger moth, *Arctia caja*, in the United Kingdom. Closed circles, sites where never captured; open circles, sites where captured in more than 5 years; triangles, sites where captured in fewer than 5 years (Source: Conrad et al. 2002. Reproduced with permission of John Wiley and Sons).

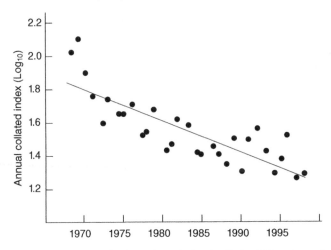

Fig. 5.5 Annual abundance trend of *Arctia caja* in the United Kingdom, from Rothamsted light trap survey data: the collated index revealing substantial decline (see text). (Line fitted by linear regression and has significantly negative slope: −0.018, P < 0.001) (Source: Conrad et al. 2002. Reproduced with permission of John Wiley & Sons).

each year following an initial 56 traps in 1968 when the scheme established. Abundance declined slowly over this period, as shown by a 'collated index' calculated (following Moss and Pollard, 1993, who had used this index for butterflies), as the ratio of total captures between successive years using only those sites sampled in both years and indicating relative change (Fig. 5.5). Until 1987 the proportion of sites occupied remained high (around 0.60) but fell in 1988 (to 0.46) and continued to decline over the remaining survey years. Mean annual abundance/site fell considerably in 1984, and thereafter remained around the lower level. Conrad et al. postulated that within 1984–1987, *A. caja* populations may have been near some critical threshold, leading to a large number of local extinctions Later, Anderson et al. (2008) summarised that the Garden tiger underwent an 85% decline in abundance since 1968, and disappeared from about 30% of sites it previously inhabited.

The 'TRIM' index (for 'TRends and Indices for Monitoring data': Pannekoek & Van Strien 2001) was used by Conrad et al. (2004) to appraise declines of the larger United Kingdom moths. Across a span of 35 years of the Rothamsted light trap records, the annual total number of all macromoths decreased by 35% (Fig. 5.6), with this trend most pronounced in the southern half of the region. Two-thirds of the 337 individual species declined, with 71 entering highly threatened categories based on percentage declines over 10 years: 5-year trends are summarised in Fig. 5.7. Many different biotopes were represented amongst the preferred habitats of declining moths, pointing to widespread deterioration of suitability – in the south perhaps linked with rapid intensification of forestry and agriculture.

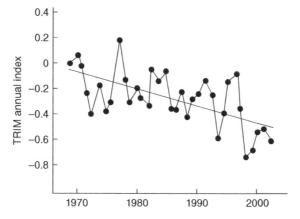

Fig. 5.6 Decrease in total annual trap catches for all macromoths (337 species) in the United Kingdom, indicated by the TRIM index (see text) (Source: Conrad et al. 2006. Reproduced with permission of John Wiley & Sons).

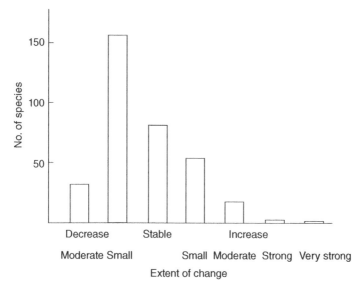

Fig. 5.7 Frequency distribution of extent and change of British macromoth populations estimated from the TRIM index, over 5 years. Stable, changes of <10%; Changes of 10–25% (small), 25–50% (moderate), 50–75% (strong) and >75% (very strong) indicated. (Source: © Rothamsted Research Ltd. Conrad et al. 2004. With permission of Rothamsted Research Ltd).

Priority for conservation

In practice, two rather different levels of decision on taxa for conservation priority have to be made: (1) whether any species needs conservation attention at present or is likely to do so in the near future, and (2) which amongst the deserving species should be given priority over others. The second implicitly includes a hierarchy amongst species, and implies some form of selection or triage whereby some of them may be neglected through lack of resources. The first is the basis for compilation of 'Red Lists' and 'Red Data Books' whose length signals extent of conservation need, and most of which rank taxa in some way to designate hierarchical level of threat. Other criteria may be superimposed on the more usual 'extent of threat' or 'risk of extinction' to partition amongst the candidates. Extent of taxonomic isolation was considered for butterflies by Vane-Wright et al. (1991), with two points of enduring importance emerging: (1) promotion of taxonomically isolated species or species groups as representing significant genetic diversity not occurring elsewhere, and (2) the converse of having to consider finer taxonomic levels (including status inflation, such as raising of putative but possibly spurious subspecies to full species status) within groups amongst which some closely related species are secure. Particularly distinctive and low richness lineages, usually highly localised endemics, may thus gain priority on both local and wider scales. The Mexican endemic swallowtail *Baronia brevicornis* is the only member of the Baroniinae and considered by many experts as perhaps the most ancestral taxon in the Papilionidae – and, even, as a living fossil representing the oldest ancestral butterfly lineage, with important implications for understanding butterfly evolution. It was signalled as a conservation priority by Collins and Morris (1985, as 'Rare'), followed by New and Collins (1991), both recommending continuing monitoring surveys of the small areas where it occurs to safeguard it against declines. Its isolated taxonomic position was emphasised also by Leon-Cortes et al. (2004) in noting that the conservation of *Baronia* 'represents much more than simply the conservation of "another butterfly species"'. Caterpillars feed on several species of *Acacia*, with the food plants more widely distributed than *Baronia*, but the butterfly does not seem to be threatened at present.

Various ways to prioritise amongst species, some drawing largely on criteria devised for better known vertebrates, are core elements of conservation legislation and policy (Chapter 11) devised with the broad general aim of ensuring that the 'most deserving' or 'most at risk' species gain preference on objective and sensible, defensible, grounds. Any such criteria become very difficult to apply to poorly known taxa, including most Lepidoptera throughout the tropics. The major exceptions are the rarer species of swallowtails and birdwings (Papilionidae) highly sought by collectors and for some of which long-term recognition of conservation need has been apparent, not least through listing of birdwings on CITES schedules (Chapter 11, p. 204).

The world's largest butterfly and one of the most charismatic of all insects is Queen Alexandra's birdwing (*Ornithoptera alexandrae*), which has a very restricted distribution in Papua New Guinea, where it depends on tropical rain forest habitats. Listed on CITES schedule 1, it has become increasingly restricted and scarce within its two main range areas, due to pressures on primary forest (for oil palm plantations around Popondetta, and for timber extraction on the Managalese Plateau). *O. alexandrae* exemplifies well the problems of conserving even a very high-profile, spectacular and relatively well documented species in a remote tropical region where resident interest is scarce, and its habitats are threatened. It has become an important flagship and umbrella species for tropical rainforests in the south-east Asia/western Pacific region. As one of Papua New

Guinea's designated 'National Butterflies' (from 1966), conservation notice is assured, but the practicalities of conserving the primary forests in the region are immensely complex, with much of the external zeal for *O. alexandrae* conservation coming from expatriates unfamiliar with life priorities within its range (Parsons 1992b). Conservation capability within a tropical nation is rare in comparison with much of the temperate regions, even to aid such species acknowledged as of global concern and importance. *O. alexandrae* has been the focus of imaginative and broad-based international conservation efforts to reduce people's dependence on rainforest clearance and provide alternative livelihoods and sources of income, involving Australian foreign aid, but the outcomes have not been wholly successful.

O. *alexandrae* brings to the fore the dichotomy (or complementarity) between 'intrinsic values' (those of the species itself) and wider conservation values – here as a remarkable and appreciated flagship taxon acting as an umbrella for sustainability of complex forest biotopes on which numerous less heralded taxa also depend. Simply that many individual Lepidoptera species of conservation significance occur in restricted or specialised biotopes has the potential to accord them 'umbrella species' status, through which their conservation contributes to wellbeing of co-occurring but often undocumented and poorly known taxa. However, whilst this role may be common, it has been specifically investigated only infrequently by detailed surveys of the other taxa present. The Karner blue (*Lycaeides melissa samuelis*) (Chapter 10, p. 189) in Ontario occurs on two sites that yielded several very rare Hymenoptera, including a bee new to Canada (Packer 1991). Conservation of all sites on which the Bay checkerspot (*Euphydryas editha bayensis*) occurs would contribute to conservation of nearly all spring-flowering native plants (Launer & Murphy 1994). Any such documented case may contribute to conservation priority rankings, but the information is usually not available.

For many Lepidoptera with lower public profiles than the few well publicised flagship species, such broad consensus over conservation priority is not easy to obtain. Many strongly held, and sometimes opposing, opinions on the most useful criteria incorporate both ideals and feasibility, often based on incomplete biological and distributional knowledge. Thus the IUCN Red List criteria of threat categorisation used widely to estimate or rank 'risk of extinction' emphasise priority for those species of greatest risk, as 'Critically Endangered' (Table 5.5). Other commentators have suggested the alternative that such taxa may be virtually impossible to conserve, so that triage application dictates that the very limited support available be devoted to cases for which chances of success are greater. Other criteria advanced for priority, in which Lepidoptera have played key roles in the ensuing debates, include (1) isolated lineages – where the 'only species' of the higher taxon, rather than 'another species' of a richer group in which some species are secure; (2) species with a small range, either narrow habitat/biotope requirements and/or local endemics, rather than eurytopic or more widespread species; (3) ecologically specialised ('K-strategist') species, rather than more generalised ('r-strategist') species; and (4) species known to have declined, rather than those that are simply rare. Many other criteria may be advanced, and some depend on local conditions and capabilities (see van Tol

Table 5.5 The five quantitative criteria used to (1) determine whether a taxon is threatened and (2) if so, which category of threat applies. These statements simply demonstrate the range of features that draw attention; the complexities of applying these criteria are discussed by IUCN (2010), and each is ranked or categorised further against stated criteria. Note that quantitative information to assess risk of extinction based on population size are extremely difficult to obtain for Lepidoptera.

Criterion

A. Declining population (past, present and/or projected)
B. Geographic range size, and fragmentation, decline or fluctuations
C. Small population size and fragmentation, decline or fluctuations
D. Very small population or very restricted distribution
E. Quantitative analysis of extinction risk (e.g. by Population Viability Analysis)

and Verdonk, 1988, for parallels amongst European Odonata). All of these, however, are information-based and can be applied more effectively to Lepidoptera than to most other terrestrial insects. Nevertheless, the quantitative information on which to assess rates of decline and estimate population sizes, implicit in the broad IUCN categories, is almost wholly lacking; even the number of populations (Chapter 6) may be very unclear. Much status assessment for Lepidoptera must rely more on distributional information. Absence of systematic surveys may render this highly serendipitous.

Some of the principles are evident in discussion of the 'SPEC' (Species of European Conservation Concern) categories for European butterflies (van Swaay et al. 2009). Butterflies were divided into four categories, depending on their global conservation status, the threat status in Europe, and proportion of the world range within Europe (Table 5.6). 'Spec 1 species' (19, van Swaay et al.) are of the highest priority through being globally threatened and meeting one of the major IUCN threat categories, for example. However, any such categorisation is 'post hoc' in that it is based on evaluations on other criteria – but emphasises the acceptance of endemism and restricted range for priority. These may apply at any spatial scale, but most priority designations are for global, national or other range state/county/province divisions – so that some commonsense approach is needed for contiguous areas in relation to distribution patterns. In Europe, for example, it is not uncommon for range edge or 'political outlier' butterfly populations in one country to be accorded high conservation status, whilst the same taxa are widespread or common elsewhere; these conditions are implicit in the SPEC hierarchy. Attributes of 'place' and attributes of 'species' coincide in various ways, ranging from specific sites as habitats for individual species to concentrations of species richness, species endemism and threats of extinction.

A parallel classification of conservation categories proposed for moths in Scotland (Bland & Young 1996) recognised both biological and distributional knowledge available and the practical aspect of whether conservation action could provide any benefit (Table 5.7). The highest-priority species are those most restricted, most threatened and so most evidently in need. Notably, even species

Table 5.6 Criteria for categories of European butterflies as Species of European Conservation Concern (SPECs). Data from van Swaay et al. (2009).

Category		Status
SPEC 1		Species of global conservation concern because restricted to Europe and considered globally threatened (Critically Endangered, Endangered, Vulnerable)
SPEC 2		Species whose global distribution is concentrated in Europe and that are considered threatened in Europe
SPEC 3		Species whose global distribution is not concentrated in Europe, but that are considered threatened in Europe
SPEC 4	4a	Species whose global distribution is restricted to Europe, but that are not considered threatened either globally or in Europe
	4b	Species whose global distribution is concentrated in Europe, but that are not threatened either globally or in Europe

Table 5.7 Categories used to indicate need for conservation attention amongst Scottish moths (Bland & Young 1996).

Category	Status
X	Species for which it is probably too late!
1A	Scottish species or subspecies with a very restricted distribution and potentially or already in need of protection
1B	Scottish species in urgent need of research into biology and/or distribution: potentially in need of protection
1C	Scottish species for which better distribution data are needed but which seem to be reasonably widespread and secure
2A	British species of very restricted distribution with colonies in Scotland in urgent need of protection
2B	British species with colonies in Scotland, in urgent need of research into biology and/or distribution
3	Species on the edge of their range in Scotland, not needing immediate protection

from the first category ('Species for which any action is probably too late', mostly because they are presumed extinct) is important, because further focused exploration may lead to discovery of previously unknown populations or sites – and, if it does not, the status might be deemed more reliable. Also from Britain, criteria qualifying a species to be treated under the United Kingdom Biodiversity Action Plan heed both local status and any listings under a range of wider Acts or Codes (Table 5.8); despite inadequate knowledge of many taxa (Parsons 2004), 53 moth species were listed according to these qualifying criteria. The species were designated according to several criteria, including (1) being threatened endemic or other globally threatened species; (2) species where the United Kingdom has more than 25% of the world or other appropriate biogeographical population; (3) species where numbers or range have declined by more than

Table 5.8 Parameters and criteria used to assess priority species under the United Kingdom Biodiversity Action Plan listings.

Species assessed under the criteria

---

1. Threatened internationally?
2. International responsibility and 25% decline in the UK
3. More than 50% decline in the UK
4. Other important factor(s), such as the species is declining and is a good 'indicator' or 'flagship' that highlights a conservation issue

---

Categories

Priority Species:
1. Globally threatened species
2. Species whose numbers or range in the UK has declined by >50% in the last 25 years

Species of Conservation Concern:
1. Those endemic or other globally threatened species
2. Species whose numbers or range has declined by >25% in the last 25 years

---

25% in the last 25 years; (4) in some instances, where the species is found in fewer than 15 10 km squares in the UK; and (5) species listed under other policy schedules, namely the European Union Birds or Habitats Directives, the Bern, Bonn or CITES Conventions, or under the Wildlife and Countryside Act 1981 and the Nature Conservation and Amenity Lands (Northern Ireland) Act 1985. The long list is reviewed constantly, but species treated as lower priority within it include (1) species not seen for several years and which may have become extinct; (2) species whose ecology is poorly known and for which no reliable site is known as a conservation focus; and (3) a few for which the resources needed for conservation are high and not currently available. Parallel priority lists of Lepidoptera exist for many parts of the world, with varying levels of accuracy and completeness (Chapter 11).

Species to areas

Fattorini (2009), again from European butterflies, discussed whether areas with large numbers of threatened species should be of high conservation interest, countering this popular advocacy principle with two widespread conservation arguments, namely (1) that in addition to targeting highly threatened species, conservation should prevent the many other taxa from becoming threatened, and (2) the point noted earlier (p. 66) that the small chances of success for highly threatened taxa might represent 'false economy'. He noted that such areas as hotspots of threatened species might achieve this status because threats are so severe that any conservation effort might prove ineffective. Concentrations of threatened species may indicate places that merit special conservation attention. In Brazil's Atlantic Forests, rare butterflies often co-occur with threatened species of other groups, leading to what Brown (1991, see also Brown & Freitas 2000) referred to as 'palaeoenvironments', in reference to presence of these rare and specialised species that had largely disappeared from other areas and anthropogenic environments. The species of concern 'merit special attention for their extreme geographical restriction, diminishing population levels, vanishing habitats, reduced adaptability, or apparent disappearance' (Brown & Freitas, 2000: 952).

'Area of occupancy' is one of the most frequently used estimators of extinction risk, and is expressed commonly as the numbers of 'grid cells' in maps of the area of interest occupied by the species within its 'extent of occurrence' (Fig. 5.8). Fattorini (2006) devised an Index of 'Biodiversity Conservation Concern' (BCC) that incorporated information on the status of each species present in a country, based on available Red List data, and the total richness of taxa (here, butterflies) present – so assigning a biodiversity value based on many species to

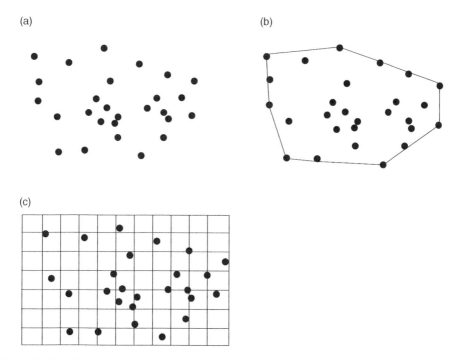

Fig. 5.8 The distinction between 'extent of occurrence' and 'area of occupancy'. (a) The 'spatial distribution of known, inferred or projected sites of occurrence; (b) a possible boundary to the extent of occurrence, as the area contained within the shortest imaginary boundary that encompasses all sites shown in 'a'; (c) a measure of area of occupancy, the sum of occupied grid squares, as the occupied area within the extent of occurrence (Source: IUCN 2010. Reproduced with permission of IUCN Publications).

an area and facilitating assessments of ecological effects based on those species. His index emphasised that high species richness does not necessarily imply high conservation priority, and that a threat analysis is also needed to determine this. A hypothetical example discussed by Fattorini compared two communities: (1) 10 species, five threatened, and (2) 20 species, five threatened. On one approach, the latter could gain priority because of higher richness, but the first community could be considered the more threatened because of the higher proportion (half) of species threatened. As with many interpretations of diversity and conservation priority, the importance of hotspots will continue to be debated – with different criteria giving different rankings but unified in a general sentiment that the greatest diversity must be conserved most effectively. As Fattorini (2009) commented, aspects of conservation concern include total species richness, and levels of endemicity and threat, which may or may not be inter-related.

Distributional and endemicity knowledge for butterflies, and some other Lepidoptera, is relatively sound, but information on risk of extinction is far less well documented. The basic criteria for species status assessments central to indices such as those noted above and used to compile advisory red lists that are subsequently influential may commonly be taken as reasonably sound; that acceptance is critical in subsequent uses. Assessing communities and assemblages by summed properties of constituent species manifestly depends on the accuracy of the species-level information incorporated.

In contrast to reliance on restricted and relatively well documented taxonomic segregates (Chapter 2, p. 26), priority areas within the archipelago for the whole of the rich butterfly fauna of the Philippines (namely a total of 915 species and 910 widely recognised subspecies, with 365 species and a total 1079 taxa not found elsewhere and, so, endemic: Treadaway 1995) were investigated by Danielsen and Treadaway (2004). As in some other parts of the Old World tropics, the number of officially designated protected areas in the Philippines is considerable – but in practice many are not acknowledged properly 'on the ground' and remain largely unprotected and subject to exploitation. One hundred and thirty-three Philippines butterfly taxa were assessed as globally threatened or conservation dependent – 100 of these butterflies are species endemic to the Philippines and 33 are endemic subspecies, collectively comprising 64 species and 69 subspecies. The entire terrestrial fauna of the Philippines depends heavily for its protection on a series of 14 sites (totalling about 800,000 ha of reserved or potentially reserved land) designated as 'Priority Sites'. Representation of the 133 priority butterflies in these is low (Figs. 5.9, 5.10). Only 29 taxa are definitely represented in one or more Priority Sites, and a further 36 are likely to be found there. Around half (68, 51%) do not occur in any Priority Site, and Danielsen and Treadaway nominated a suite of additional critical areas needed for protection to increase butterfly taxon representation. They noted that 94 (71%) of the taxa do not have a resident population within any existing Priority Site (68, above) or that any existing population there is not stable (a further 26 taxa). With their new site suggestions, there are 29 'totally irreplaceable areas' for threatened or conservation-dependent butterflies and, if all are to be conserved, a total of 25 areas additional to the currently recognised 14 will be needed. Table 5.9 lists some of the caveats that result from incomplete knowledge, and – despite the

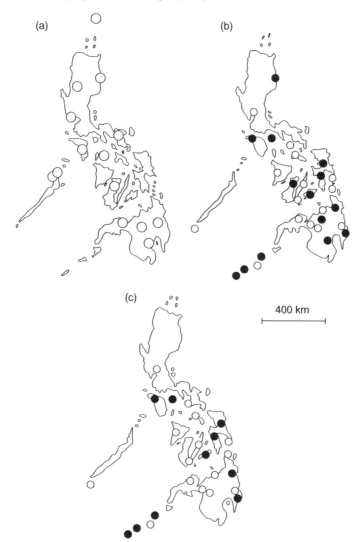

Fig. 5.9 Adequacy of protected areas for representation of butterflies in the Philippines. (a) Locations of the 14 terrestrial Priority sites, totaling ca 800,000 ha of reserved or proposed reserved land as major foci for conservation of terrestrial biota. (b) Pattern of irreplaceability irrespective of Priority Sites for threatened/conservation dependent species/subspecies of butterflies; (c) same, but on patches other than Priority Sites. (b, c: open circles, one taxon; solid circles, more than one taxon not covered without that site) (Source: Danielsen & Treadaway 2004. Reproduced with permission of John Wiley & Sons).

disadvantages of relying on single-site representation of all taxa (as likely to be inadequate in view of future changes) – Danielsen and Treadaway (2004) noted that many Philippine endemic butterflies are currently known only from single sites, and that the practicalities of protecting sites within existing pressures and levels of forest degradation are very limited. However, innovatively, the sites

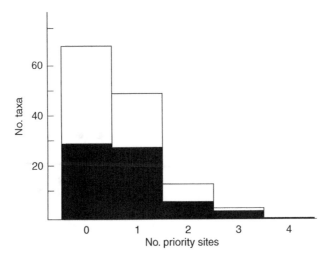

Fig. 5.10 Representation of threatened and conservation-dependent butterfly species (black) and subspecies (open) in the 14 priority areas of the Philippines (Source: Danielsen & Treadaway 2004. Reproduced with permission of John Wiley & Sons).

Table 5.9 Caveats noted by Danielsen & Treadaway (2004) for the designations of priority/important areas and priority sites for butterflies in the Philippines.

1	Current knowledge of distribution, status and biogeography of Philippines butterflies is very limited. Many taxa still poorly recorded, and new taxa continue to be discovered and described. Survey sites are thus in general very incompletely recorded, and some presumed local endemics and disjunct populations may have wider distribution that currently documented
2	Unknown if records and assigned conservation status accurately reflect true distributions and status. Consistent application of criteria difficult, and much inference is inevitable, particularly when information collated/evaluated by different people
3	Butterflies were used as surrogate information for overall biodiversity value of sites, but such patterns remain to be investigated or validated, and cross-taxon congruence between complementary sets of priority areas not yet established

were compared for the important parameters of vulnerability (the proportion of the areas deforested, indicating risk of habitat loss and change) and irreplaceability (the number of threatened and conservation-dependent taxa confined, or possibly confined, to each site, indicating extent to which loss of area will influence total taxon conservation). The outcome helps to demonstrate aspects of priority – but in some cases, as exemplified well in this case, no known replacement sites are available, so that site conservation is the only option for conserving, in this case, 21 distinctive taxa.

The Philippines butterflies illustrate another important general principle in emphasising the importance of island endemic Lepidoptera. On the Comoro Islands (Indian Ocean) the rich endemic butterfly fauna is restricted largely to upland forest areas that now remain only at those higher elevations, above about

500 m (Lewis et al. 1998). In common with many islands, including most throughout the Pacific, lowland areas have undergone dramatic changes for human needs, with the twin major impacts of losses of native vegetation and biotopes and facilitation of alien invasive species (Chapter 12). The modified lowland forests of the Comoros support very few endemic taxa, whilst the upland forests contain few non-endemic species. The islands harbour about 32 endemic butterfly taxa, of which four are globally threatened. Lewis et al. noted that these endemic taxa are taxonomically diverse, representing many different genera, so would collectively receive 'high value' in conservation rankings that draw on issues of taxonomic distinctiveness.

The pioneering IUCN Red Data Book for the notable flagship group of swallowtail butterflies (Collins & Morris 1985) remains one of the most influential and thorough compilations for any insect group, and the introduction encapsulates many of the 'flagship values' of these insects. They were selected for treatment because (1) they are relatively well known, and familiar to people in many walks of life; (2) they are large, spectacular, and relatively easy to see, identify and monitor; and (3) they have a worldwide distribution, and occur in a variety of habitats across many different food plant families, but with many very localised species. Collins and Morris recognised 573 species of Papilionidae, and the family is clearly at its richest within the tropics (Fig. 5.11). However, even for these, arguably the best known of all Lepidoptera, about 100 species could not

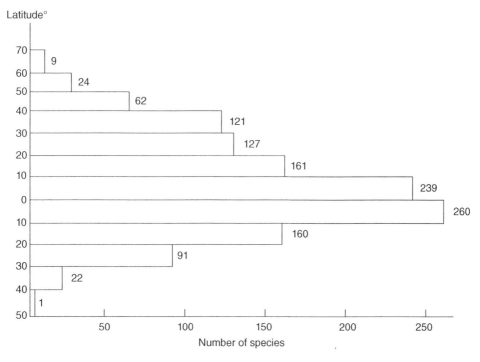

Fig. 5.11 Global distribution profile of swallowtail butterflies (Papilionidae). Latitudinal distribution (10° intervals north and south from the Equator) with number of species recorded from each zone (Source: Collins & Morris 1985. Reproduced with permission of IUCN Publications.).

be assigned to any Red Data Book category because of being poorly known. For the 78 taxa for which some degree of threat was then known or suspected, Collins and Morris emphasised that status attribution was dynamic and open to change as conditions change or new information accrues. Individual dossiers were provided for endangered species (4), endangered subspecies (3), vulnerable species (24), rare species (20), indeterminate species (13) and those insufficiently known (14). Knowledge of many of these taxa has not increased much since that account was published, and the variety of concerns and the wide geographical distribution of species of concern remains.

Critical faunas

One broader application to establishing priorities is that of 'critical faunas', in which patterns of richness and incidence are used to determine a minimum set of priority areas that collectively harbour at least one population of every species. The principle was initiated through analyses of two well known butterfly groups, Papilionidae (Collins & Morris 1985) and Nymphalidae: Danainae (Ackery & Vane-Wright 1984), in both of which species-level phylogenies enable sound critical interpretation of distribution and speciation in relation to food plants.

The development of the 'critical faunas' approach to assessing butterfly conservation priority has been a notable advance from single species assessments, but based fundamentally on those. Units in which to detect and evaluate endemism and richness can either be habitat-based (as in Brown's 1991 Neotropical rainforest refugia), biogeographical provinces (as by Lamas, 1982, who used butterflies to distinguish 48 provinces in Peru, many of them with endemic species or subspecies), or political entities. The last of these was adopted for swallowtails (Collins & Morris 1985), on the pragmatic grounds that these are the units within which conservation is administered. They set out to determine the smallest number of countries that could collectively include all 573 taxa. The principles are summarised in Table 5.10, which shows only the first part of a published table of 51 countries, in descending order of endemic species or species not represented in countries higher in the list, so adding most taxa to a cumulative total. The 51 countries complete representation of all taxa in at least one country. Forty-three countries include one or more endemic species. Where two or more countries had the same number of endemic species, they are ranked in descending order of total species richness.

The list commences with Indonesia, a complex archipelago with far more endemic Papilionidae than any other country and a total of 121 species. The 'top five' countries (Indonesia, Philippines, China, Brazil, Madagascar) together account for more than half the world total (309 species, 54%). The system indicates well the relative contributions of each country to global faunal representation, but having such background information notionally so valuable in suggesting priority is very far from practical conservation, and even defining what that conservation should comprise. Indonesia, for example, was compared across the seven provinces (Collins & Morris 1985), helping to indicate within-country priorities. That analysis suggested strongly where needs may be greatest,

Table 5.10 The 'critical faunas' approach to setting conservation priorities, as outlined by Collins & Morris (1985) for swallowtail butterflies. Reproduced with permission from IUCN Publications.

Objective:

To find the smallest number of countries that include all recognized species, here 573 species of Papilionidae

---

Approach:

---

1. Determine number of endemic species in each country (or other analysis unit) (A, below)
2. List countries (or other) in decreasing order of numbers of endemics
3. If two or more have same number of endemics, list in decreasing order of total number of species
4. For each entry on list, list also also the number of non-endemic species in two classes as (B) not found in any country higher on the list and (C) also found in previously listed countries
5. The 'newly accountable species' for each country are thus 'endemics' plus 'non-endemics not previously listed' (A + B), to give a cumulative representation of species with each list entry (E)
6. 'Richness' can also be listed, as 'total species found in the country' (endemics plus both categories of non-endemics: A + B + C)
7. List countries in sequence with all species accounted – if no endemics, list in decreasing order of additional representation. In this example, 43 countries contained one or more endemic swallowtails, with the remaining non-accounted species (12) distributed across a maximum of eight further areas, so that 51 countries contributed to the total world fauna

---

Example: the top five countries for Papilionidae, to illustrate above.

	Status (numbers of species)				
	Endemic	Non-endemic		Newly accounted	Cumulative
Country	A	B	C	(A + B)	E
Indonesia	53	68	0	121	121
Philippines	21	4	24	25	146
China	15	61	28	76	222
Brazil	11	63	0	74	296
Malaysia	10	3	0	13	309

and could imply a relative duty of conservation responsibility – but applies mostly to countries where practical measures are largely impracticable, and in which even high-ranking designated protected areas continue to be degraded. Country-level analysis is only the first step in a reducing hierarchy to demonstrate areas and biotopes that have within-country priority. Those further analyses have only rarely been undertaken, but local 'hotspots' of richness are sometimes well known to local enthusiasts. Other considerations, of course, may apply – Collins and Morris exemplified some of these through Myanmar (country number 44 in the list), which has no endemic swallowtails but, with 68 species, has the fifth largest natural richness total, so clearly merits high priority on that basis. In contrast they noted also Italy (country 51) with no endemics and only eight species – but with two of those noted as threatened, so becoming possible flagships in a conservation-aware country.

In contrast to the widely distributed Papilionidae, most of the 157 'macrospecies' of Danainae (milkweed butterflies) are concentrated in the Indo-Pacific region (Ackery & Vane-Wright 1984), so that a stronger regional focus is possible in assessing area priorities through critical faunas. Thus, the four countries/ islands with greatest endemism (Sulawesi, nine endemic species; Biak, four; Mindanao, four; New Guinea, three) have combined representation of 69 Danainae – over 40% of the global fauna and more than half of the regional pool. Further sequential adding of regional areas with endemics brings the regional total to 115 (or 92%). 'Only' 31 selected territories are needed to represent populations of all 157 species (Fig. 5.12), but the considerable evolutionary interest of danaines led Ackery and Vane-Wright to list other 'outstanding danaine faunas' under the major themes of evolution of mimicry, ecology and biogeography.

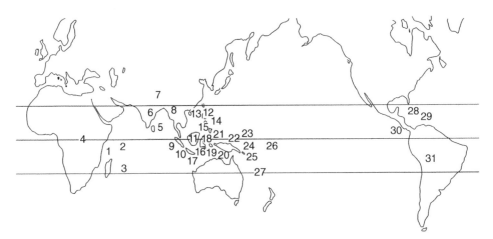

Fig. 5.12 The critical regional faunas of milkweed butterflies (Nymphalidae: Danainae) as complementary conservation priorities to assure representation of at least one population of each species (numbered areas are: 1, Comoro Islands; 2, Seychelles; 3, Mauritius; 4, Zaire; 5, Sri Lanka; 6, southern India; 7, Nepal, 8, Burma (Myanmar); 9, Sumatra; 10, Java; 11, Borneo; 12, Luzon; 13, Negros; 14, Mindanao; 15, Sulawesi; 16, Sumbawa; 17, Sumba; 18, Flores; 19, Timor; 20, Seram; 21, Biak; 22, New Guinea; 23, New Ireland; 24, New Britain; 25, Guadalcanal; 26, San Cristobal; 27, New Caledonia; 28, Hispaniola; 29, Cuba; 30, Costa Rica; 31, Bolivia) (Source: Ackery & Vane-Wright 1984. Reproduced with permission of The Natural History Museum.).

Related approaches

A related but somewhat different approach, based on a considerably greater level of information, was involved in selecting 'Prime Butterfly Areas' (PBAs) for Europe (van Swaay & Warren 2003, 2006). Selection was based on sites within each country and focusing on species that are agreed conservation priorities. Selection of the taxa drew on the SPEC framework (p. 75), with 34 species selected on at least two of three criteria as (1) global distribution restricted to Europe (pool of 189 species); (2) threatened in Europe (71); and (3) listed in Appendix 2 of the Bern Convention, or on the European Habitats Directive (23). Site selection aimed to select the three 'best sites' or wider areas for each species, or larger regions for more widely distributed species, with areas for different target taxa combined wherever possible. Altogether, 431 priority Butterfly Areas were identified across Europe, encompassing 37 countries and three archipelagos and comprising 1.8% of Europe's land area. These PBAs should progressively be recognised 'officially' as appropriate targets for conservation. At present, many are not protected under any international laws, and actions recommended from this document included increasing and assuring protection, with sound and sympathetic management within and around the areas, in conjunction with systematic monitoring of target species.

This exercise has stimulated a number of penetrating attempts to define important areas for butterflies in individual European and adjacent countries, capitalising on the amount of information available to explore ways in which this selection can be approached to distinguish areas that merit high priority for conservation. Each of these efforts differs from simply setting a national list of threatened species or Red List (see later in this section) to providing spatially explicit information of value for practical management. Thus, for Italy, Girardello et al. (2009) envisaged three ways in which the outcomes of their analysis could be used, as (1) selecting new protected areas for butterflies; (2) modifying the approach for use at regional scales by local authorities, should sufficient fine-scale information be available; and (3) selecting areas where special emphasis is placed on long-term planning that integrates wider conservation measures. The dual outcome was detection of areas of high biological value and those of interest for species of conservation concern, and the Italian approach drew on a zonation study that provided hierarchical prioritisation of the landscape based on values of sites for 232 butterfly species distributed across 670 $10 \times 10\,$km recording squares. Likewise, the butterflies of Turkey have been used in a site prioritisation exercise (Zeydanli et al. 2012) based on complementarity but drawing on distributional information from 358 species.

Site rankings for priority can draw on many detailed approaches incorporating Lepidoptera, but most fall into one of two major groups, as (1) score-based, with some numerical value accorded for selected criteria such as species richness, endemism, threat status of species present or others that can be summed to imply relative value; and (2) complementarity-based approaches involving selecting sites to progressively complement each other in the biodiversity they support, toward the ideal of at least one representation of every species. The first approach is by far the more common but the second – sometimes referred to as 'systematic conservation planning' or 'SCP' – is receiving increasing attention.

For the Aegean Islands butterflies, Fattorini (2006) ranked areas based on conservation value using a combination of the conservation status of each species within an assemblage and the total richness, an approach he termed the Index of Biodiversity Conservation Concern. This was claimed to have two important features, namely utilising many species to assign a biodiversity value, and facilitating multispecies assessments of ecological effects because of its emphasis on degree of threat. Fattorini noted also that it includes information on endemicity, simply because most endemic butterflies are considered threatened.

Level of threat to the individual species present is a frequent component of any score-based system for appraising sites, with much of the requisite information derived from various forms of national or other listing of threatened species. 'Red Listing' is a critical aspect of increasing awareness of conservation need and estimating the extent of that need. The lists are advisory, but many have formed the basis for legal schedules that confer conservation obligations on the enacting bodies (Chapter 11). However, as Lewis and Senior (2011) forcefully put it 'We do not know what fraction of butterflies globally is threatened with extinction'. Within the IUCN (most recently IUCN 2010) criteria (Table 5.5), criterion B is the most usable in butterfly assessments, reflecting extent and rate of decline of geographical range. The 'Data Deficient' category is sometimes very difficult to apply, and may be confused with 'Critically Endangered' unless historical information is available. However, the vast majority of the many species known from single specimens must clearly fall into this category. Lewis and Senior (2011) discussed the example of the lycaenid *Pseudaletis arrhon*, known from a single male collected in Cameroon about a century ago. Recent intensive collecting in the region has led to many other Theclinae associated with *Crematogaster* ants being accumulated, but no further *P. arrhon*. Many members of this group of butterflies are very localised, and it is possible that *P. arrhon* may still occur in the large remaining areas of forest – but the ambiguity over its continued existence is clear.

More broadly, extinctions among West African butterflies may in fact be far fewer than might be expected in view of the massive loss and degradation of forests in the region. Less than 13% of forest cover remains, and this substantial habitat loss for forest-dependent species suggests correlative parallel losses of butterflies. Larsen (2008) noted the likelihood of local butterfly extinctions, but explored whether all such butterflies would 'survive somewhere in West Africa'. His persistent field studies and scholarship (see Larsen 2005) revealed the regional fauna to comprise 972 forest species recorded from Africa west of the Dahomey Gap – 842 essentially forest species (120 endemic to the region), 50 ubiquitous species and 80 adventive savanna species (Table 5.11). Almost all (938 species) were recorded during the period 1990–2006, with those few not caught personally by Larsen verified by him, so that 97% of all butterflies known from the region were positively confirmed. Examining further, all but three species were present sometime during 1960–2006. Of the 34 species not recorded most recently, 22 were found elsewhere in Africa, so that only 12 were deemed 'missing'. All of these are rare, seasonal and highly local – Larsen (2008) referred

Table 5.11 Butterflies of West Africa, to the west of the Dahomey Gap; broad biogeographical and ecological categories (Source: Larsen 2008. With kind permission from Springer Science+Business Media).

Total butterfly species	1042
Endemic to both subregions west of Dahomey Gap	70
Endemic to Liberian subregion	43
Endemic to Ghanaian subregion	33
Wide distribution	896
Species occurring in forests:	972
Essentially forest species	842
Adventive savanna species	80
Ubiquitous species	50
Strict savanna species:	70
Centred on Guinea savanna	22
Centred on Sudan savanna	48
Forest species	
Present in West Africa 1960–2006 969 1990–2006 938	

to some as 'once-in-a-lifetime butterflies', very unlikely to be encountered during any individual survey visit. Six were eventually recorded during 2007–2008 and a few others may be 'phantom species' confused with others or wrongly labelled. Following from the above discussion, several of the missing species remain 'Data Deficient'.

The nine categories amongst which Red List assessed species are now allocated (Table 5.12) provide a 'snapshot' of a current situation (as does Collins & Morris 1985) that can help to indicate taxa, habitats and political entities for which heightened concerns exist, in addition to showing patterns of richness and ende-micity. The more recently adopted 'Red List Index' compares the number of threatened species across repeated re-assessments – and Lewis and Senior (2011) noted that (in March 2010) only 425 Lepidoptera in total had any global Red List assessment; 372 are true butterflies (Papilionoidea) and many appeared to have been given a status that exaggerated their risk of extinction, as a precautionary approach to reflect lack of knowledge. Many assessments are also out-of-date, and there is clear geographical bias toward the better known temperate faunas. For many parts of the world, assessment of conservation status of individual species is largely speculative and, at most, very preliminary as not backed by any substantial objective information. In such cases, conservation need might, rather, be guided initially by richness, so that putative 'hotspots' of Lepidoptera diversity gain notice as a future focus for more detailed investigation and comparison. Thus, for China, the records of 184 butterfly species (more than 10% of those recorded from China) from the Bifeng Valley within a Natural Reserve in Gansu Province – and from which the reserve total of 464 butterfly species was far higher than from other parts of the region – led Li et al. (2011) to claim that the valley had very high conservation importance. It also represented a transitional zone between the temperate Palaearctic fauna and the subtropical Oriental region.

Table 5.12 The IUCN Red List categories: summary of notation and meanings. (Source: IUCN 2010. Reproduced with permission from IUCN Publications).

Category	Abbreviation	Description
Extinct	EX	No individuals remaining
Extinct in the wild	EW	Known only to survive in captivity or as a naturalised population outside its historical range
Critically Endangered	CR*	Extremely high risk of extinction in the wild
Endangered	EN*	High risk of extinction in the wild
Vulnerable	V*	High risk of endangerment in the wild
Near threatened	NT	Likely to become threatened in the near future
Least concern	LC	Lowest risk. Does not qualify for a more at risk category. Widespread and abundant taxa are included here
Data Deficient	DD	Insufficient data to make an assessment of risk of extinction
Not Evaluated	NE	Has not yet been evaluated against the criteria

*These three categories constitute 'Threatened'

Table 5.13 Criteria used to allocate Noctuidae species from Serbia to the IUCN categories of extinction risk (Source: Stojanovic et al. 2012. With kind permission from Springer Science+Business Media).

Category (no. spp)	Criteria used for allocation
Regionally extinct (RE) (2)	Not recorded in last 50 years (since 1960) at former localities in Serbia
CR (28)	Species with fragmented distribution, found at few locations; often single or two populations at 1–2 localities. Populations of low abundance, estimated to have declined by >80% over 10 years
EN (49)	Fragmented distribution, in low number of localities; often five or fewer populations at 3–5 localities. Populations of low abundance, estimated to have declined by 50–79% over 10 years
V (58)	Fragmented distribution, but more widely distributed; often 20 or fewer populations at 6–20 localities. Populations relatively abundant and stable, declines estimated at 30–49% over 10 years
NT (64)	Area of distribution across whole territory of Serbia; often 30 or fewer populations across 21–30 localities. Populations of low abundance but stable, declines of at least 30% in area of occupancy, so occupancy values meet threshold for VU but only one of the subcriteria satisfied
DD (22)	Area of distribution over whole territory of Serbia; often 40 or fewer populations at 31–40 localities. Populations of great abundance
LC (342)	Area of distribution across whole territory of Serbia; more than 40 populations at more than 41 localities. Populations characterised by great abundance of individuals

Applying the IUCN criteria consistently across a regional fauna and comparing the outcomes with other regions requires a 'commonsense approach' to assure credibility. For assessments of the Serbian Noctuidae, for which 223 of the 556 resident species were placed on the country's Red List, with many of these threatened (Stojanovic et al. 2012) according to criteria summarised in Table 5.13, incorporating scale of distribution, numbers of populations and wider abundance. The status allocated was based predominantly on assessment against two important IUCN criteria – reduction of population size (A2) and area of occupancy (B2), with the first of these commonly not reliable in other places because of lack of historical information on trends, and insufficient habitat data. The Serbian surveys included about 350 sites and more than 230,000 specimens, and the background given emphasises the cautious approach needed in exercises of this nature.

References

Ackery, P. & Vane-Wright, R.I. (1984) Milkweed Butterflies: their Cladistics and Biology. British Museum (Natural History), London and Cornell University Press, Ithaca, New York.

Anderson, S.J., Conrad, K.F., Gillman, M.P., Woiwod, I.P. & Freeland, J.R. (2008) Phenotypic changes and reduced genetic diversity have accompanied the rapid decline of the garden tiger moth (*Arctia caja*) in the U.K. *Ecological Entomology* **33**, 638–645.

Asher, J., Warren, M., Fox, R., Harding, P., Jeffcoate, G. & Jeffcoate, S. (2001) The Millennium Atlas of Butterflies in Britain and Ireland. Oxford University Press, Oxford.

Axmacher, J.C. & Fiedler, K. (2004) Manual versus automatic moth sampling at equal light sources – a comparison of catches from Mt. Kilimanjaro. *Journal of the Lepidopterists' Society* **58**, 196–202.

Barlow, H.S. & Woiwod, I.P. (1989) Moth diversity of a tropical forest in Peninsular Malaysia. *Journal of Tropical Ecology* **5**, 37–50.

Beck, J., Schulze, C.H., Linsenmair, K.E. & Fiedler, K. (2002) From forest to farmland: diversity of geometrid moths along two habitat gradients on Borneo. *Journal of Tropical Ecology* **18**, 33–51.

Birkinshaw, N. & Thomas, C.D. (1999) Torch-light transects for moths. *Journal of Insect Conservation* **3**, 15–24.

Bland, K.P. & Young, M.R. (1996) Priorities for conserving Scottish moths. pp. 27–36 in Rotheray, G.E. & MacGowan, I. (eds) Conserving Scottish Insects. Edinburgh Entomological Club, Edinburgh.

Bodner, F., Mahal, S., Reuter, M. & Fiedler, K. (2010) Feasibility of a combined sampling approach for studying caterpillar assemblages – a case study from shrubs in the Andean montane forest zone. *Journal of Research on the Lepidoptera* **43**, 27–35.

Brehm, G. & Axmacher, J.C. (2006) A comparison of manual and automatic moth sampling methods (Lepidoptera: Arctiidae, Geometridae) in a rain forest in Costa Rica. *Environmental Entomology* **35**, 757–764.

Brown, K.S. Jr (1991) Conservation of Neotropical environments: insects as indicators. pp. 349–404 in Collins, N.M. & Thomas, J.A. (eds) The Conservation of Insects and their Habitats. Academic Press, London.

Brown, K.S. Jr & Freitas, A.V.L. (2000) Atlantic Forest butterflies: indicators for landscape conservation. *Biotropica* **32**, 934–956.

Bucheli, S.R., Horn, D.J. & Wenzel, J.W. (2006) Sampling to assess a re-established Appalachian forest in Ohio based on gelechioid moths (Lepidoptera: Gelechioidea). *Biodiversity and Conservation* **15**, 489–502.

Collins, N.M. & Morris, M.G. (1985) Threatened Swallowtail Butterflies of the World. IUCN, Gland and Cambridge.

Conrad, K.F., Woiwod, I.P. & Perry, J.N. (2002) Long-term decline in abundance and distribution of the garden tiger moth (*Arctia caja*) in Britain. *Biological Conservation* **106**, 329–337.

Conrad, K.F., Woiwod, I.P., Parsons, M., Fox, R. & Warren, M.S. (2004) Long-term population trends in widespread British moths. *Journal of Insect Conservation* **8**, 119–136.

Conrad, K.F., Warren, M.S., Fox, R., Parsons, M.S. & Woiwod, I.P. (2006) Rapid declines of common, widespread British moths provide evidence of an insect biodiversity crisis. *Biological Conservation* **132**, 279–292.

Danielsen F. & Treadaway, C.G. (2004) Priority conservation areas for butterflies (Lepidoptera: Rhopalocera) in the Philippines islands. *Animal Conservation* **7**, 79–92.

Ehrlich, P.R. & Hanski, I. (eds) (2004) On the Wings of Checkerspots. A Model System for Population Biology. Oxford University Press, Oxford.

Fattorini, S. (2006) A new method to identify important conservation areas applied to the butterflies of the Aegean Islands (Greece). *Animal Conservation* **9**, 75–83.

Fattorini, S. (2009) Assessing priority areas by imperiled species: insights from the European butterflies. *Animal Conservation* **12**, 313–320.

Fiedler, K. & Truxa, C. (2012) Species richness measures fail in resolving diversity patterns of speciose forest moth assemblages. *Biodiversity and Conservation* **21**, 2499–2508.

Fisher, R.A., Corbet, A.S. & Williams, C.B. (1943) The relation between the number of individuals in a random sample of an animal population. *Journal of Animal Ecology* **12**, 42–58.

Fry, R. & Waring, P. (2001) A Guide to Moth Traps and their Uses. Amateur Entomologists' Society, London.

Gibson, L. & New, T.R. (2007) Problems in studying populations of the golden sun moth, *Synemon plana* (Lepidoptera: Castniidae) in south-eastern Australia. *Journal of Insect Conservation* **11**, 309–313.

Girardello, M., Griggio, M., Whittingham, M.J. & Rushton, S.P. (2009) Identifying important areas for butterfly conservation in Italy. *Animal Conservation* **12**, 20–28.

Groenendijk, D. & van der Meulen, J. (2004) Conservation of moths in The Netherlands: population trends, distribution patterns and monitoring techniques of day-flying moths. *Journal of Insect Conservation* **8**, 109–118.

Grotan, V., Lande, R., Engen, S., Saether, B.E. & DeVries, P.J. (2012) Seasonal cycles of species diversity and similarity in a tropical butterfly community. *Journal of Animal Ecology* **81**, 714–723.

Haddad, N.M., Hudgens, B., Damiani, C., Gross, K., Kuefler, D. & Pollock, K. (2008) Determining optimal population monitoring for rare butterflies. *Conservation Biology* **22**, 929–940.

Harding, P.T., Asher, J. & Yates, T.J. (1995) Butterfly monitoring 1 – recording the changes. pp. 3–22 in Pullin A.S. (ed.) Ecology and Conservation of Butterflies. Chapman & Hall, London.

Hebert, P.D.N. (1980) Moth communities in montane Papua New Guinea. *Journal of Animal Ecology* **49**, 593–602.

Holloway, J.D. (1977) The Lepidoptera of Norfolk Island. Their Biogeography and Ecology. W. Junk, The Hague.

Holloway, J.D. & Nielsen, E.S. (1999) Biogeography of the Lepidoptera. pp. 423–462 in Kristensen, N.P. (ed.) Lepidoptera: Moths and Butterflies. Vol.1. Evolution, systematics and biogeography. Handbuch der Zoologie, Vol. 4, part 35. W. de Gruyter, Berlin and New York.

Hudgins, R.M., Norment, C. & Schlesinger, M.D. (2011) Assessing detectability for monitoring of rare species: a case study of the cobblestone tiger beetle (*Cicindela marginipennis* Dejean). *Journal of Insect Conservation* **16**, 447–455.

Intachat, J. & Woiwod, I.P. (1999) Trap design for monitoring moth biodiversity in tropical rainforests. *Bulletin of Entomological Research* **89**, 153–163.

Isaac, N.J.B., Girardello, M., Brereton, T.M. & Roy, D.B. (2011) Butterfly abundance in a warming climate: patterns in space and time are not congruent. *Journal of Insect Conservation* **15**, 141–148.

Iserhard, C.A., Brown, K.S. & Freitas, A.V.L. (2013) Maximised sampling of butterflies to detect temporal changes in tropical communities. *Journal of Insect Conservation* DOI 10-1007/s10841-013-9546-z.

IUCN (2010) Guidelines for using the IUCN Red List Categories and Criteria, version 8.1 (August 2010). IUCN Species Survival Commission, Gland (http://intranet.iucn.org/webfiles/doc/SSC/RedListGuidleines.pdf)

Janzen, D.H. (1993) Caterpillar seasonality in a Costa Rican dry forest. pp. 448–477 in Stamp, N.E. & Casey, T.M. (eds) Caterpillars. Ecological and Evolutionary Constraints on Foraging. Chapman & Hall, New York.

Kadlec, T., Tropek, R. & Konvicka, M. (2012) Timed surveys and transect walks as comparable methods for monitoring butterflies in small plots. *Journal of Insect Conservation* **16**, 275–280.

Kitching, R.L., Orr, A.G., Thalib, L., Mitchell, H., Hopkins, M.S. & Graham, A.W. (2000) Moth assemblages as indicators of environmental quality in remnants of upland Australian rain forest. *Journal of Applied Ecology* **37**, 284–297.

Kremen, C. (1994) Biological inventory using target taxa; a case study of the butterflies of Madagascar. *Ecological Applications* **4**, 407–422.

Kuussaari, M., van Nouhuys, S., Hellmann, J.J. & Singer, M.C. (2004) Larval biology of checkerspots. pp. 138–160 in Ehrlich, P.R. & Hanski, I. (eds) On the Wings of Checkerspots. A Model System for Population Biology. Oxford University Press, Oxford.

Lamas, G. (1982) A preliminary zoogeographical division of Peru based on butterfly distributions. pp. 336–357 in Prance, G.T. (ed.) Biological Diversification in the Tropics. Columbia University Press, New York.

Landau, D., Prowell, D. & Carlton, C.E. (1999) Intensive versus long-term sampling to assess Lepidoptera diversity in a southern mixed mesophytic forest. *Annals of the Entomological Society of America* **125**, 435–441.

Larsen, T.B. (2005) The Butterflies of West Africa. Apollo Books, Svendborg.

Larsen, T.B. (2008) Forest butterflies in West Africa have resisted extinction . . . so far (Lepidoptera: Papilionoidea and Hesperioidea). *Biodiversity and Conservation* **17**, 2833–2847.

Launer, A.E. & Murphy, D.D. (1994) Umbrella species and the conservation of habitat fragments: a case of a threatened butterfly and a vanishing grassland ecosystem. *Biological Conservation* **69**, 145–153.

Leinonen, R., Soderman, G., Itamies, J., Rytkonen, S. & Rutanen, I. (1998) Intercalibration of different light-traps and bulbs used in moth monitoring in northern Europe. *Entomologica Fennica* **9**, 37–51.

Leon-Cortes, J.L., Perez-Espinosa, F., Marin, L. & Molina-Martinez, A. (2004) Complex habitat requirements and conservation needs of the only extant Baroniinae swallowtail butterfly. *Animal Conservation* **7**, 241–250.

Lewis, O.T. & Hurford, C. (1997) Assessing the status of the marsh fritillary butterfly (*Eurodryas aurinia*): an example from Glamorgan, UK. *Journal of Insect Conservation* **1**, 159–166.

Lewis, O.T. & Senior, M.J.M. (2011) Assessing conservation status and trends for the world's butterflies: the Sampled Red List Index approach. *Journal of Insect Conservation* **15**, 67–76.

Lewis, O.T., Wilson, R.J. & Harper, M.C. (1998) Endemic butterflies on Grande Comore: habitat preferences and conservation priorities. *Biological Conservation* **85**, 113–121.

Li, X., Zhang, Y., Fang. J., Schweiger, O. & Settele, J. (2011) A butterfly hotspot in western China, its environmental threats and conservation. *Journal of Insect Conservation* **15**, 617–632.

Maes, D., Vanreusel, W., Jacobs, I., Berwaerts, K. & Van Dyck, H. (2012) Applying IUCN Red List criteria at a small regional level: a test case with butterflies in Flanders (north Belgium). *Biological Conservation* **145**, 258–266.

Miller, D.G., Lane, J. & Senock, R. (2011) Butterflies as potential bioindicators of primary rainforest and oil palm plantation habitats on New Britain, Papua New Guinea. *Pacific Conservation Biology* **17**, 149–159.

Moss, D. & Pollard, E. (1993) Calibration of collated indices of abundance of butterflies based on monitored sites. *Ecological Entomology* **18**, 77–83.

Muirhead-Thomson, R.C. (1991) Trap Responses of Flying Insects. Academic Press, London.

Murphy, D.D. (1987) Are we studying our endangered butterflies to death? *Journal of Research on the Lepidoptera* **26**, 236–239.

Murphy, D.D. & Weiss, S.B. (1988) A long-term monitoring plan for a threatened butterfly. *Conservation Biology* **2**, 367–374.

New, T.R. (2009) Insect Species Conservation. Cambridge University Press, Cambridge.

New, T.R. & Britton, D.R. (1997) Refining a recovery plan for an endangered lycaenid butterfly, *Acrodipsas myrmecophila*, in Victoria. *Journal of Insect Conservation* **1**, 65–72.

New, T.R. & Collins, N.M. (1991) Swallowtail Butterflies. An Action Plan for Their Conservation. IUCN, Gland.

Nieminen, M., Siljander, M. & Hanski, I. (2004) Structure and dynamics of *Melitaea cinxia* populations. pp. 63–91 in Ehrlich, P.R. & Hanski, I. (eds) On the Wings of Checkerspots. A Model System for Population Biology. Oxford University Press, Oxford.

Novotny, V. & Basset, Y. (2000) Rare species in communities of tropical insect herbivores: pondering the mystery of singletons. *Oikos* **89**, 564–572.

Orr, A.G. & Haeuser, C.L. (1996) Temporal and spatial patterns of butterfly diversity in a lowland tropical rainforest. pp. 125–138 in Edwards, D.S., Booth, W.E. & Choy, S.C. (eds) Tropical Rainforest Research – Current Issues. Kluwer Academic Publishers, Dordrecht.

Packer, L. (1991) The status of two butterflies, Karner blue (*Lycaeides melissa samuelis*) and Frosted elfin (*Incisalia irus*), restricted to oak savannah in Ontario. pp. 327–331 in Allen, G.W., Eagles, P.F.J. & Price, S.W. (eds) Conserving Carolinian Canada. University of Waterloo Press, Waterloo, Ontario.

Pannekoek, J. & van Strien, A.J. (2001) Trim 3 Manual. Statistics Netherlands, Voorburg.

Parsons, M.J. (1992b) The world's largest butterfly endangered: the ecology, status and conservation of *Ornithoptera alexandrae* (Lepidoptera: Papilionidae). *Tropical Lepidoptera* 3, Supplement 1, 35–62.

Parsons, M.S. (2004) The United Kingdom Biodiversity Action Plan moths – selection, status and progress in conservation. *Journal of Insect Conservation* 8, 95–107.

Pollard, E. & Yates, T.J. (1993) Monitoring Butterflies for Ecology and Conservation. Chapman & Hall, London.

Powell, A.F.L.A., Busby, W.H. & Kindscher, K. (2007) Status of the regal fritillary (*Speyeria idalia*) and effects of fire management on its abundance in northeastern Kansas, USA. *Journal of Insect Conservation* 11, 299–308.

Powell, J.A. (2004) Assessment of inventory effort for Lepidoptera (Insecta) and the status of endemic species on Santa Barbara Island, California. pp. 351–370 in Garcelon, D.K. & Schwenn, C.A. (eds) Proceedings of the Sixth California Islands Symposium 2003, Ventura, California. Institute for Wildlife Studies, Arcata, CA.

Richter, A., Weingold, D., Robertson, G. et al. (2012) More than an empty case: a non invasive technique for monitoring the Australian critically endangered golden sun moth, *Synemon plana* (Lepidoptera: Castniidae). *Journal of Insect Conservation* DOI 10.1007/s 10841-012-9537-5.

Robertson, I.A.D. (1997a) Records of insects taken at light traps in Tanzania. V. Seasonal changes in catches and effect of the lunar cycle on hawkmoths of the subfamily Asemanophorinae (Lepidoptera: Sphingidae). *Centre for Overseas Pest Research, Miscellaneous Report* 30, 1–14.

Robertson, I.A.D. (1997b) Records of insects taken at light traps in Tanzania. VI. Seasonal changes in catches and effect of the lunar cycle on hawkmoths of the subfamily Semanophorinae (Lepidoptera: Sphingidae). *Centre for Overseas Pest Research, Miscellaneous Report* 37, 1–20.

Robinson, G.S. & Tuck, K.R. (1996) Describing and comparing high invertebrate diversity in tropical forest – a case study of small moths in Borneo. pp. 29–42 in Edwards, D.S., Booth, W.E. & Choy, S.C. (eds) Tropical Rainforest Research – Current Issues. Kluwer Academic Publishers, Dordrecht.

Roy, D.B., Rothery, P. & Brereton, T. (2007) Reduced effort schemes for monitoring butterfly populations. *Journal of Applied Ecology* 44, 993–1000.

Samways, M.J., McGeoch, M.A. & New, T.R. (2010) Insect Conservation. A Handbook of Approaches and Methods. Oxford University Press, Oxford.

Sands, D.P.A. & New, T.R. (2002) The Action Plan for Australian Butterflies. Environment Australia, Canberra.

Schtickzelle, N. & Baguette, M. (2009) (Meta)population viability analysis: a crystal ball for the conservation of endangered butterflies. pp. 339–352 in Settele, J., Shreeve, T., Konvicka, M. & Van Dyck, H. (eds) Ecology of Butterflies in Europe. Cambridge University Press, Cambridge.

Schulze, C.H., Linsenmair, K.E. & Fiedler, K. (2001) Understorey versus canopy: patterns of vertical stratification and diversity among Lepidoptera in a Bornean rain forest. *Plant Ecology* 153, 133–152.

Spalding, A. (1997) The use of the butterfly transect method for the study of the nocturnal moth *Luperina nickerlii leechi* Goater (Lepidoptera: Noctuidae) and its possible application to other species. *Biological Conservation* 80, 147–152.

Stojanovic, D.V., Curcic, S.B., Curcic, B.P.M. & Makano, S.E. (2012) The application of IUCN Red List criteria to assess the conservation status of moths at the regional level: a case of provisional Red List of Noctuidae (Lepidoptera) in Serbia. *Journal of Insect Conservation* DOI 10.1007/s10841-012-9527-7

Summerville, K.S. & Crist, T.O. (2008) Structure and conservation of lepidopteran communities in managed forests of northeastern North America; a review. *Canadian Entomologist* 140, 475–494.

Thomas, C.D. & Abery, J.C.G. (1995) Estimating rates of butterfly decline from distribution maps: the effects of scale. *Biological Conservation* 73, 59–65.

Thomas, J.A. (1993) Holocene climate changes and warm man-made refugia may explain why a sixth of British butterflies possess unnatural early-successional habitats. *Ecography* 16, 278–284.

Thomas, J.A. & Simcox D. (1982) A quick method for estimating larval populations of *Melitaea cinxia* L. during surveys. *Biological Conservation* 22, 315–322.

Treadaway, C.G. (1995) Checklist of the butterflies of the Philippine islands (Lepidoptera: Rhopalocera). *Apollo*, Supplement 14, 7–118.

Tribert, P. & Braggio, S. (2011) Remarks on some families of leaf-mining Microlepidoptera from central-southern Sardinia, with some ecological considerations (Lepidoptera: Nepticulidae, Bucculatricidae, Gracillariidae). *Conservazione Habitat Invertebrati* 5, 567–581.

van Dyck, H., van Strien, A.J. & van Swaay, C.A.M. (2009) Declines in common widespread butterflies in a landscape under intense human use. *Conservation Biology* 23, 957–965.

van Swaay, C.A.M., Maes, D. &Warren, M.S. (2009) Conservation status of European butterflies. pp. 322–339 in Settele, J., Shreeve, T., Konvicka, M. & Van Dyck, H. (eds) Ecology of Butterflies in Europe. Cambridge University Press, Cambridge.

van Swaay, C.A.M. & Warren, M.S. (2003) Prime butterfly areas of Europe: priority sites for conservation. National Reference Centre for Agriculture, Nature and Fisheries; Ministry of Agriculture, Nature Management and Fisheries, The Netherlands.

van Swaay, C.A.M. & Warren, M.S. (2006) Prime Butterfly Areas of Europe: an initial selection of priority sites for conservation. *Journal of Insect Conservation* 10, 5–11.

van Tol, J. & Verdonk, M.J. (1988) The protection of dragonflies (Odonata) and their biotopes. *Nature and Environment Series, Council of Europe* 38, 1–181.

Vane-Wright, R.I., Humphries, C.J. & Williams, P.H. (1991) What to protect? Systematics and the agony of choice. *Biological Conservation* 55, 235–254.

Wagner, D.L. (2006). A precautionary tale: on the larva and life history of *Lithophane joannis* (Lepidoptera: Noctuidae). *Journal of the Lepidopterists' Society* 60, 174–176.

Young, M.R. & Barbour, D.A. (2004) Conserving the New Forest burnet moth (*Zygaena viciae* [Denis and Schiffermueller]) in Scotland; responses to grazing reduction and consequent vegetation changes. *Journal of Insect Conservation* 8, 137–148.

Zeydanli, U.S., Turak, A.S., Balkiz, O. et al. (2012) Identification of Prime Butterfly Areas in Turkey using systematic conservation planning: challenges and opportunities. *Biological Conservation* 150, 86–93.

6

Population Structures and Dynamics

Introduction: Distinguishing populations

Defining and delimiting a population of a moth or butterfly is not always easy, but incorporates considerations of numbers of individuals and residential permanency in a spatial arena. Loss or decline of populations, usually manifest by decrease in number and/or distribution, is the most frequent harbinger of conservation need, so that population size and population structure are both central to understanding that need. The more detailed survey methods noted in the last chapter have potential to identify both extinctions and colonisations based on 'presence' or 'absence' but – as Pollard and Yates (1992, 1993) noted – it is necessary to define those states clearly, in itself not always straightforward in relation to short-term variations in abundance and detectability. Based on the United Kingdom Butterfly Monitoring Scheme transect walk approach, Pollard and Yates adopted definitions as follows, because 'they seemed reasonable': (1) it was assumed that a breeding population was present if a species was recorded in four successive flight periods and absent if there were no such records for four successive seasons, with other combinations of records considered inconclusive; (2) extinction at a site was assumed if 'presence' was followed by 'absence' at some later time; and (3) colonisation was assumed if 'absence' was followed by by 'presence'. A sequence of eight flight periods is thus needed to determine either extinction or foundation. The latter may represent range expansion (mostly near a current range edge) or recolonisation of previously occupied sites, and either may be confused by mobility of some species leading to 'casual occurrence'. Many Lepidoptera are highly mobile, with extensive adult migrations a regular feature of their life cycles (Chapter 4, p. 45), whilst many others are relatively sedentary and disperse rather little, so tend to be more restricted in distribution. Adult activity is a major focus of surveys, with them either viewed

Lepidoptera and Conservation, First Edition. T.R. New.
© 2014 John Wiley & Sons, Ltd. Published 2014 by John Wiley & Sons, Ltd.

with little disturbance or concentrated by attraction to traps by baits or lights. Populations are dynamic in time and space but, however they are defined, are a primary focus of conservation, with changes by natural or anthropogenic causes interpreted variously as normal or creating vulnerability. That interpretation involves interactions between dispersal need/capability and the distribution of resources (Chapter 7, p. 119) at a variety of scales from local to landscape. External influences differ across the distribution of any species, as also may the life cycle – numbers of annual generations of some widely distributed moths and butterflies, for example, may be influenced strongly by climate and, hence, differ across a latitudinal range. The causes of change in any individual population may differ in detail from any other. Except for taxa for which very few populations exist, some level of imprecision is almost always inevitable.

Both during surveys and in later interpretation of populations and their structures, the terms used in any study should be defined clearly, to avoid ambiguity. For the Sand-verbena moth (*Copablepharon fuscum*, Noctuidae) the term 'location' has been used to define an isolated metapopulation (discussed later in this chapter), and so comprises a variable number of population units defined by the patchy distribution of the sole larval host plant (sand verbena, *Abronia latifolia*) and which is disjunct from any other location (BCIRT 2008). This moth is confined to small areas of coastal sand dunes in British Columbia (four locations) and Washington State (five locations).

It is thus one of numerous highly localised, and ecologically specialised, species that may be restricted to discrete patches of suitable terrain, and separated from other populations by inhospitable areas that they are unable to traverse. They may be essentially sedentary and 'self-contained', isolated as 'closed populations' in which the influences of immigration and emigration on numbers are negligible. Numerical changes then result almost wholly from internal demographic processes, of births and deaths. Population size is linked strongly with the carrying capacity of the residential site, as the supply and quality of food and other critical resources. However, because each population is discrete, the number of populations can then be assessed, and their distributions plotted to determine the extent and possible impacts of that isolation (Chapter 8).

A contrasting population structure involves greater levels of dispersal, so that many Lepidoptera are not bound as tightly to individual sites and, despite concentrations of numbers in particular areas, range more freely across the local landscape. Changes in numbers in a population must then incorporate arrivals and departures, so adding dispersal effects to those of internal demographics. In practice these two categories intergrade to form a continuum of structures. The extremes link to categories defined initially for British butterflies as 'habitat specialists' and 'wider countryside specialists' (Pollard & Eversham 1995, Asher et al. 2001, Table 6.1), in relation to dispersal. Whilst the extremes of open or closed populations are generally clear, the development of the metapopulation concept, arising largely from studies on butterflies, provides considerable insight with the realisation that many local losses of populations occur naturally. If these losses were of closed populations these would cause conservation concern, and often have done so in the past, but metapopulations work through sequences of extinction–colonisation that represent normal population dynamics across the

Table 6.1 Comparative attributes of two main ecological categories of butterflies, based on the British taxa and reflecting need for different approaches to practical conservation (after Pollard & Eversham 1995, as presented by Asher et al. 2001.).

Habitat specialists
Confined to specific, usually discrete, habitat areas that are patchy in the landscape
Rarely or never use linear habitats such as hedgerows or road verges
Usually only one or two species of larval food plants
Relatively sedentary
Mostly with single generation each year

Wider countryside specialists
Broad habitat requirements, or use habitats that are distributed widely in the farmed
    countryside: generalists
Can use linear habitats such as hedgerows or road verges
Often with several species of larval food plants
Relatively mobile
Often with multiple generations each year

array of accessible habitat patches so that, at any time, some patches are likely to be vacant, although suitable for habitation, and others are occupied. Individual habitat patches may differ greatly in extent, quality, internal dynamics and the characteristics of the surrounding area (the 'matrix') and can become difficult to define. Some of the problems of definition were elucidated by a mark–release–recapture study of Europe's most threatened butterfly (the Scarce fritillary, *Euphydryas maturna*) in the Czech Republic, and at its last remaining site in the country (Cizek & Konvicka 2005). *E. maturna* occupies open areas, such as clearings and glades, within a matrix of thick forest in which good habitat may develop into non-habitat as natural vegetational changes occur – so that the distinction between habitat and non-habitat becomes blurred. From surveys and models of migration, Cizek and Konvicka defined eight categories where at least one butterfly was captured (either sex, male or female, one or more than one individual, clusters of adjacent open areas separated from other clusters by >50 m of forest, openings with larval nests, and merged patches with nests into larger areas separated by wide (>100 m) areas of forest). Dispersal differed between the sexes. Males exhibited higher dispersal mortality when dispersing amongst small patches and dispersed only rarely in patchworks of large patches – probably reflecting reluctance to leave after establishing territorial perches. Females moved more readily and tended to leave even the largest patches. Interpretation is complex, but Cizek and Konvicka (2005) noted three possible explanations as (1) small patches may not contain all possible resources, so that wider searches become necessary; (2) low population density decreases chances of finding mates, again inducing dispersal; and (3) because of succession rendering every patch progressively unsuitable, females may benefit by dispersal after laying some eggs as the colonised patch might be a younger stage and, so, more hospitable. However, this case exemplifies well that defining patch suitability on presence of the insect alone may be overly simplistic.

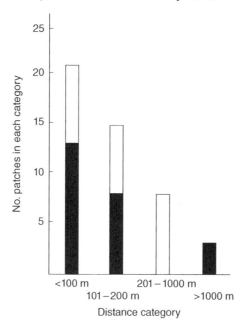

Fig. 6.1 The spatial separation of habitat patches of *Notoreas* sp. in New Zealand: the number of patches in each distance class (measured to nearest neighbouring patch) and the number inhabited by *Notoreas* (black) (Sinclair 2002. Reproduced with permission of Taylor & Francis Ltd.).

Habitat patches need not be necessarily either large or resource-rich in order to support notable species. The unnamed *Notoreas* sp. (Geometridae) surveyed by Sinclair (2002) is of notable conservation concern (as a species in 'serious decline') in New Zealand, and three-quarters of the 47 habitat patches detected each contained fewer than 15 host plants. Small patches occupied by the moth were usually within 200 m of another occupied patch and, whilst large patches were generally occupied, many very small patches also yielded moths (Fig. 6.1). Collectively, moths were found on 22 patches over the 5 years of survey, but occupation varied considerably: no site revealed moths on every occasion visited. Conservation was recommended to focus on larger patches and small neighbouring patches presumed to be within normal colonisation range, with isolated patches given lower priority unless their size and connectivity to inhabited patches could be enhanced.

Metapopulation biology

The historical development and understanding of metapopulation biology, discussed by Nieminen et al. (2004), owes much to extensive long-term studies on two nymphaline butterflies, the checkerspots *Melitaea cinxia* (the Glanville fritillary) in Europe and *Euphydryas editha bayensis* (the Bay checkerspot) in California (Ehrlich & Hanski 2004).

The classic population studies of *M. cinxia* in the Aland archipelago, Finland, involved surveys of 1452 habitat patches monitored by larval surveys (Chapter 5, p. 62) annually since 1993. From 1993–2001, only 33 patches were occupied continually, and 842 patches have been occupied at least once. Several levels of spatial arrangements were differentiated in a hierarchy. Thus larval groups move little but may be located in different parts of a habitat patch in different years – but some plants or groups of plants receive eggs more often or abundantly than others, so that patchiness occurs even at this very local scale, possibly reflecting local microclimate differences. The habitat patches, in turn, are grouped into 'patch networks' each supporting a more-or-less independent meta-population not reaching other patch networks. In the Aland Islands 25% of networks were occupied every year from 1993–2001, with 38–55% occupied in any given year. Finally, networks are grouped within 'regions' with clear boundaries such as extensive forests or water barriers. Rapid changes in butterfly occupancy occur at any of these spatial scales but tend to be smaller at the larger scales.

The key features for a classic metapopulation (Hanski et al. 1995; Hanski 1999) are essentially fourfold: (1) suitable habitats occur as small discrete patches in a landscape, and can support local breeding populations of the butterfly; (2) all local populations have a high risk of extinction, so are not permanent, and any patch may be occupied or vacant at any given time; (3) the patches are not too isolated to prevent recolonisation; and (4) local and regional influences on population dynamics are not synchronous, so that simultaneous extinction of all local populations is unlikely. Without appreciating the complex dynamics of such systems it becomes easy to exaggerate the consequences of loss of a focal species from an individual patch – but in reality, such concern may be justified by ignorance as part of the precautionary principle so important in practical conservation.

Closely related species can exhibit very different population structures, but some form of metapopulation structure is very common in butterflies – to the extent that many species presumed earlier to have closed populations actually comprise metapopulations. Hanski and Kuussaari (1995), for example, estimated that 65% of Finland's 94 butterfly species existed in this form. Major factors fostering this structure are, perhaps, the patchiness of suitable habitat dictating occurrence of discrete local populations and the extent of adult movements between those populations. Particularly if depending on successional habitats, it is necessary to continually track and colonise suitable patches as they appear. In other instances, habitats are not as transient, but often small and isolated, comprising networks of more persistent patches. The studies on checkerspots noted earlier (and summarised comprehensively by Ehrlich and Hanski 2004) were central in developing models that help to clarify many aspects of metapopulation dynamics and structures. Some of the key issues have considerable relevance in both fundamental ecology and conservation planning (Table 6.2, Hanski et al. 2004). A clear understanding of 'habitat' is fundamental (Chapter 7), together with relationships between habitat and dispersal capability and behaviour.

Metapopulations can have various spatial structures (p. 105), again elucidated largely from butterfly studies (Harrison 1994, Hanski 1999), but one universal

Table 6.2 Lessons for the conservation of metapopulations drawn from long-term studies on checkerspot butterflies, as listed and discussed by Hanski et al. (2004).

1. Metapopulation theory can be used to predict how the structure of a fragmented landscape influences metapopulation size and persistence
2. Models developed for well studied common species may provide insight to the dynamics of ecologically related species that are rare and threatened
3. Metapopulation models combined with reserve-site selection algorithms provide a robust framework for conservation of metapopulations
4. An intermediate level of connectivity is generally most beneficial for long-term persistence
5. Species inhabiting successional habitats often exhibit metapopulation dynamics
6. Single large high-quality patches may play an important role in maintaining regional persistence
7. The presence of a metapopulation in the current landscape may be deceptive because metapopulation dynamics track environmental changes with a substantial delay
8. Many causes of local extinctions may operate in a large metapopulation
9. Inbreeding depression may be an important cause of local extinction in species with a fragmented population structure
10. Habitat fragmentation may select for increased migration rate
11. Local adaptations may influence the persistence of populations in a metapopulation

consequence of increased habitat fragmentation is that previously functional metapopulations may be disrupted and transformed into a series of isolated closed populations between which dispersal and functional connectivity is lost; any local extinction then represents 'real loss'. However, this has only rarely been tested experimentally or by field observations.

Analyses of metapopulation viability have to be species-specific, because different species in the same biotope can have very different dispersal propensity, behaviour and capability, with extent of movements amongst habitat patches characteristically differing across species, as well as being influenced differentially by the local landscape (Baguette et al. 2000). Patch area seemed to be an important influence, supporting the common utterance that large patch size is associated with decreased emigration, together with distance between patches, and the individual species' dispersal prowess. In that study, of three specialist butterfly species in chalk grassland, the Small blue (*Cupido minimus*) dispersed less than either of two much larger species. Baguette et al. noted that the persistence of the less mobile species cannot be assured if habitat networks are designed on the basis of studies on dispersive taxa alone and – whilst this may seem self-evident – comparative studies on the dispersal of co-occurring specialist species remain rather sparse.

In Poland, three species of large blues (*Maculinea/Phengaris*, Lycaenidae) formed one of few such targeted investigations (Nowicki et al. 2013). All three species occupied food plant patches, and 61 such patches were surveyed. Local populations of *M. alcon*, by far the rarest of the three, were small, rarely with more than 100 individuals present – but these were also stable. *M. teleius* and *M. nausithous* typically occurred in population units of several hundred to about 1000, but a few populations of each were much larger. High colonisation by *M. nausithous* was suspected but, with all patches occupied in the first year of a 2-year survey, colonisation could not be detected. The study also drew attention to the importance of the condition of patch surrounds – here as areas habitable by the obligatory host *Myrmica* ant species.

Any such study of dispersal involves need for recognition of individual butterflies or inhabitants of a single patch and tracking these over wider scales. Mark–release–recapture techniques have been used widely in butterfly population ecology, and the large wings of Lepidoptera are very suitable for marking, using either individual coded marks or numbers, or broader colour codes to represent batches of common origin or capture occasion. The marked insects are released and recaptured at some later time or times (background on the techniques and approaches: Samways et al. 2010), demonstrating movements within and between populations/patches and, when relevant, gaining a substantial amount of related information on changes in number, sex ratio, longevity and relative extent and distance of movement at different ages and in or between different patches helping to estimate extent of genetic interchange in relation to landscape features such as distance and topography.

Evaluating the extent and frequency of migration among local populations is central to evaluating metapopulations, together with the spatial pattern of habitat patches – for Lepidoptera most commonly patches of larval foodplants. Together with estimating genetic interchange, they help determine the sustainability of the structure revealed. A recent study on another lycaenid, the Silver-studded blue (*Plebejus argus*) in Britain employed these complementary approaches to show that although relatively sedentary, the butterfly undertook migrations at most sites and even the most isolated sites received migrants regularly. Lewis et al. (1997) thus demonstrated that patch isolation was here not generally a barrier to colonisation, with the provision that dispersal over more than about 5 km was unlikely. This last provision emphasises the importance of determining the spatial status and relationships of any focal species in the landscape, with sustained connectivity between demographic units a key need in conservation. For metapopulation viability and projecting management needs, the dispersion of habitat patches in relation to dispersal capability of the species involved is therefore a critical focus, driving the needs for connectivity (Chapter 7, p. 132). Dispersal capability is difficult to assess realistically; as Franzen and Nilsson (2007) commented, even if dispersal distances are substantially underestimated, in many Lepidoptera studies on small areas, the true size of the arena needed for a valid estimate is not known. For the two species of burnet moths (*Zygaena*) they studied in southern Sweden, they suggested that a study area of at least $50 \, km^2$ is needed, with individual marked moths confirmed to fly as much as 5.6 km. Clearly, retrieval of individuals declines substantially with distance from release points, and that only small numbers can be marked for studies on many low-abundance conservation-interest species adds to the difficulties of tracking. *Zygaena* species are commonly regarded as sedentary, but the major inference from Franzen and Nilsson's study was simply that interpreting dispersal patterns from studies over small areas may need to be very cautious. Whilst mobility is highly significant in conservation studies, at the local or regional scale, the true distance over which most species are capable of moving is generally unknown. Some Lepidoptera are well known long-distance migrants, whilst many others move little, as discussed later in this chapter. Perhaps the greatest individual distance reported for a marked moth is for the cutworm *Agrotis ipsilon*, for which Showers (1997) reported a straight line distance from mark to recapture of 1818 km in China.

The probability of dispersal reflects the local population context, and also that individuals within a population can vary widely in their dispersal ability and need. In addition to patch size and connectivity, as above, Hovestadt et al. (2011) noted that the structure of the landscape may influence whether dispersal emanates from normal 'every day movements' or more distinct 'dispersal episodes'. Reinterpreting the results of an earlier study of the Dusky large blue, *Maculinea nausithous*, in Germany (Binzenhofer & Settele 2000), movements between patches were relatively frequent, found in almost half the individuals marked, in a survey arena with maximum distance between patches just over 5800 m, and in which the longest butterfly dispersal distance recorded was 3800 m (average 600 m, median 390 m, n = 145 recaptures). Modelling revealed that a small proportion of individuals followed different dispersal rules from the greater numbers pursuing more normal movement. The long-distance dispersers may be a key to understanding how butterflies track climatic and other changes occurring within a changing landscape. A more formal exploration of this phenomenon, for the Meadow brown (*Maniola jurtina*) revealed that some individuals used a different 'direct flight strategy' involving straight line flight when they left a patch and enabling them to avoid turning back into that patch (Delattre et al. 2010). This strategy may be beneficial when dispersal risks and costs are high. Females employed this flight mode more than males, and also undertook longer flights with fewer stops.

The complexity and variety of butterfly dispersal and its interpretation includes the variety that may occur within a species (Stevens et al. 2010). That review revealed the dangers in presuming that dispersal mode is invariably species-specific, and that some of the ambiguities arise from the methods used to study dispersal. Three areas for future research were suggested, namely (1) relationships between movement rates and rates of effective dispersal in the context of environmental change; (2) how habitat quality and environmental changes influence how dispersal evolves and changes; and (3) the processes that affect observed patterns in within-species variations in dispersal ability. The dispersal estimates used widely in butterfly studies are not always distinguished clearly; these are summarised in Table 6.3 to indicate the range of relevant contexts. Ambiguities may flow from difficulties in directly comparing different studies: 'In the rich literature on butterfly dispersal and movement abilities there is a high degree of heterogeneity in the methods used' (Stevens et al. 2010: 639).

Dispersal data for moths is, in general, sparser than that available for butterflies, but its assessment is clearly of equivalent importance in assessing conservation need, and several categories of dispersal capability are apparent. Data accumulated from literature sources on 119 species of day-flying moths revealed

Table 6.3 A classification of mobility of butterflies and moths based on relationship of the adult stage to its larval habitat and extent of wandering behaviour (adapted from van der Meulen & Groenendijk 2005; see also Stevens et al. 2010. With permission of Netherlands Entomological Society.).

| Mobility class | Classification criteria | | Flight range | Value |
	Relationship to larval habitat	Wandering behaviour		
Very area-restricted	Restricted to larval habitat	None	Very small	1
Area-restricted	Foraging in adjacent habitats	Limited	Small	2
Moderately restricted	Foraging in non-adjacent habitats	Rather limited/ outbreaks depending on year	Moderate	3
Wanderer	Foraging far outside larval habitat	Wandering behaviour evident	Considerable	4
Migrant	Completion of life cycle in area far removed from larval habitat	Migrant behaviour prominent	Extremely large	5

a wide lack of quantitative criteria, but allowed a more qualitative general appraisal of 'mobility' based on the whereabouts of the adult moth in relation to its larval habitat (van der Meulen & Groenendijk 2005). The preliminary inferences, evaluated and refined also by a number of moth experts, led to allocation of a 'mobility value' to almost all species, under the rationale shown in Table 6.3. As a general ranking system, with a few species not categorised because of lack of information, the numbers of taxa allocated to each category were (value 1) 6, (2) 36, (3) 60, (4) 6, (5) 10, from this combination of published record and expert opinion. The categories shown may help to guide management needs for assemblages and, although based on the fauna of north-west Europe, have clear relevance elsewhere. Whilst much of the immediate relevance of mobility is in relation to population structure and dynamics, capability to undertake range shifts as responses to climate change may also reflect this variation. Estimating the characteristic mobility level of taxa from field studies may thus have wider conservation applications.

A survey of noctuid moths on the island of Utklippan (about 16 km from the Swedish mainland and 8 km from the nearest other island in the Baltic Sea) by light trapping yielded 98 species, of which 51 were classified as non-resident, that is as highly mobile in that they must have traversed a minimum of 16 or 8 km (Betzholtz & Franzen 2011). Polyphagous and oligophagous species moved to the islands more readily than monophagous species, possibly reflecting a more general trait within Lepidoptera of wider distribution of generalists. Species with adults active in late summer (August, September) were also more strongly represented than those flying earlier in the summer, linked tentatively with higher temperatures enhancing suitable conditions for flight. That study is important in showing that dispersal can occur in a substantial proportion of taxa in a local assemblage, and that the mobility correlates also with definable species traits. Those traits may be highly relevant in considering mobility in changing and fragmented landscapes.

More generally, the mechanisms of metapopulation dynamics can be both complex and difficult to interpret. As Hanski and Singer (2001) demonstrated from an elegant experimental study of *Melitaea cinxia*, the role of 'patch quality' can be perceived differently by individuals of different butterfly phenotypes, so that particular patches become more suitable or less suitable for colonisation depending on that perception. In the Aland Islands, *M. cinxia* has two larval food plants (*Plantago lanceolata, Veronica spicata*). An empty habitat patch containing mostly one or other of these is more likely to be colonised if there is a history of higher relative use of that plant in surrounding patches: Hanski and Singer termed this the 'colonisation effect'. Understanding how dispersal capability/propensity and population structures interact is central to interpreting conservation needs. As Samways and Lu (2007) illustrated, this can be approached by comparative study of related sympatric species that differ in ecological features and conservation status. In their study, the threatened Karkloof blue (*Orachrysops ariadne*, Chapter 12, p. 232) was compared with the common sympatric Grizzled blue (*O. subravus*) in South Africa by mark–release–recapture surveys of individual movements within adjacent restricted habitat patches. Outcomes may be unexpected: in this example, the rare *O. ariadne* flew greater distances than *O. subravus*, and the species showed no significant differences in survival or longevity, so that the extreme scarcity of the host plant of the Karkloof blue appeared to be related to its scarcity. Samways and Lu suggested that selection for strong flight might be linked with extreme microhabitat specialisation, with the scarce habitat patches further isolated by anthropogenic effects. Countering the barriers imposed by recent agroforestry may be the key conservation need for this butterfly.

Common butterflies such as *M. jurtina* and the Gatekeeper, *Pyronia tithonus*, which occur together, can be studied easily by direct observations on individual insects. Flight behaviour of both species at permeable boundaries between their usual habitat (long grassland) and inhospitable mown short grassland implied that butterflies were aware of the position of such margins (Conradt & Roper 2006). Many butterflies that crossed the boundary returned to their original patch, so undertaking 'foraging loops'. For *M. jurtina*, only three of 142 movements observed out of the habitat patch resulted in movement to other patches, whilst the other 139 all returned to the patch they had left. The patterns of such loops vary, as indicated in Fig. 6.2, but Conradt and Roper concluded that dispersing butterflies moved systematically, with the rate of habitat return 'under active and effective control', with foraging loops enabling individuals to reconnoitre the local landscape. Earlier work (Conradt et al. 2000) revealed three possible suggested reasons why *M. jurtina* dispersed in such non-random ways: (1) when released in unsuitable habitat, movements were directed toward familiar habitat patches from up to 125 m away, and to novel patches from up to at least 70 m, so that the normal dispersal range (of 40–70 m) is within that in which they can orientate actively; (2) increasing release distances from a suitable

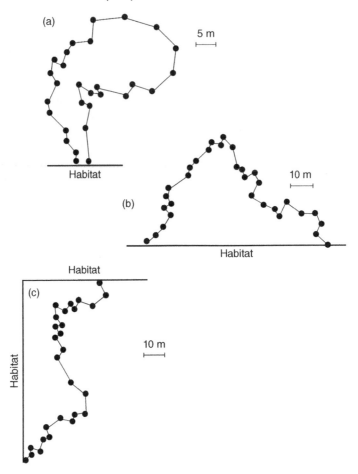

Fig. 6.2 Examples of dispersal loops recorded for individual *Maniola jurtina* butterflies in leaving and re-entering habitat patches. Each such 'foray loop' was observed directly, with each point indicating a turn in flight path recorded by dropping a marker in that spot; distances between markers were then measured directly (Source: Conradt & Roper 2006. Reproduced with permission of Ecological Society of Americal Ltd.).

habitat correlated with increasing choice of flight patterns suggesting systematic search, in large loops around the release point; and (3) butterflies preferentially returned to their familiar patch when given a choice between this and an unfamiliar one. Metapopulation models that involve presumption of random dispersal may need careful interpretation to cater for such other eventualities, with the form of dispersal system affecting flow of individuals between habitat patches and, more indirectly, influencing natality and mortality in each patch. Boughton (2000, studying *Euphydryas editha*) emphasised the importance of the scale of patchiness – so that a fine scale related to mobility means that butterflies encounter many patches, whilst at large scales, most dispersing butterflies may encounter only a single patch. Colonisation of new patches then becomes an unusual stochastic act.

Examples such as this also highlight the diversity of structures encompassed by the term 'metapopulation' and recognised as having different consequences for resisting environmental changes. Rates of local extinctions and recolonisations are highly relevant in conservation genetics, with extinctions accelerating genetic drift and colonisations increasing interchanges of genetic material through interbreeding with existing residents. Much early theoretical interpretation of metapopulations assumed that all local populations are equally prone to extinction, so that recolonisation is essential to metapopulation persistence. However, this 'classical' interpretation may be relatively rare, with several other structures more common. These alternatives to the 'classical metapopulation' fall broadly into three categories (Harrison & Hastings 1996, Fig. 6.3) as: (1) 'mainland–island' systems, in which large permanent or relatively permanent populations, often on larger or richer habitat patches, constitute the source of individuals that disperse to other sites where populations are more transient; (2) local populations are strongly interconnected by dispersal, so dispersal is frequent and local extinctions are rare – there is a single 'patchy population'; and (3) in contrast, local populations may be connected very weakly by dispersal, so that extinctions are not sufficiently countered by recolonisations and the population (sometimes termed 'non-equilibrium') is on the trajectory to extinction.

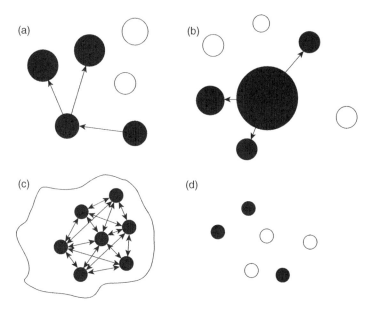

Fig. 6.3 Patterns of metapopulation structure: (a) classical, all local populations equally prone to local extinction and recolonisation; (b) mainland (source)–island (sink), local populations uneven in size and duration; (c) patchy, not truly divided so a single demographic population at patches within a greater habitat area; (d) non-equilibrium, weakly connected by dispersal, so that local extinction is not balanced by recolonisation (black, occupied patch; open, vacant patch, arrows show direction of movement) (Source: Harrison & Hastings 1996. Reproduced with permission of Elsevier.).

Mainland populations have stability, in that they rarely (mostly, never) become extinct, so that the species' survival does not depend on dispersal. Tendency for 'island populations' to differentiate genetically is countered by homogenisation from immigrants that are always from the same source population. A patchy population is genetically mixed, with a variety of subpopulation interchanges, whilst a non-equlibrium metapopulation provides, in Harrison and Hasting's words, 'the ideal conditions for population divergence, providing that local populations do not die out too quickly'. Changing levels of migration are clearly important influences on patterns of genetic variations.

However, despite the attractions of stating that a given species manifests a particular kind of population structure, this is often a considerable over-simplification – as Thomas (2001) presciently noted from the diversity of scales at which populations are considered 'we should seek a process-based framework for trying to understand population structure, rather than trying to force complex systems into descriptive categories they will rarely fit'. His studies on the Silver-spotted skipper butterfly (*Hesperia comma*) in Britain demonstrated the relevance of spatial scale – whether, for example, a 'patch' (to or from which dispersal may occur) is delineated as a single food plant (a tuft of the grass *Festuca ovina*), or a continuous grassland area containing scattered suitable tufts and separated from other such areas. The various scenarios available could lead to the butterfly fitting several different structural population categories, with variation across the species' range and individual populations subject to different structuring processes. Thus, 'It is a waste of time to attempt to force any or all *H. comma* systems into a single population category' (Thomas 2001: 327). This species may not be unusual, but considerations of scale are universal in considering habitat isolation and management need in conservation. Thus, even single tussocks of the sedge *Gahnia filum* can be managed individually in conservation of the Altona skipper (*Hesperilla flavescens*) in Australia. Burning of individual senescent tussocks with old dry foliage induces rapid growth of new foliage of this sole larval food plant, that becomes suitable for oviposition by *H. flavescens* within a few months (Chapter 12, p. 231), and the individual wetland sites on which these tussocks occur constitute another 'layer' for consideration (Relf & New 2009).

Following Thomas and Kunin (1999), who emphasised that populations vary along a continuum of demographic and dispersal events (the 'Compensation Axis', Fig. 6.4), Thomas (2001) stressed the advantages of the 'Mobility Axis' (defined, using the notation of Fig. 6.4, as '(I + E) – (B + D)') to provide a continuous approach to incorporate these processes.

The functional and practical need is to interpret the levels of isolation of supposed populations, and their consequence for survival and, possibly, conservation attention. The scale of a study may strongly influence the inferences drawn, as implied above, and the interactions between dispersal behaviour and capability and landscape structure and resource availability dominate much conservation management for Lepidoptera (Chapter 8). Comparative studies on the same species in different landscapes can be instructive. One such study, on Fender's blue butterfly (*Icaricia icarioides fenderi*), compared the responses of the butterfly to three structural components of the landscape (dense wood, open

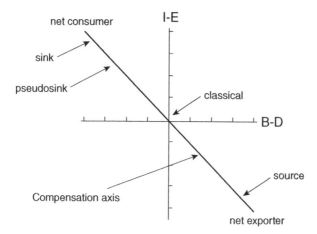

Fig. 6.4 The 'compensation axis' of population structures, in which the axis reflects the per capita rates of balance between I (immigration and E (emigration) (vertical axis), and B (births) and D (deaths) (horizontal axis) (Source: Thomas & Kunin 1999. Reproduced with permission of John Wiley & Sons.).

wood, prairie) and to presence/absence of the larval food plant (*Lupinus orega-nus*) patches (Schultz et al. 2012). Although both structure and resources influenced movements, resources appeared to have the greater effect. For example, dense and open woods are partial boundaries to movement, but the boundary response disappeared when *Lupinus* occurred near the boundary with open woods: the combined influences are not evident from responses to either factor in isolation. The two sexes also responded differently. Females tended to respond more strongly to resources (returning to or staying in lupin patches), whilst males explored more widely and moved more evidently across structural boundaries.

More generally, many habitually migratory Lepidoptera, and others that are normally highly mobile, are unlikely to be affected severely by most landscape discontinuities. Far more conservation concern devolves on those that are towards the 'closed' pole of the 'open–closed population continuum' (Dover & Settele 2009), for which even relatively small interpatch distances across alienated terrain may constitute severe barriers to movement. Rather than the conventional view of landscapes as polarised 'hospitable patch' and 'inhospitable matrix', the more integrated picture advocated by Dennis and his colleagues (Chapter 7) and which emphasises levels of matrix permeability, ensures that for any particular species within a particular landscape, the effective limits of dispersal are almost always unknown. An additional practical point noted by Dover and Settele is that many published accounts cite (or imply tacitly) distance travelled as 'straight line distance'; this might not reflect the real distance travelled (p. 100), but is often taken as a 'reasonable indication' of minimum distance capability.

Colonisation rates may generally decline with increased distances between patches, with this trend enhanced in more sedentary species. Dispersion of suitable patches at and near the leading edge for colonisation (Chapter 7) of the population may thus become critical. Background to this (for example, Wilson et al. 2009) emphasised that metapopulation maintenance will be enhanced by greater connectivity within the landscape, and expansion beyond current range margins, and inhibited by landscapes that enable only low colonisation rates and linking in some cases with high extinction rates in temporarily established populations. As described by simulation modelling for *Hesperia comma* (p. 113), habitat fragmentation can thwart expansion of range boundaries as climate warms, with many species able to accomplish this only when suitable habitat is sufficiently dense. Active conservation might be necessary to assure this. The important theme of 'assisted colonisation' whereby range extensions are undertaken deliberately by translocations or introductions is dealt with in Chapter 10.

The importance of conservation modules of interacting or mutualistic species (sensu Mouquet et al. 2005) is highlighted by possible differential responses of the constituent species to climate or other change. Populations within a species participate in two kinds of networks (Bergerot et al. 2010), namely (1) food webs within local communities and (2) local linkages amongst populations, as metapopulations, by dispersal. The recent concept of 'metacommunity' draws on both of these to emphasise local communities (including equivalents to modules) linked by dispersal of the interacting taxa and so leading to understanding how functional relationships may persist across a fragmented or otherwise changing landscape. Bergerot et al. discussed the example of a braconid parasite (*Cotesia glomerata*) and a host, the Large white butterfly (*Pieris brassicae*), to show that butterfly densities were not affected by habitat fragmentation along an urbanisation gradient in France, whereas parasitisation rate decreased strongly with increasing urbanisation. This difference was attributed to contrasted dispersal of the participants, as several kilometres (*Pieris*) and only several hundred metres (*Cotesia*). Parallels could be implied for almost any combination of interacting species across patchy landscapes, but both the colonisation in relation to climatic tolerances, and the ability to track patchy resources in a changed range become important considerations in attempting to model or predict outcomes. Novel combinations may also arise – such as for the Brown argus butterfly (*Aricia agestis*) in Britain, for which climatically facilitated northward spread has enabled it to increasingly use a widespread plant that was previously unused, as a new interspecific association (Pateman et al. 2012). *A. agestis* was previously restricted to the single larval food plant, rockrose (*Helianthemum nummularium*, Cistaceae), in southern England, but its northward spread (of about 80 km over 20 years) has encompassed many areas where this food plant is absent. There, it has exploited dovesfoot cranesbill (*Geranium molle*, Geraniaceae), which is very widely distributed. This novel interaction has facilitated rapid range expansion of the butterfly, with transition from a highly localised distribution to one that incorporates almost any grassland containing either host species. As Pateman et al. (2012) claimed, such flexibility – even if exceptional – counters the more widely advocated view that interactions between species constrain responses to climate change.

Vulnerability

As for other taxa, but illustrated well through Lepidoptera, vulnerability of populations arises from two rather different, but interacting, scenarios (Caugh-

Table 6.4 Processes influencing extinction in metapopulations (from Hanski 2004).

Scale of extinction	Intrinsic causes (type of stochasticity)	Extrinsic causes
Local extinctions	Demographic Environmental Genetic	Habitat loss Generalist enemies and competitors Persecution by humans
Metapopulation extinctions	Migration (in small populations)	Specialist enemies and competitors
	Extinction–colonisation	Habitat loss and fragmentation (extinction typically delayed)
	Regional	

ley 1994) involving the condition of the population. Caughley differentiated the 'declining population paradigm', involving factors extrinsic to the population (including most of the familiar threats normally anticipated for management in conservation: habitat changes, alien species impacts, exploitation and others) from the 'small population paradigm' (representing the array of processes intrinsic to the populations and with impacts increasing as populations become smaller, and including stochastic effects). Hanski (2004) emphasised that these two may not operate in isolation, and that risks of extinction generally increase as populations become smaller, whether entire populations or single metapopulation units are involved. However, whilst the site of an extinct metapopulation unit may be recolonised, loss of a closed population has finality. Extinction of entire metapopulations may occur from a variety of causes (Table 6.4, Hanski 2004).

In this table, 'stochasticity' includes three distinct categories of impacts: (1) 'demographic stochasticity' refers to the inherent unreliability in births and deaths; (2) 'environmental stochasticity' refers to variations in birth and death rates over time when correlated among individuals that are affected by the same environmental conditions to cause variation beyond that of normal demographics – the extreme case sometimes differentiated as 'catastrophic stochasticity' with impacts affecting all (or nearly all) individuals falls into this broad category; and (3) 'genetic stochasticity', refers to inbreeding and genetic drift leading to changes in birth and death rates. Genetic deterioration, particularly attributed to inbreeding depression, is a widespread concern in declining populations suffering increased isolation and decreased numbers, so that sib-mating increases. For some taxa, however, this seems relatively normal. Haikola et al. (2001, 2004) noted that many metapopulation units of *Melitaea cinxia* in the Aland Islands can consist of only single larval groups that are almost certainly siblings. The suggestion that this usual situation may lead to loss of inbreeding depression (by 'purging' of deleterious recessive alleles) is not supported – earlier studies on this fritillary cited by Haikola et al. described the substantial effects of inbreeding depression on egg hatching, larval survival and local population extinctions. Nevertheless, *M. cinxia* do not actively prefer non-sibling mates when presented with a choice.

Habitat dynamics (Chapter 7) are often considered more important than stochastic factors in allowing many real populations and metapopulations to persist (Thomas 1994), with lack of knowledge of precise habitat requirements of some species leading to local extinctions being attributed (simply through lack of understanding) to stochasticity. Even unusual weather – at sub-catastrophic levels – cited as causing extinction might be open to alternative interpretation. Thomas (1994) suggested that population stochasticity may become important only in the final generations when the population's fate has already been sealed.

Individual unanticipated catastrophic events can have major impacts on small or concentrated populations. One of the most spectacular documented examples for butterflies is of a single severe winter storm (from 12–14 January 2002) affecting overwintering *D. plexippus* colonies in Mexico (Chapter 4, p. 45, Brower et al. 2004). Systematic quadrat counts of monarchs on the forest floor implied, conservatively, that 5000 individuals/m² were killed across the entire overwintering region, for an estimated 467.5 million monarchs killed by this single storm. Mortality for two colonies studied most intensively was estimated at 75%. These high levels were in part the outcome of previous forest clearing and thinning, increasing exposure of resting butterflies to extreme microclimates and disturbance.

However, it remains very difficult to assess the long-term impacts even of such major losses. Brower et al. (2012) noted the long-term (15 year) downward trend, with the overwintering monarch population in Mexico in 2009–2010 at 'an all-time low', and attributed the decline to severe weather, the loss of forests in the region and also to losses of breeding habitat in the United States resulting from genetically modified herbicide-resistant crops with associated losses of milkweed plants and increased land development. Difficulties of appraising such declines were emphasised by Davis (2012), who suggested that the population remains essentially stable with high fecundity enabling recovery from low winter numbers. The population issues for even such intensively studied and popular taxa remain very complex and difficult to investigate.

In 1992, Hurricane Andrew in Florida nearly eliminated the Schaus swallowtail butterfly (*Papilio aristodemus ponceanus*), reducing the already small population to a reported 73 individuals.

Direct mortality, such as in the above cases, is perhaps the most obvious outcome of such catastrophic events. A rather different process ensued from severe frosts in 1992 killing all larval host plants used by a population of *Euphydryas editha* (Thomas et al. 1996). Again, prior disturbance contributed to the impact – with a history of logging in the late 1960s creating clear-cut areas that supported butterfly populations that functioned as source populations for more transient populations on nearby outcrops. With frost kills of the food plant *Collinsia torreyi*, caterpillars starved to death, leaving very few individuals. Only two egg batches were found the following year, and neither developed, so that the population became extinct. In a rather different example of habitat loss, severe storm damage in 2005–2006 was a major impact on the coastal sand dunes inhabited by the Sand-verbena moth (*Copablepharon fuscum*, Noctuidae), causing losses of patches of the larval food plant over a 200 m stretch of beach – a substantial part of the overall habitat occupied by this threatened and highly localised moth in western North America (BCIRT 2008).

Perhaps even more insidious, the impacts of catastrophes such as the – thankfully very infrequent – nuclear accidents on insects through increased mutation rates are even more difficult to assess. Following the meltdown and explosion of the Fukushima Dai-ichi nuclear power plant in Japan (12 March 2011) the common local lycaenid butterfly *Zizeeria maha* suffered physiological and genetic change, evident in morphological abnormalities that were inherited through generations, and sampled from May 2011 (Hiyama et al. 2012). Similar abnormalities were induced by laboratory radiation exposures, and it was postulated that the field changes may result from low-level internal exposure from caterpillars eating contaminated foliage, leading to increased random mutations from the radiation.

Population structure, in addition to spatial arrangement, also reflects genetic connectedness. Reduced dispersal opportunity through habitat losses (Chapter 7) and increased fragmentation of remnant populations in the landscape, isolates populations. Increasingly, genetic approaches, utilising DNA information, are being employed in analysing these effects – with a frequent aim being to determine whether a species occurs as populations that are genetically distinct (isolated, as possible ESUs, Chapter 2, p. 23), or whether the population is panmictic, with interchanges between units. If the former, the features leading to isolation may become clearer, as they are indeed 'barriers'. Many fragmented populations have lost genetic variation, and the present-day genetic 'architecture' is likely to be a consequence of that fragmentation. Such inbreeding depression is of concern in small populations, and may also manifest in captive-bred stock used for introductions or translocations (Chapter 10).

Historical demographics, considering the past effects of inter-population movements, can be related to geographical location and past dispersal movements and opportunities.

The Australian Golden sun-moth (*Synemon plana*, Chapter 9, p. 174) occurs in isolated populations on grasslands, and dispersal is poor; males are reluctant to cross any unhospitable terrain, and females are largely sedentary, so that recolonisation effects are very unlikely to occur (Clarke & Whyte 2003). Findings from both allozyme and mtDNA markers confirmed that grouped populations from Victoria and from the New South Wales/Australian Capital Territory region (Fig. 6.5) represented distinct ESUs, so that – as for the Karner blue butterfly discussed earlier – translocations between those regions should be avoided in order to conserve this evolutionary heritage. The variation found in *S. plana* led to recognition of five 'management units' that Clarke and Whyte recommended for separate management.

Implications of genetic variation for conservation are complex, as discussed by Habel et al. (2012b), and reflect properties of any given species' dispersal, abundance and ecology, with differences inferred widely to be general between 'specialists' and 'generalists'. The latter are typically presumed to be more homogeneous, with the three parameters of dispersal (ranging from sedentary to highly

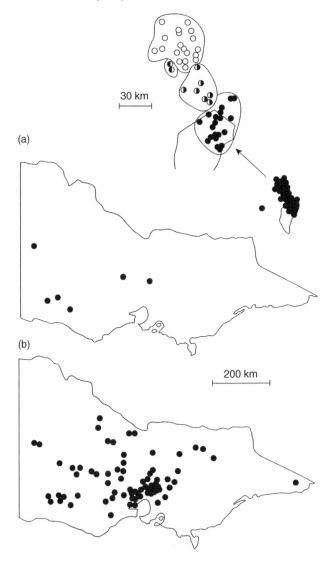

Fig. 6.5 The Golden sun-moth (*Synemon plana*, Castniidae) in south eastern Australia. (a) Extant populations known in the late 1990s, with paucity of records in Victoria (bounded) and concentration of populations within the Australian Capital Territory and adjacent parts of New South Wales; enlargement of the latter represents the groups of populations for which genetic analyses revealed four distinct categories (indicated by different spots) with the Victorian populations a further, possibly basal, lineage; (b) increased number of populations known in Victoria by early 2012, as consequence of detailed surveys for presence, largely on grassland patches near Melbourne (centre of map) investigated for possible development (a, after Clarke & Whyte 2003; b, after Brown et al. 2012. Reproduced with permission of Field Naturalists Club of Victoria.).

mobile, Chapter 7, p. 128), habitat size requirements (very small to very large) and extent of ecological specialisation (such as monophagy to polyphagy) influential. Habel et al. showed that many generalists (in their survey spanning 22 grassland species: 18 butterflies, four Zygaenidae) have complex genetic structures, and conservation of these may depend on maintaining habitat connectivity across the landscape, so promoting continuous gene flow. If broad-scale habitat degradation and loss occurs, such taxa might indeed suffer more severely than more specialised taxa with 'simpler' genetic constitution. Conservation of specialists, in contrast, may need focus on sustaining limited but high-quality habitat.

Patterns of genetic variation across a species' range are, of course, widespread amongst Lepidoptera but have been investigated most in ecologically specialised taxa for which restricted or fragmented habitats promote population isolation. The southern North American moth *Exyra semicrocea* (Noctuidae) is obligately associated with *Sarracenia* pitcher plants in bogs, and its life cycle occurs wholly within the plant leaves (Stephens et al. 2011). Western and eastern populations differed considerably in genetic makeup, with clear implication that the Mississippi River alluvial plain may constitute an effective break between these major groups – three each to the east and west of the river. Each of those groups was considered worthy of independent tailored individual management. Such distribution of genetic diversity among populations may become important if the principle of retaining (conserving) a species' full capacity for future evolution and adaptation is to be fulfilled, when the delineation of 'evolutionary significant units' is useful. The functional need is then to harmonise taxonomic consensus (Chapter 2) with practicalities of dispersal or transfer of individuals so not unwittingly destroying these units that may be recognisable only with the aid of DNA analyses across the taxon range. Maintenance of genetic diversity may be a key to long-term survival, and can raise complex issues. For *Lycaena dispar* in Europe (Chapter 9, p. 170), for example, it is important for conservation measures at this scale of consideration to transcend political boundaries and for the butterfly to be treated as an international concern with distinctive populations and sites accorded proper conservation (Lai & Pullin 2004). Concerns are expressed commonly over genetic distinctiveness of stocks used for recolonisation or reinforcement exercises (Chapter 10), especially movements to or from long-isolated populations that may be distinctive. However, relatively few practical studies have considered this theme. Considering possible origins of stocks of the Chequered skipper butterfly (*Carterocephalus palaemon*) for reintroducing it to England, Joyce and Pullin (2004) found no deep genetic differences between candidate donor populations from Scotland and continental Europe, and suggested that it was preferable to make the choice on ecological grounds, so giving preference to Belgian stocks over Scottish populations. Parallel problems will continue to arise as considerations of assisted colonisation (Chapter 10, p. 192) recur.

The related topic of 'population refuges' was discussed by Thomas and Jones (1993) for *Hesperia comma* in Britain. The needs of the butterfly include close-cropped calcareous grassland, where eggs are laid on the grass *Festuca ovina*. From the mid-1950s, and due largely to loss of rabbits from myxomatosis, overgrowing of grasslands led to declines of *H. comma* to only some 46 isolated 'refuge habitats' where conditions remained suitable, and on which the species depended for the next 20–30 years. Expansion of habitat area, aided by restoration, led to increase of recorded populations to 257 (in 109 2 × 2 km sampling tetrads) in 2000, compared with only 68 populations (in 30 tetrads) in 1982 (Davies et al. 2005). Many, however, were very small – 100 of them with fewer than 100 individuals seen on the peak sampling day. Recovery reflected (1) recovery of rabbit populations, (2) increased conservation management of the grasslands, and (3) warming, increasing area of potentially suitable habitat.

References

Asher, J., Warren, M., Fox, R., Harding, P., Jeffcoate, G. & Jeffcoate, S. (2001) The Millennium Atlas of Butterflies in Britain and Ireland. Oxford University Press, Oxford.

Baguette, M., Petit, S. & Queva, F. (2000) Population spatial structure and migration of three butterfly species within the same habitat network: consequences for conservation. *Journal of Applied Ecology* 37, 100–108.

BCIRT (British Columbia Invertebrates Recovery Team) (2008) Recovery strategy for the Sand-verbena moth (*Copablepharon fuscum*) in British Columbia. British Columbia Ministry of Environment, Victoria, British Columbia.

Bergerot, B., Julliard, R. & Baguette, M. (2010) Metacommunity dynamics: decline of functional relationship along a habitat fragmentation gradient. *PLoS ONE* 5, e11294. doi:10.1371/journal.pone.0011294.

Betzholtz, P.-E. & Franzen, M. (2011) Mobility is related to species traits in noctuid moths. *Ecological Entomology* 36, 369–376.

Binzenhofer, B. & Settele, J. (2000) Vergleichende autokologische Untersuchungen an *Glaucopsyche* (*Maculinea*) *nausithous* Bergstr. und *G.* (*M.*) *teleius* Bergstr. (Lepidoptera, Lycaenidae) im nordlichen Steigewald. pp. 1–98 in Settele, J. & Kleinwietfeld, S. (eds) Populationokologische Studien an Tagfalten. UFZ-Bericht 2/2000.

Boughton, D.A. (2000) The dispersal system of a butterfly: a test of source-sink theory suggests the intermediate-scale hypothesis. *American Naturalist* 156, 131–144.

Brower, L.P., Kust, D.R., Salimas, E.R. et al. (2004) Catastrophic winter storm mortality of monarch butterflies in Mexico during January 2002. pp. 151–166 in Oberhauser, K.S. & Solensky, M.J. (eds) The Monarch Butterfly. Biology and Conservation. Cornell University Press, Ithaca, New York.

Brower, L.P., Taylor, O.R., Williams, E.H., Slayback, D.A., Zubieta, R.R. & Ramirez, M.I. (2012) Decline of monarch butterflies overwintering in Mexico: is the migratory phenomenon at risk? *Insect Conservation and Diversity* 5, 95–100.

Brown, G., Tolsma, A. & McNabb, E. (2012) Ecological aspects of new populations of the threatened Golden sun moth *Synemon plana* on the Victorian Volcanic Plains. *Victorian Naturalist* 129, 77–85.

Caughley, G. (1994) Directions in conservation biology. *Journal of Animal Ecology* 63, 215–244.

Cizek, O. & Konvicka, M. (2005) What is a patch in a dynamic metapopulation? Mobility of an endangered woodland butterfly, *Euphydryas maturna*. *Ecography* 28, 791–800.

Clarke, G.M. & Whyte, L.S. (2003) Phylogeography and population history of the endangered golden sun moth (*Synemon plana*) revealed by allozymes and mitochondrial DNA analysis. *Conservation Genetics* 4, 719–734.

Conradt, L. & Roper, T.J. (2006) Nonrandom movement behavior at habitat boundaries in two butterfly species: implications for dispersal. *Ecology* 87, 125–132.

Conradt, L., Bodsworth, E.J., Roper, T.J. & Thomas, C.D. (2000) Non-random dispersal of the butterfly *Maniola jurtina*: implications for metapopulation models. *Proceedings of the Royal Society, series B* 267, 1505–1510.

Davies, Z.G., Wilson, R.J., Brereton, T.M. & Thomas, C.D. (2005) The re-expansion and improving status of the silver-spotted skipper butterfly (*Hesperia comma*) in Britain: a metapopulation success story. *Biological Conservation* 124, 189–198.

Davis, T. (2012) A review of the status of Microlepidoptera in Britain. Butterfly Conservation Report No. S 12-02. Butterfly Conservation, Wareham, Dorset.

Delattre, T., Burel, F., Humero, A., Stevens, V.M., Vernon, P. & Baguette, M. (2010). Dispersal mood revealed by shifts from routine to direct flights in the meadow brown butterfly *Maniola jurtina*. *Oikos* **119**, 1900–1908.

Dover, J.W. & Settele, J. (2009) The influence of landscape structure on butterfly distribution and movement: a review. *Journal of Insect Conservation* **13**, 3–27.

Ehrlich, P.R. & Hanski, I. (eds) (2004) On the Wings of Checkerspots. A Model System for Population Biology. Oxford University Press, Oxford.

Franzen, M. & Nilsson, S.G. (2007) What is the required minimum landscape size for dispersal studies? *Journal of Animal Ecology* **76**, 1224–1236.

Habel, J.C., Rodder, D., Lens, L. & Schmitt, T. (2012b) The genetic signature of ecologically different grassland Lepidoptera. *Biodiversity and Conservation*. DOI 10.1007/s 10531-012-0407-y

Haikola, S., Fortelius, W., O'Hara, R.B. et al. (2001) Inbreeding depression and the maintenance of genetic load in *Melitaea cinxia* metapopulations. *Conservation Genetics* **2**, 323–335.

Haikola, S., Singer, M.C. & Pen, I. (2004) Has inbreeding depression led to avoidance of sib mating in the Glanville fritillary butterfly (*Melitaea cinxia*)? *Evolutionary Ecology* **18**, 113–120.

Hanski, I. (1999) Metapopulation Ecology. Oxford University Press, Oxford.

Hanski, I. (2004) Biology of extinctions in butterfly metapopulations. pp. 577–602 in Boggs, C.L., Watt, W.B. & Ehrlich, P.R. (eds) Butterflies. Ecology and Evolution Taking Flight. University of Chicago Press, Chicago and London.

Hanski, I. & Kuussaari, M. (1995) Butterfly metapopulation dynamics. pp. 149–171 in Cappuccino, N. & Price, P.W. (eds) Population Dynamics. New Approaches and Syntheses. Academic Press, San Diego.

Hanski, I. & Singer, M.C. (2001) Extinction-colonization dynamics and host-plant choice in butterfly metapopulations. *The American Naturalist* **158**, 341–353.

Hanski, I., Pakkala, T., Kuussaarii, M. & Lei, G. (1995) Metapopulation persistence of an endangered butterfly in a fragmented landscape. *Oikos* **72**, 21–28.

Hanski, I., Ehrlich, P.R., Nieminen, M. et al. (2004) Checkerspots and conservation biology. pp. 264–287 in Ehrlich, P.R. & Hanski, I. (eds) On the Wings of Checkerspots. A Model System for Population Biology. Oxford University Press, Oxford.

Harrison, S. (1994) Metapopulations and conservation. pp. 111–128 in Edwards, P.J., May, R.M. & Webb, N.R. (eds) Large-Scale Ecology and Conservation Biology. Blackwell Publishing, Oxford.

Harrison, S. & Hastings, A. (1996) Genetic and evolutionary consequences of metapopulation structure. *Trends in Ecology and Evolution* **11**, 180–183.

Hiyama, A., Nohara, C., Kinjo, S. et al. (2012) The biological impacts of the Fukushima nuclear accident on the pale grass blue butterfly. *Scientific Reports* **2**, 570 DOI 10.1038/srep00570.

Hovestadt, T., Binzenhofer, B., Nowicki, P. & Settele, J. (2011) Do all inter-patch movements represent dispersal? A mixed kernel study of butterfly mobility in fragmented landscapes. *Journal of Animal Ecology* **80**, 1070–1077.

Joyce, D.A. & Pullin, A.S. (2004) Using genetics to inform re-introduction strategies for the Chequered skipper butterfly (*Carterocaphalus palaemon* Pallas) in England. *Journal of Insect Conservation* **8**, 69–74.

Lai, B.-C.G. & Pullin, A.S. (2004) Phylogeography, genetic diversity and conservation of the large copper butterfly *Lycaena dispar* in Europe. *Journal of Insect Conservation* **8**, 27–35.

Lewis, O.T., Thomas, C.D., Hill, J.K. et al.(1997) Three ways of assessing metapopulation structure in the butterfly *Plebejus argus*. *Ecological Entomology* **22**, 283–293.

Mouquet, N., Belrose, V., Thomas, J.A., Elmes, G.W., Clarke, R.T. & Hochberg, M.E. (2005) Conserving community modules: a case study of the endangered lycaenid butterfly *Maculinea alcon*. *Ecology* **86**, 3160–3173.

Nieminen, M., Siljander, M. & Hanski, I. (2004) Structure and dynamics of *Melitaea cinxia* populations. pp. 63–91 in Ehrlich, P.R. & Hanski, I. (eds) On the Wings of Checkerspots. A Model System for Population Biology. Oxford University Press, Oxford.

Nowicki, P., Halecki, W. & Kalarus, K. (2013) All natural habitat edges matter equally for endangered *Maculinea* butterflies. *Journal of Insect Conservation* **17**, 139–146.

Pateman, R.M., Hill, J.K., Roy, D.B., Fox, R. & Thomas, C.D. (2012) Temperature-dependent alterations in host use drive rapid range expansion in a butterfly. *Science* **336**, 1028–1030.

Pollard, E. & Eversham, B.C. (1995) Butterfly monitoring 2 – interpreting the changes. pp. 23–36 in Pullin A.S. (ed.) Ecology and Conservation of Butterflies. Chapman & Hall, London.

Pollard, E. & Yates, T.J. (1992) The extinction and foundation of local butterfly populations in relation to population variability and other factors. *Ecological Entomology* 17, 249–256.

Pollard, E. & Yates, T.J. (1993) Monitoring Butterflies for Ecology and Conservation. Chapman & Hall, London.

Relf, M. & New T.R. (2009) Conservation needs of the Altona skipper butterfly, *Hesperilla flavescens flavescens* Waterhouse (Lepidoptera: Hesperiidae) near Melbourne, Victoria. *Journal of Insect Conservation* 13, 143–149.

Samways, M.J. & Lu, S.-S. (2007) Key traits in a threatened butterfly and its common sibling: implications for conservation. *Biodiversity and Conservation* 16, 4095–4107.

Samways, M.J., McGeoch, M.A. & New, T.R. (2010) Insect Conservation. A Handbook of Approaches and Methods. Oxford University Press, Oxford.

Schultz. C.B., Franco, A.M.A. & Crone, E.E. (2012) Response of butterflies to structural and resource boundaries. *Journal of Animal Ecology* 81, 724–734.

Showers, W.B. (1997). Migratory ecology of the black cutworm. *Annual Review of Entomology* 42, 393–425.

Sinclair, L.J. (2002) Distribution and conservation requirements of *Notoreas* sp., an unnamed geometrid moth on the Taranaki coast, North Island, New Zealand. *New Zealand Journal of Zoology* 29, 311–322.

Stephens, J.D., Santos, S.R. & Folkerts, D.R. (2011) Genetic differentiation, structure and a transition zone among populations of the Pitcher plant moth *Exyra semicrocea*: implications for conservation. *PLoS ONE* 6(7): e22658, doi: 101371/journal.pone.0022658.

Stevens, V.M., Turlure, C. & Baguette, M. (2010) A meta-analysis of dispersal in butterflies. *Biological Reviews* 85, 625–642.

Thomas, C.D. (2001) Scale, dispersal and population structure. pp. 321–335 in Woiwod, I.P., Reynolds, D.R. & Thomas, C.D. (eds) Insect Movement; Mechanisms and Consequences. CAB International, Wallingford.

Thomas, C.D. (1994) Extinction, colonization and metapopulations; environmental tracking by rare species. *Conservation Biology* 8, 373–378.

Thomas, C.D. & Jones, T.H. (1993) Partial recovery of a skipper butterfly (*Hesperia comma*) from population refuges: lessons for conservation in a fragmented landscape. *Journal of Animal Ecology* 62, 472–481.

Thomas, C.D. & Kunin, W.E. (1999) The spatial structure of populations. *Journal of Animal Ecology* 68, 647–657.

Thomas, C.D., Singer, M.C. & Boughton, D.A. (1996) Catastrophic extinction of population sources in a butterfly metapopulation. *American Naturalist* 148, 957–975.

van der Meulen, J. & Groenendijk, D. (2005) Assessment of the mobility of day-flying moths: an ecological approach. *Proceedings of the Netherlands Entomological Society Meetings* 16, 37–50.

Wilson, R.J., Davies, Z.G. & Thomas, C.D. (2009) Modelling the effect of habitat fragmentation on range expansion in a butterfly. *Philosophical Transactions of the Royal Society, series B* 276, 1421–1427.

7

Understanding Habitats

Introduction: The meaning of 'habitat'

A suitable 'habitat' is the feature cited universally as a primary need for conservation, and habitats deemed unsuitable in some way are, correspondingly, the most frequently cited and pervasive causes of losses and declines. Loss and degradation of 'habitat' is the paramount threat to Lepidoptera, as to most other taxa. However, defining optimal habitat for any given species is far from easy. Comparison of a supposedly poor area with those on which equivalent losses or declines have not occurred is a common approach. Interpreting such comparisons, often with the belief that the 'more pristine' (or less disturbed) sites are in some way better than others has underpinned much conservation management for Lepidoptera, and discerning the needs of focal species is the major key to conservation management, reflecting that many species are rare or becoming rarer because of loss of key food plants or other resources. In managed environments, such as agroecosystems and much of the world's forests, promoting diversity of habitat management and spatial heterogeneity through management mosaics across the regional landscape may benefit many species by preventing the greater homogenisation of local structure through large-scale monoculture or other uniform aged plantations. The central role of 'habitat' in practical conservation is the subject of this chapter. Understanding how to characterise habitats, and how to interpret changes and disturbances is the core foundation of much conservation management.

Historically, the term has been equated most commonly with 'place', often synonymised with site or biotope as 'a place to live', with the term 'patch' applied widely to signal a discrete suitable and, often, occupied, area. However they are defined formally, much discussion of habitats has devolved on the separation of habitat patch (occupied or able to be occupied) and inhospitable surrounding

Lepidoptera and Conservation, First Edition. T.R. New.
© 2014 John Wiley & Sons, Ltd. Published 2014 by John Wiley & Sons, Ltd.

matrix in the adjacent landscape. Reserve systems, such as the United Kingdom Sites of Special Scientific Interest, may reflect these characters, but are often designated on the presence of species of conservation interest, as well as broader contexts. In many instances, insect 'habitat' has been defined by vegetation features or other broad 'biotope characters'. Many examples, such as that of *Euphydryas maturna* (Chapter 9, p. 166), confirm that this abrupt contrast is misleading, with the matrix (1) providing at least some of the facilities needed by the species and (2) being 'permeable' to dispersal to varying extents. Whilst habitat can be denoted very simplistically in this way, such exclusive and polarised contrasts are very misleading for real-life interpretations.

As Dennis et al. (2003, 2006) have emphasised, a resource-based concept, whereby landscapes are viewed as a continuum of overlapping and intergrading resource distributions, may have considerable value in understanding lepidopteran biology as a basis for conservation. Indeed, Dennis' magisterial book (Dennis 2010) encapsulates numerous original insights and ideas on integrating butterfly biology with environmental features. It demonstrates repeatedly the subtleties of such connections, and further comment pales by comparison; it is 'required reading' for all conservation biologists working with Lepidoptera, with the key points raised for each section invaluable pointers to management. With increasing awareness of the nuances of species' needs, arising substantially from studies on British butterflies, Dennis (2009, 2010) emphasised that recent approaches to modelling 'require a new interpretation of habitats involving a shift from simple notions of habitat equating with vegetation units to species resources, and that a biotope is not itself a habitat, but an environment in which the critical resources needed come together in space and time in ways sufficient to foster the species' wellbeing'. A 'woodland butterfly' or a 'grassland moth' is an initial, useful but bland, characterisation indicating where it may thrive, but within which numerous complex variables affect that existence. Again from Dennis, 'Unless we isolate the elements of habitat – resources and conditioners [see later in this chapter] – and understand how they impact on individuals, and amalgamate to integrate populations, our grasp of butterfly biology will remain deficient, artificial' – and carefully founded management and selection of priority sites for treatment will not be optimal. The balance between resource dynamics and spatial dynamics (as emphasised primarily in much metapopulation investigation) thus changes away from the extreme classical scenario of contrast between patch and matrix, to incorporate a much wider picture of the insects' needs. The 'landscape', as the context for any site, biotope, or patch of conservation interest, may influence the integrity of most conservation measures. Thus, for woodland moths in Britain, the percentage cover of woodlands in surrounding areas (within about 500 m) may be an important consideration for moth conservation management (Fuentes-Montemayor et al. 2012), but with macromoth management influenced by cover at larger scales, of about 1500 m.

Holometabolous insects are complex to conserve, simply because each species is in essence two very different animals! The resources needed by a relatively sedentary herbivorous caterpillar with chewing mouthparts are very different from those critical for a mobile nectarivorous adult with suctorial mouthparts – but either or both may be very specialised and restricted. Those resources must coincide and be accessible for a resident species, whose life cycle may be tuned finely to their seasonal availability and suitability. 'Habitat' must include these resources where and when they are needed, and their supply is perhaps the most central need of conservation management. Definition of these 'critical resources' is thereby important. Habitat use essentially devolves on patterns of resources, and the population structure(s) of the species needing those resources. The amount, quality and seasonal supply and availability of critical resources, and those of lesser prominence, influence site carrying capacity (and, so, maximum population size) and the needs for dispersal as migration to other patches. That species richness increases with patch size is a paradigm well entrenched in ecology, and has been illustrated for Lepidoptera on numerous occasions – however, as with the alpine butterflies of central Europe appraised by Bila et al. (2012), most or all available sites may be small, so that even small proportional losses are of concern, together with any resulting increase in site isolation.

In large patches, 'resource quantity' may be sufficient to (at least partially) offset 'resource quality' (Dennis et al. 2006), but a resource-based definition of habitat is based on a given species requiring a definable suite of resources within an environment suitable for it to function and thrive. The array of those resources is large and must be documented separately for larval and adult stages and subsequently integrated to provide a profile for the species. Of the major resource categories differentiated by Dennis and his colleagues, 'consumables' are the easier to define (most simply as larval food plants, often specific or a closely related taxonomic group, and adult nectar or other food) and also those most easily manipulated in conservation management. Often less obvious than consumables, and much more difficult to evaluate comprehensively, are the 'utilities', the wider environmental needs and conditions for sustaining normal behavioural performance and access to consumables. A very long list of attributes may be needed for this category, but representative utilities may include suitable pupation, oviposition, hibernation or aestivation sites, bare ground for basking as thermoregulation or display, vegetation edges with perches for territorial flights, refuges from natural enemies, mating assembly sites – such as hilltops – and so on. Many of these do not necessarily coincide in space with consumables, and for some species may be provided by the matrix separating patches of consumables. The critical importance of structural components as resources is often underestimated, or understudied, but many Lepidoptera may depend substantially on suitable structures for mate location, hibernation or roosting, for examples. Investigation on the Silver-studded blue butterfly (*Plebejus argus*) in North Wales revealed the importance of tall herb/shrub vegetation for roosting and mating – with fewer than 3% of butterflies roosting within larval host plant areas (Dennis 2004). The taller vegetation (>15 cm tall) is used as perching and territorial 'observation points' in mate location. Shelter from wind leads to concentrations on the lee side of shrubs outside the breeding area – scrub cover may be important for shelter and, for many Lepidoptera, be particularly significant in exposed areas such as wind-prone coasts and upland sites. 'Shelter belts' are important habitat components for butterflies in open landscapes (Dover et al. 1997). Windiness may also affect the outcomes of transect counts (Kuussaari et al. 2007a), again emphasising the needs for standardising sampling conditions.

Microclimate, sometimes considered as a separate category (as a 'conditioner'), may have important influences and ultimately govern access to other resources. In particular, temperature may both determine the range over which those resources occur and the activity of the searching butterfly or moth. Relatively simple features of site topography, for example, may restrict occupancy if insolation is insufficient to keep that site warm enough. The classic study of the Adonis blue butterfly (*Lysandra bellargus*) in southern Britain by Thomas (1983) clearly demonstrated the need for south-facing slopes for the butterfly, then at the northern fringe of its distribution in Britain and in a climatically marginal environment. In a related context, 'shelter' from winds or extreme weather may be a key need – for example, hedgerow trees in exposed agricultural landscapes are recognised widely as beneficial to moths, with increased numbers even of species not associated with the tree species as larval foods (Merckx et al. 2010), and leading to advocacy for retention and establishment of such trees in agri-environment schemes.

More generally, Dennis et al. (2006) regarded information on thermal resources as a vital component of habitat definition then largely absent from their accumulating database. They anticipated that the approach used by Bryant and Shreeve (2002) could lead to microclimate modelling at relevant scales, drawing on use of neural networks leading to plotting thermal resources over landscapes by combining weather data with topographic and vegetational features as overlays.

Accumulation of all information on resources needed by Lepidoptera species into centralised Resource Data Bases, as initiated for the British butterflies (Dennis et al. 2008), would assemble a tool of immense potential value in conservation management, in providing much of the background needed for focused planning. Not least, it would help greatly in reducing the extent of the primary basic biological research almost invariably needed as each additional species becomes of conservation significance. Knowledge of such exact requirements is likely to reduce the failure rate of conservation management, and practical benefits include (1) anticipating how the insect may respond to future changes in its effective environment; (2) understanding why co-occurring species respond in different ways to change or management of the occupied area and its surrounds; and (3) understanding the potential for the species on different sites, as an aid in setting priorities or defining conservation needs. Whilst currently impracticable to follow equivalent detail in many other regions, the approach demonstrates the universal critical relevance of resources in sound conservation planning and management.

Resource issues are also the basis for assessing the relative position of species along the 'specialist–generalist continuum', with integrated information also critical in indicating functional groups or guilds (Chapter 1, p. 9) and which species may co-occur within assemblages. Two major sources of information have contributed to such databases. For Britain, Dennis et al. noted these as (1) the independently gained and mostly published natural history observations accumulated over many years, and (2) the more formal (and, generally, more recent) 'scientific' data from autecological studies (many of them site-specific) such as doctoral theses and agency projects. The primary subsequent need was for effective assembly and inventory of all this information and of the sites studied, to indicate where significant gaps remain to be addressed. Even for the British butterflies, those gaps are considerable. Site-based conservation measures or strategies can generally preserve threatened habitat specialists only by heeding their specific resource needs, which may differ substantially across sites, as for *Maculinea alcon* in Belgium (Maes et al. 2004, Chapter 9, p. 162).

Ensuing calls for this approach to be extended to cover the more accessible groups of moths complement and enhance database values considerably. Resource-based habitat associations for central European macromoths (other than Geometridae and Noctuidae) led to recognition of five broad-based habitat groups, paralleling those based on trap samples for North America shown by Summerville (2004, Chapter 1, p. 9) (Pavlikova & Konvicka 2012, Table 7.1). This grouping was modelled for a large array of life-history and resource uses, drawing extensively on the pioneering listing for British butterflies used by Shreeve et al. (2001) to delimit four habitat-based groups there – namely butterflies of woodland, open tall-herb formations, short-sward grasslands and ruderal environments – but expanded the scope considerably from the broader array of 174 moth species evaluated. For example, woodland butterflies are relatively few in Britain, whereas the far greater richness of European forest moths enabled further categorisation within this major biotope category. For many moth species, however, boundaries between habitats along a gradient were not wholly discrete. As Fig. 7.1 summarises, with the analyses based in correlative ordinations, closed canopy moths (n = 52 species) contain relatively few threatened species, whereas threatened species were also over-represented amongst grassland taxa (39 species). As elsewhere in the world, European grasslands are highly vulnerable to land use changes, including agricultural intensification, and specialised resident taxa are often susceptible to loss. In contrast, closed canopy woodlands have increased in

Table 7.1 The broad-based habitat association groups of European macromoths distinguished from a resource-based classification of habitats founded on 178 life history attributes across 174 species (Source: Pavlikova & Konvicka 2012. With kind permission from Springer Science+Business Media.).

Habitat association group	Comments
Closed canopy moths	(n = 53) Develop on trees; pupae overwinter; early stages tend to occur in groups
Open canopy moths	(n = 47) Develop on woody shrubs; adults not nectar feeders; caterpillars often hairy
Grassland moths	(n = 39) Develop on herbs/grasses; most oviposit in large batches, are polyphagous; larvae hibernate on ground surface
Herb-feeding hawk moths	(n = 14) Larvae almost all on herbs; overwinter as pupae; adults nectar feeders; some (7) are long-distance migrants
Lichen feeders	(n = 21) All are Actiidae: Lithosiinae; univoltine development; overwintering larvae

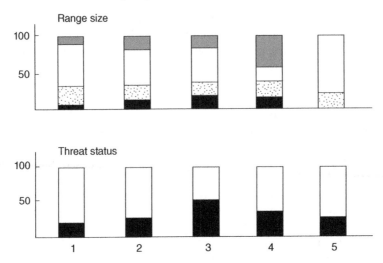

Fig. 7.1 Larger moths in central Europe: indication of range size (a) and threat status (b) of species in five main habitat associations. Proportional representation obtained by ordination analyses of life-history attributes. Range (scale to 100%), top to bottom: huge, large, intermediate, small. Threat status: black, threatened; open, unthreatened (Pavlikova & Konvicka 2012. With kind permission from Springer Science+Business Media.).

the region through afforestation for timber supply and wider amenity uses, so that this resource suite has apparently also fostered moth wellbeing.

Notwithstanding the greater information that accrues from resource considerations in this way, it is almost inevitable that the near future will be dominated by more general considerations of habitat based largely on biotopes, simply because of the urgency of conservation and the incomplete understanding of the more detailed needs, particularly beyond the northern temperate regions. Thus, habitat loss and degradation (reflecting loss and lowered amount, quality and availability of critical resources) is the paramount theme in Lepidoptera conservation, and is interwoven intricately with population structure and dynamics. Those changes reflect natural ones, such as vegetational succession on sites, and the array of human disturbances inflicted, and discussed in Chapter 12, with increased habitat (patch) isolation and decreased size fostering vulnerability to stochastic causes (Chapter 6). Nevertheless, the most effective habitat restoration and management depends on sound biological understanding of the focal species' needs.

A single informative biological response may provide information of fundamental importance to conserving a rare species. Thus for the North American Mardon skipper (*Polites mardon*), which has declined markedly in recent decades and for which prairie restoration is seen as a key conservation need, practical management was hampered by lack of knowledge of host plant(s) and key habitat needs by the early stages – information clearly relevant to considering re-introductions and resource enhancement (Henry & Schultz 2013). Direct observations of skippers, tracking individual females to detect oviposition in the field, were instrumental in remedying this: 86 of 88 observed ovipositions were

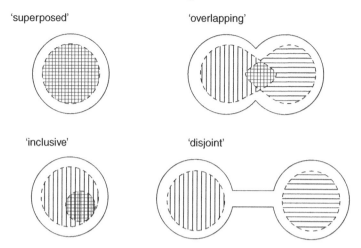

Fig. 7.2 Patterns of spatial arrangement of resources for butterflies in habitat patches: 'superposed', larval food plants and adult nectar resources coincide in distribution; 'overlapping', these coincide over only part of the habitat, with each occurring without the other elsewhere in the patch; 'inclusive', nectar plants occur within part of a more extensive larval food plant range; 'disjoint', the two entirely separate but not separated by barriers (habitat patch boundary, solid line; larval food plant distribution, vertical hatching; adult nectar sources, horizontal hatching) (after Turlure et al. 2005. Reproduced with permission of Pensoft Publishers.).

on a single native bunchgrass species (*Festuca roemeri*, Roemer's fescue), mostly on small tufts in sparsely vegetated areas. The skipper was confirmed as a host specialist in this study, in marked contrast with some earlier oviposition records across 23 grass species, so that further discrimination beyond species, to condition and structure was evident; changes in vegetation structure contributed to population declines, with regional differences also reflecting local climate differences. In any such case, local behaviour and habitat use patterns are critical components in planning local conservation management.

Even when two (or more) species use the same resource, the spatial and temporal distribution of those species may lead to markedly differing modes of exploitation. Turlure et al. (2005) noted four different basic patterns of resources within a bounded habitat (Fig. 7.2) and considered these in relation to comparative use of the same host plant by two butterflies in peat bogs (namely, *Lycaena helle*, *Proclossiana eunomia*, on *Polygonum bistorta*). They differed: adult *L. helle* required other nectar plant species, leading to habitat boundaries varying in time; and distances used by *P. eunomia* were greater, so that the two species were partially separated in space. This example, of differing distribution of just two critical resources, larval food plants and adult nectar plants, may reflect much more complex systems in real life as more resources enter the pattern.

Habitat loss

The Dakota skipper (*Hesperia dakotae*) has declined considerably from occupying a once continuous prairie habitat over much of central North America. Less than 2% of the pre-European settlement extent of prairie now remains (Britten & Glasford 2002), with the remnants largely occurring in small fragments within large areas converted for agricultural production. The butterfly has declined in parallel, with genetic studies implying that effective immigration rates across populations are low, and effective population sizes small – so that each population is effectively isolated on a prairie fragment (but with historical affinity between nearby groups of sites) and subject to genetic drift. Management recommendations include maximising effective population size within each *H. dakotae* population, and avoiding any measures likely to cause harm to adults – so that seasonal management that avoids the peak flight season may be dictated (Britten & Glasford 2002).

Analogous scenarios are widespread, and the core of many conservation programmes, so that issues of habitat extent and population connectivity are central themes dictating needs and scales of consideration. In addition to smaller areas of suitable habitat, changes in land use may also enforce isolation by rendering the matrix increasingly less permeable and more inhospitable. This has apparently been the case for a European burnet moth (*Zygaena loti*, Zygaenidae) which is now restricted to isolated remnants of formerly much more extensive seminatural meadows, and which has now declined considerably in abundance. The moth disperses little and, with polyphagous caterpillars, was formerly regarded as common on xerothermic meadows and related habitats. A study of the genetic diversity across seven populations selected to represent different population sizes and degrees of supposed isolation, in a region spanning western Germany and adjacent parts of France and Luxembourg led Habel et al. (2012a,b) to suggest that populations with high genetic diversity represent the general structure of previously interconnected units, whilst low genetic diversity was the outcome of reduced connectivity. As for the Dakota skipper, management to reconnect fragmented sites to promote genetic interchange is needed. The few remaining genetically diverse populations of *Z. loti* in the region merit strong protection as possible sources for future translocations (Chapter 10, p. 190). This study, significantly, revealed the twin states of (1) high genetic diversity representative of common species and (2) low genetic diversity more typical for restricted specialists. It suggested the important scenario that common taxa with originally extensive population networks and high genetic diversity may suffer more from sudden habitat fragmentation than may highly specialised species with low genetic diversity. The latter may have persisted in isolated patches for some time. Habel et al. suggested that in the absence of conservation measures *Z. loti* might become extinct, not directly through loss of habitat but by the rapid rate of habitat change initiating rapid genetic change. However, correlation of genetic diversity with habitat structure alone may be over-simplistic: as representatives among the *Z. loti* populations were evaluated, for example, Habel et al. (2012a) found (1) a small population close to others but with low genetic diversity, and (2) a large isolated habitat with a high genetic diversity population.

Despite the widely held generalisation that habitat fragmentation and genetic diversity are correlated closely, such exceptions may be quite frequent. Genetic 'bottlenecks' for tiny populations are frequently postulated for Lepidoptera, with recovery associated with regaining variety. An unusual butterfly example is for a local subspecies of the Apollo (*Parnasius apollo vinningensis*, Papilionidae) in Germany. Strongly isolated populations, including post-bottleneck populations, showed complete lack of genetic variability, with all sampling sites belonging to one homogeneous gene pool (Habel et al. 2009) and the historical interpretation suggesting long isolation of the populations.

Habitat fragmentation is claimed commonly to be more detrimental to specialised than to generalised species, with the principle flowing in part from theoretical models predicting that 'stacked specialists' (a term used by Holt et al., 1999, to designate monophagous species at higher trophic levels) are more influenced by habitat area than are lower trophic levels, including the plants on which they feed. Trends in diversity of butterflies on fragmented calcareous grasslands in Germany were used to test several relevant predictions relating influences of habitat area (Steffan-Dewenter & Tscharntke 2000) (Table 7.2). Densities of monophagous species (n = 4) increased with habitat area, whereas densities of oligophagous (20, plus an additional 27 less restricted taxa) and polyphagous species (18) decreased. Plotted slopes of species–area relationship (the 'z-values') increased with food plant specialisation – revealing differential responses to habitat area by these trophic groups (Fig. 7.3), both in richness and

Table 7.2 Predictions used to explore influences of habitat fragmentation on butterflies. (Source: Steffan Dewenter & Tscharntke 2000. Reproduced with permission of John Wiley & Sons).

1. Species diversity of butterfly communities increases with habitat area and decreases with habitat isolation
2. Habitat fragmentation affects particularly species at higher trophic levels, food plant specialists and species with limited dispersal abilities
3. Habitat quality may add to the predictive value of habitat area and thereby modify species–area relationships
4. Species richness per se, based on a sample size that increases with area, has larger z-values than species richness per plot, based on equal sample size
5. Total butterfly densities increase with habitat area

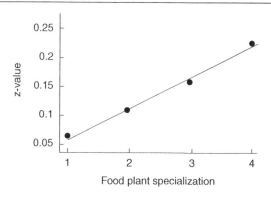

Fig. 7.3 The z-values of species–area curves of butterfly species in relation to extent of food plant specialisation: (1) polyphagous; (2) oligophagous; (3) strictly oligophagous; 4, monophagous (Source: Steffan-Dewenter & Tscharntke 2000. Reproduced with permission of John Wiley & Sons.).

density. Many butterflies in agricultural landscapes depend on warm, protected microclimates, which are largely confined to non-cultivated areas.

Difficulties of generalisation are demonstrated well by comparative studies of closely related but ecologically different taxa. Although such investigations are few, they help to emphasise that different species may respond differently to the same general landscape pattern. The principle involved appears rather simple, although very difficult to investigate properly: that a given landscape may be fragmented for a habitat specialist with poor dispersal powers and much less so for dispersal-capable generalist species. Generalists that are poor dispersers, and specialists that can disperse well, should show some intermediate levels of occupancy. Three species of *Thymelicus* (Hesperiidae) in central Europe were compared by Louy et al. (2007), and their relationships between ecology and genetic structure are summarised in Fig. 7.4. In summary:

1 the Essex skipper, *T. lineola*, is a habitat generalist, but with higher dispersal ability, and the high availability of habitat fosters a panmictic population structure (Fig. 7.4a);
2 the Small skipper, *T. sylvestris*, is also a habitat generalist with low dispersal ability, fitting the principle of progressive isolation with distance (Fig. 7.4b); and
3 the Lulworth skipper, *Thymelicus acteon*, is a habitat specialist with low dispersal ability, and the limited availability of habitats results in complete isolation; genetic drift then acts independently in each population (Fig. 7.4c).

Dispersing Essex skippers fly sufficiently strongly to reach more distant habitat patches, as well as those nearby, so that regional gene flow occurs. In contrast, the Lulworth skipper is likely to prove vulnerable (and, indeed, is declining strongly over much of Europe, in contrast to the other two species) because

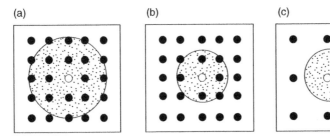

Fig. 7.4 Schematic depiction of dispersal ability and habitat availability and its influences on genetic structure of butterfly species, using three species of *Thymelicus* (Hesperiidae) as comparative examples. (a) high dispersal ability (circle) from an origin (central point) and high availability of habitat patches (dark spots) led to intensive gene flow and panmictic population structure as in *T. lineola*; (b) lower dispersal ability in landscape with high habitat availability reduces gene flow, so an 'isolation by distance' system will establish, as in *T. sylvestris*; (c) low dispersal ability in combination with limited habitat availability results in complete isolation, with genetic drift acting independently in each population, as in *T. actaeon* (Source: Louy et al. 2007. With kind permission from Springer Science+Business Media.).

losses from habitat fragmentation are unlikely to be countered by recolonisations. In *T. acteon*, genetic diversity equivalent to that of the other two species is not a driver of loss – rather, habitat loss *per se* or stochastic loss of any individual population on an isolated patch may constitute local extinction.

Examples noted in this chapter and the preceding one emphasise that co-occurring specialist species may have very different population structures and dispersal patterns within the same arena of habitat patches. Comparison of three unrelated butterflies in Belgium, as a further instructive example, confirmed that patterns of patch occupancy and migration can differ strongly (Baguette et al. 2000), with the design of a suitable habitat network for conservation thus needing to consider the most susceptible species of concern amongst the candidates of interest. Patch areas, patch quality, dispersal propensity and distance between patches all reflect the scale needed for an appropriate network.

For each species, an individual factor may predominantly induce decline. Most populations of the Adonis blue butterfly (*Lysandra* or *Polyommatus bellargus*) in Britain suffered severe declines in the late 1970s, because a severe drought in 1976 caused the host plant, the vetch *Hippocrepis comosa*, to wilt. More than 90% of the known populations of *P. bellargus* in the United Kingdom had become extinct by 1981. Since then, gradual recovery has occurred, with populations varying considerably in size and the extent of isolation (Harper et al. 2003), but the butterflies disperse little (Thomas, 1983, noted dispersal distances of generally less than 25 m), and even nearby populations may be functionally isolated, essentially forming closed populations. Larger populations supported considerably higher levels of genetic diversity, with the corollary that small populations may lose genetic diversity through drift and – with the widely supported 'general rule' of such losses increasing probability of extinction – population size itself becomes a critical conservation concern, with need to counter declines. Harper et al. concluded that patch size may be more important to this species than connectivity, and that the conservation of large habitat patches should be given high priority.

Issues around the relative importance of site (resource) quality and size and of promoting effective connectivity in the landscape are widespread in Lepidoptera conservation planning, and subtle ecological and behavioural differences between even very closely related species can lead to very different emphases in management. The *Thymelicus* skippers, discussed earlier, are one such example. Another, dealing also with congeneric species of very considerable conservation interest, is of two species of large blue butterflies, *Maculinea teleius* and *M. nausithous*, in the Netherlands. Both had become extinct in the country and were re-introduced successfully in 1990; they were monitored for the next decade (van Langevelde & Wynhoff 2009). Both re-introductions thrived, and both species appeared to be still restricted to a few sites after this period, despite other apparently suitable habitat patches being available. *M. teleius* remained only on the meadows to which it was re-introduced, and *M. nausithous* was slightly more dispersive and had formed a new colony about 600 m away. The surveys were undertaken at an unusually fine and detailed scale, using 587 randomly selected square metre plots that each contained one or more larval food plants (*Sanguisorba officinalis*), on which adult presence or absence was recorded each year. *M. teleius* constantly undertook short distance colonisation within the meadow release site, with a dynamic 'expansion–retraction' pattern – attributed by van Langerveld and Wynhoff to slow expansion due to limited movement countered by stochastic impacts of weather conditions, but together confirming very high site fidelity by this butterfly, with little likelihood of movement to unoccupied sites, however suitable those might be. On average, *M. nausithous* undertook longer-distance colonisations, leading to the suggested contrast between species that (1) *M. teleius* is confined to using habitat near to where adults emerge, even if it is of poor quality, whilst (2) *M. nausithous* has greater ability to disperse and seek patches of higher quality. This difference contributes to need for differing management protocols, as (1) improving local habitat quality for *M. teleius* through a range of highly targeted management actions (Chapter 11), augmented by creation of stepping stones where possible to enhance limited connectivity, and (2) creating networks of habitat patches to take advantage of *M. nausithous*' greater dispersal capability (of up to about 450 m). Growth and spread after reintroduction can thereby be fostered in different ways for the two species.

However, simply assessing dispersal can become complex. Possible generalisations appeal, so that support for the general concept that butterfly species occupying early seral habitats are necessarily able dispersers – or relatively more mobile than many species restricted to long-lived habitats (Shreeve 1995) – in order to track short-lived habitats in patchy landscapes is strong. Very generally, again, high-mobility species tend to be 'survivors', and sedentary species, vulnerable – but, more surprisingly, Thomas (2000) demonstrated that 'intermediate mobility' species of butterflies have declined in Britain more than sedentary species. Dispersal rates and behaviour determine species distributions and the relationships between populations and may also influence how a habitat patch is delineated. Short-term weather features can influence dispersal considerably. From studies on burnet moths (Zygaenidae) in southern Sweden, Franzen and Nilsson (2012) found that in an unusually warm year, dispersal rates of both *Zygaena lonicerae* and *Z. viciae* between patches increased markedly, and suggested that prolonged warm weather could have a more general influence on metapopulation dynamics – perhaps particularly in areas toward the poleward edge of the range. In contrast, warmer weather at the trailing edge of the range might be disadvantageous if the habitat provides insufficient buffer. Variation in dispersal of Lepidoptera between years has been considered only rather rarely in relation to temperature (Hill et al. 2002) and, in Sweden (where all six resident Zygaenidae are red-listed), a higher frequency of warm summers may increase moth survival, as long as sufficient habitat remains. Climate warming can influence both the extent and rate of colonisation as taxa shift range, with implications of differential interactions with other taxa in different years.

Such changes may be both rapid and subtle. The habitat boundaries of the Silver-studded blue (*Plebejus argus*) in North Wales, for example, appeared to change within hours or days in relation to weather. This species commonly disperses rather little (p. 119), but much greater distances are traversed during hot, clear and calm conditions (Dennis & Sparks 2006), when butterflies can move out into exposed countryside beyond their more usual sheltered shrubby areas on which the host plant occurs. In this example, probably representing numerous other taxa, definition of 'habitat' depends on the conditions in which surveys are undertaken. Answering the apparently simple question of 'what is a habitat patch?' is far more complicated than initially appears. Implications for conservation include addressing the definition of 'patches' for assessing metapopulation status, with the need to monitor range spatial changes related to weather rather than just those sites that are discerned by presence of core residency.

As Parmesan et al. (1999) noted, range shifts of butterflies (in response to climate change, Chapter 12, p. 210) occur at the level of the population, fundamentally from changes in the ratio of extinctions to colonisations at the northern and southern range boundaries. A northward shift then occurs when there is either or both of net colonisation at the northern boundary and net extinction at the southern boundary – with these conditions apparent for many butterflies in the northern hemisphere. Within any species' range, populations can differ widely in genetic constitution. Collectively, population individuality and genetics add considerable complication to more general predictions that a species as a whole will shift range. Studies on two butterflies in western North America induced Zakharov and Hellmann (2007) to urge caution in assuming that populations do not vary along latitude, and that responses by species as single ecotypes to climate probably do not occur. An additional complication (Davis et al. 1998) in predicting responses to change by the most familiar approach of matching the species' 'climate envelope' (through which the current plotted climate–space distribution is used as a template to predict future incidence) is that not only the species' range but also its interactions with other species change. Thus, more realistic prediction should ideally include both dispersal and interactions. But, as Crozier (2004a,b) emphasised, species respond to specific environmental changes in individualistic ways and 'we are far from a detailed understanding of range-limiting factors for most species'.

That species differ in their responses was shown clearly by Parmesan et al.'s (1999) survey of European butterflies, which also demonstrated the clear trend of northward range shifts. Of 35 non-migratory species with data from both northern and southern range boundaries available for analysis, 63% had shifted north by 35–240 km during the twentieth century, and only 3% southward. More broadly, the northern boundaries of 34 (of 52) species had extended with any trend consistent if different countries were examined separately. Southern boundaries (data from 40 species) showed 22% contracting northward, 72% remaining similar and 5% extending further south. The most common pattern, then, was for northward extension with the southern boundary stable (about two-thirds of the 22 species) or contracting northward.

Northward range extensions seem to have arisen from sequential colonisations of new populations along latitude, as the primary gradient of climatic variation, with the poleward edge progressively presenting suitable conditions for residency.

A survey by Zakharov and Hellman (2007) was amongst the first to examine butterfly genetic diversity in relation to population differentiation and climate changes, so challenging the broader concept of more uniform population regimes enabling selective advantage and characterising adaptations at opposing range ends, and anticipating possible genetic differences with divergence of peripheral populations correlating with climate. Interactions of climate warming with habitat favourability are complex – the increase in thermally hospitable space is simply the possible template for more intricate changes to develop, and in which increased survival may occur in species restricted by temperature regimes (Crozier 2004a,b) in either or both of space and season. The North American Sachem skipper, *Atalopedes campestris*, has recently colonised areas that have become 2–4 °C warmer over the last 50 years or so, with winter warming shown to be a prerequisite for this expansion (Fig. 7.5), reflecting that the overwintering caterpillars survive better; cold stress was probably the major cause of mortality, and survivorship was higher within the core range than at the range edge. Likewise, requirements of the Silver-spotted skipper, *Hesperia comma*, in Britain may change in very subtle ways with warming (Davies et al. 2006). Thus, the space available for oviposition increased, leading to increased oviposition rate, so potentially increasing the rate of population increase. The changes documented by Davies et al. from earlier appraisal of this species included marked changes in its presence on habitat patches of different aspects (Table 7.3), with increased population numbers established on northerly-facing slopes, and the proportion of bare ground optimal for oviposition halving over the 20 years between surveys. The former 'prescription' for habitat management had become deficient in relation to recent expansion and temperature-induced changes in resources. Rather than sustaining the intensive habitat management needed for the earlier-documented scenario, greater emphasis on habitat heterogeneity, with wider benefits also to other resident taxa, was suggested: habitat needs can change very subtly over time, and management regimes must be adapted accordingly.

One evolutionary dilemma that arises is the need to modify or impose changes on life history traits that are not continuously variable in response to continuous thermal gradients postulated from climate changes. Fundamentally, temperature influences (1) the length of the developmental period (often defined or delimited in large part by diapause) and (2) the speed of development possible during that period. If more than one generation a year may occur, whether or not to do this involves a binary 'choice' between diapause or continuing development – there is no intermediate state, as discussed by Burke et al. (2005) for the Brown argus, *Aricia agestis*. For bivoltinism, rather than univoltinism, the traits found were (1) speeding up development; (2) consequent reduction in adult body size; and (3) shifting the threshold for diapause induction. The univoltine population needed longer day length for development, and caterpillars entered diapause in July/early August, whereas bivoltine populations did not do so until late August/September. Univoltine larvae took considerably longer (25% longer at 25 °C, and 29% longer at 20 °C) to develop.

Seasonal temperatures, either of winter or summer, may each prove influential and limit distribution. Experimental translocations of the skipper *Atalopedes campestris* demonstrated the crucial interaction between these in defining current range limits in north-western North America (Crozier 2003, 2004a,b) (Fig. 7.5). That range is wholly within the –4 °C January average isotherm, and cold tolerance trials revealed –4 to –7 °C a critical thermal limit within which high

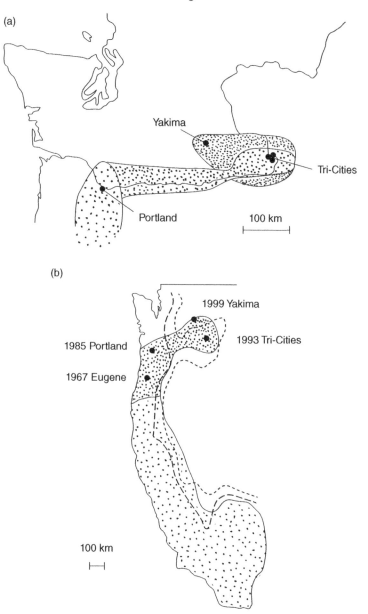

Fig. 7.5 Distribution and range parameters and change of the skipper *Atalopedes campestris* in the north-western United States. (a) Expansion of range from 1990 (pale), through 1993 (intermediate shading) and 1999 (dark); (b) overwintering range indicating change in range with climate change as paler to darker shading (dashed line: January average minimum –4 °C isotherm from 1950–1969; dotted line, same for 1990–1998 (Adapted from Crozier 2003, 2004a,b).

Table 7.3 Changing aspects of occupied habitat patches in 1982 and colonised over 1983–2000 by *Hesperia comma* in southern England (Source: Davies et al. 2006. Reproduced with permission of John Wiley & Sons.).

	Proportion occupied at each aspect							
	N	NE	E	SE	S	SW	W	NW
1982 (n = 64)	0.00	0.00	0.01	0.09	0.41	0.39	0.05	0.05
2000 (n = 179)*	0.14	0.09	0.12	0.23	0.06	0.10	0.14	0.12

*Excluding sites occupied in 1982

mortality occurred. Whist this opportunistic species is not restricted by habitat or dispersal prowess, the minimum temperature is an important constraint, and future winter warming is likely to be critical for the skipper's range expansion. Summer temperature constraints were studied through experimental field translocations of caterpillars along a natural thermal gradient of 3 °C across the range edge. Crozier (2004a) found the developmental time significantly slower beyond the current range, reflecting cooler temperatures, and this was sufficient to reduce the number of possible generations, reducing the chances of population survival. However, winter temperature alone may be the more important influence – but Crozier's (2004a: 148) comment that 'asymmetry in seasonal warming trends will have biological consequences' is a salutary indication of little-documented complexities of much wider relevance.

The foregoing reflects that both habitat quality and habitat (patch) isolation are critical in conservation. However, they are often treated as alternatives, rather than having joint impacts and, as Thomas et al. (2001) emphasised, problems in interpretation can then easily arise. Very broadly, decrease in 'quality' reflects both extent and suitability of resources and, so, carrying capacity and can culminate in declines and losses of populations, whilst isolation leads to breakdown of population structure through hampering colonisation and with habitat patches becoming progressively too few, too small and too separated for persistence. The management consequences of this polarisation in emphasis may be profound. Emphasis on habitat quality gives priority to conserving optimal habitat within patches, sometimes on few sites; emphasis on consequences of isolation, rather, advocates conservation of more numerous (10–20) biotope patches within the species' dispersal range, even if some of these are low quality. Whilst it may seem self-evident that both practices are important, the recommendation (initially by Harrison and Bruna, 1999, and recapitulated by Thomas et al., 2001) that metapopulation theory 'should not be employed as a substitute for within-site habitat management' still needs to be adopted more widely.

However, and linking with the difficulties of defining critical resource needs, the full array of factors contributing to high habitat quality – as a prerequisite for optimal management – are usually unknown. This caveat is evident in many individual species' conservation management plans (Chapter 9), in which it is not unusual for a butterfly or moth to be known from single or few sites that become the focus of management to emulate the conditions deemed desirable there and, in some cases, to replicate them elsewhere through active restoration. It is often not clear whether those conditions are indeed optimal, or whether the species is simply 'hanging on' at those sites as the last possible refuges but within stressed or impoverished resource environments. Carefully designed management regimes might then only be emulating marginal conditions for survival, rather than encouraging more suitable resource regimes, and so compromise their aims unknowingly.

The roles of 'suboptimal habitats' in conservation may become complex, and difficult to determine. In another such context, the intriguing concept of 'ecological traps' was raised for forest moths by Summerville and Crist (2008). The broad principle, derived largely from studies on birds and reviewed by Battin (2004), recognises an ecological trap habitat as a low-quality habitat in which population growth is negative (so, a 'sink habitat') that is actively chosen rather than avoided. Higher-quality habitat, which would enable higher fecundity, is ignored, and presence of such traps in a landscape may usually drive a local population to extinction. Summerville and Crist ventured that degraded or restored habitats might function in this way for some moths, with the burden of proof relying on demonstrating that (1) high-quality habitat exists in a region and (2) forest moths occur in and respond to lower-quality habitat. They used the example that a forest management technique might be considered benign to moth communities if the same suite of species is present in managed and unmanaged habitats without reference to any changes in fecundity in managed sites; comparative surveys based solely on adult moths, as the most usual approach, could not reveal the lowered density of caterpillars that might indicate a trap situation. Despite the rather theoretical nature of the concept, Battin (2004) warned that 'we cannot afford to ignore the possibility of ecological traps or fail to take them into account in the study, management and conservation of animal populations'.

Additional management complications may arise if the two sexes have differing resource needs to accommodate their behaviour, sometimes resulting in subtle differences in habitat use, such as need for territoriality by males and specific oviposition sites for females. These may sometimes vary across the species' range in response to local environment. Near the northernmost limit of its distribution in Europe, *Lycaena tityrus* selected the warmest microhabitats available (Pradel & Fischer 2011), so preferring sites with a higher proportion of bare ground within the continuous habitat, and females selected young host plants within low vegetation or near bare ground for oviposition. Nectar sources may be exploited differently between the two sexes of the same species. *Parnassius apollo* males frequent flowers on top of long stems, whilst females seek nectar near the ground, reflecting the cryptic behaviour flowing from forced courtship (Baz 2002); as a result, different nectar-providing species are important for the two sexes.

Understanding the intricate and often individual linkages between resources and space ('habitat'), population structure and climate changes is perhaps the most important theme in designing conservation management for Lepidoptera and most other terrestrial invertebrates. In considering the effects of habitat disturbance on tropical hawkmoths (Sphingidae), Beck et al. (2006) distinguished three classes, as (1) primary habitats without significant human disturbance – usually primary rainforest; (2) secondary habitats, ranging from selectively logged forests to partially forested sites; and (3) heavily disturbed anthropogenic sites, as useful broad divisions for comparison. On gross comparison, no major overall influence of habitat disturbance on sphingids in south-east Asia was found – but closer examination showed a rather more complex situation. Smerinthinae decreased with disturbance, whilst Macroglossinae increased, and seemed to compensate for the lost diversity of the former group. This might reflect their different life-history strategies: Macroglossinae are 'income breeders', depending on nectar sources to mature their eggs, so that they can thrive in disturbed areas with floral resources. In contrast, many Smerinthinae do not feed as adults, so depend wholly on larval resources, as 'capital breeders', for which loss of larval habitats, including suitable primary habitats, is critical, and is not compensated by nectar supplies in disturbed areas. This difference in strategy is relevant also to considering the 'minimum area requirement' (MAR) of habitat patches for Lepidoptera. The MAR, that smallest patch size needed for population persistence, has been related to several life history traits and body size (wing size) in a survey of European butterflies (Baguette & Stevens 2013). Four traits, of 17 measured, were significantly associated with MAR. Species that are myrmecophilous or are capital breeders have smaller area thresholds than non-myrmecophilous species or those with income-breeding females. Similarly, species with territorial or patrolling males need larger areas than those with less concentrated mate-seeking behaviour. Lastly from this survey, species with adults indifferent to temperature or with very precise temperature requirements need smaller areas than those with intermediate thermal tolerances.

Many attempts have been made to predictively model the various scenarios of resource patterns and usage, and to assess the relative influences of habitat (location) and climate on species incidence and community change. At higher elevations, climate has been suggested to be the major driver (or 'filter': Pellissier et al. 2012) of community change, in selecting for those species that can cope with harsher conditions (Boggs & Murphy 1997). At lower elevations, where land use is commonly more intensive, complications in prediction arise from those very varied changes and impacts. Models might have potential to predict aspects of alpine assemblages, by indicating those species that could not cope with extreme conditions, and those models used by Pellissier et al. (2012), using Swiss alpine butterflies in different elevational zones, predicted some aspects of species richness and community composition along elevational and plant species richness gradients, with the 'filter effect' supported by increased ability to predict absence at higher elevations (Fig. 7.6). A central consideration is that all montane insect distributions are dynamic, so that loss or fragmentation of sites may necessitate close attention to the conservation of the more sedentary species (Boggs & Murphy 1997). Montane species are necessarily adapted to short growing seasons, in which vagaries of weather or seasonal nectar availability may induce local extinctions. With climate changes, it is expected that the most vagile species of Lepidoptera may have the best chances of regional persistence, with the more sedentary species becoming the more vulnerable (Wilcox et al. 1986, Boggs & Murphy 1997). A broad categorisation of dispersal trends amongst the species

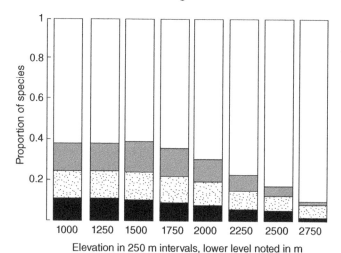

Fig. 7.6 The components of predicting communities, compared along an elevational gradient (intervals of 250 m, lower level for each bar indicated). Each such bar has four components summing to the total community; from top to bottom these are (1) the proportion of species both observed and predicted to be present; (2) the proportion of species that were not observed but predicted to be present (overprediction); (3) the proportion of species that were observed but predicted to be absent (underprediction); (4) the proportion of species that were both absent and predicted to be absent (Source: Pellissier et al. 2012. Reproduced with permission of John Wiley & Sons.).

in an assemblage may have some value in predicting vulnerability (Chapter 12). The three 'mobility classes' for British butterflies (Pollard & Yates 1993) include an 'intermediate mobility' group of species between the more familiar 'sedentary' and 'wide-ranging'. Thomas (2000) suggested that this group has probably fared poorly because of a combination of metapopulation dynamics and mortality of migrating individuals failing to colonise new patches in a fragmented landscape. Both sedentary and intermediate mobility groups tend to breed in discrete habitats that characteristically become fragmented, whilst highly mobile species are less affected by this trend.

Attempting to predict, and generalise on, how different species may react to habitat disturbance is a key area of concern for Lepidoptera conservation. Whilst extent of change or disturbance is the most common focus, Aviron et al. (2007) implied that the rate of changes in the landscape might become of even greater importance in determining population survival, because dispersal of many species may be insufficient to keep up with rates of local extinctions. That intensively changed areas (such as for agricultural cultivation) are often associated with 'homogenisation' of Lepidoptera assemblages, as specialists are lost in favour of relative ecological generalists, is a widespread concept, but has only rarely been investigated thoroughly or quantitatively. Samples of butterflies (60 species, 21,695 individuals) and diurnal geometrid moths (39 species, 10,858 individuals) across 134 fragmented landscapes in Finland by transect walks (Ekroos et al. 2010) provided convincing evidence that diversity was influenced by agricultural intensification. Diversity was assessed as (1) alpha diversity, α, the mean species richness observed in 10 sample transects from each landscape; (2) gamma diversity, γ, as the total number of species observed within the landscape ; and (3) beta diversity, β, calculated by ' $\beta = \gamma - \alpha$'. Beta diversity was the largest fraction, as 67% of gamma diversity (butterflies) and 73% (geometrid moths). Both alpha and beta diversity decreased with increased cover of agricultural fields, with accelerated declines at greater than 60% of cover by these – indicating a threshold value beyond which agricultural intensification may have increased impacts. Increased homogenisation was thus confirmed, with diversity decreases representing losses of poor dispersers and habitat specialists from highly altered landscapes.

Historical implications can be informative in suggesting long-term changes in habitat use. Thus, 10 British butterfly species (18% of residents) were noted by Thomas (1993) as restricted to early seral stages, such as open grasslands or woodland clearings – whilst in central Europe (with warmer spring and summer temperatures) the same species occur in later seral stages. Thomas suggested that the British habitat distribution, reflecting the warmest accessible microclimates, reflected use of anthropogenic habitats (downs, heaths, early successional stages in woodlands) to which the species retreated as temperatures cooled in the past and rendered UK habitats paralleling those now occupied in central Europe too cool: these habitats were essentially anthropogenic refugia for the butterflies, enabling them to avoid extinction as cooling occurred. With warming conditions as at present, Thomas' hypothesis would be supported by habitat range expansion in Britain.

The emphasis on understanding dispersal becomes of practical value only when placed into the context of landscape structure and geometry, the pattern of how resources (patches) are distributed, and how these may be managed or otherwise included in conservation planning. Amongst other factors, successful colonisation reflects the numbers, sizes and separation distances of suitable patches, as well as the 'permeability' of intermediate areas that may variously impose barriers or facilitate movements. Management to increase this 'connectivity' involves enhancing chances of this occurring, most commonly by planning for resource enhancement in some way, either by enriching existing patches or constructing new ones. Position of new patches may be idealised, planned to link those already available, but in practice may be constrained by opportunity, such as land tenure and availability. As Haddad (2000) commented 'Because corridors remain one of few tangible management options, they have become one of the most popular strategies proposed for habitat conservation in fragmented landscapes', but their roles in influencing movements – and, hence, population sizes and persistence – remain complex to interpret. Many linear features in landscapes, such as highways, can act as either corridors or barriers to enhance or impede movements (Munguira & Thomas 1992). In particular, the relative roles of corridors and a more open matrix between patches have only rarely been compared experimentally. Haddad used the butterfly *Junonia coenia* to investigate whether corridors increase colonisation of open patches within pine forests, irrespective of distance, and having determined previously that corridors indeed increased movements between patches. He released butterflies at

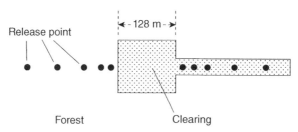

Fig. 7.7 Estimating dispersal and patch colonisation by *Junonia coenia* in cleared and forest areas. Butterflies were released at one of five sites (marked) along transects in forest (left of diagram, unmarked) and cleared areas (right of diagram, shaded) 16–192 m from a cleared experimental patch (centre) (Source: Haddad 2000. Reproduced with permission of John Wiley & Sons.).

sites in forest and open areas (Fig. 7.7), knowing also that they are capable of moving up to hundreds of metres. At the larger distances (128–192 m), the number of butterflies colonising a patch from corridors was twice as high as from forest – while there was a broad trend toward lower colonisation with increasing distances from forest, colonisation from the open corridor remained constant over distance. This study helped to indicate that distance between patches is important in any decision to opt for either a corridor or stepping stones in management. More generally, if inter-patch distances are short relative to the lepidopteran's dispersal capability, stepping stones may be a more effective strategy. Stepping stone efficacy may be influenced by (1) detection from a source patch; (2) the insects not restricted or diverted by habitat boundaries; (3) reluctance to enter corridors with extensive edge habitats; and (4) preference by species characteristic of naturally fragmented landscapes (as in the case of Fender's blue, Chapter 12, p. 224, and in the following paragraph: Schultz 1998). The lupin patches needed by this butterfly are isolated, and it is unlikely that the weakly flying butterflies will encounter new patches if they leave their parental one so that, following the United Kingdom examples discussed by Webb and Thomas (1994), stepping stones may be supplied to constitute 'terrestrial archipelagos' of habitat patches.

Countering the impacts of habitat change and loss involves resource management and restoration, with restoration of habitat networks also central to management of any species surviving in highly fragmented environments. As McIntyre et al. (2007) emphasised, any such effort necessitates combining biological information on the species involved with 'the landscape, economic and social realities of the restoration effort'. Within the context of the landscape, restoration commonly involves providing new potential resources and habitat area that can aid connectivity and persistence. Much of any such exercise is necessarily experimental, but modelling exercises can sometimes provide insight, with its roles complex, as described for Fender's blue, *Icaricia icarioides fenderi*, on prairie remnants in North America (Schultz & Crone 2001) in a study continued over 14 years (McIntyre et al. 2007). Nevertheless, characterising habitat in terms of resource needs and dynamics of supply is the most urgent consideration on Lepidoptera species and assemblage conservation. As Early et al. (2008) demonstrated for the Marsh fritillary (*Euphydryas aurinia*), such information can be used to predict distribution of suitable habitat and infer presence or absence of the butterfly. However, presence of defined critical resources does not necessarily imply presence of the user species – and there are many cases where this is so and for which the factors restricting the distribution of the lepidopteran remain unclear. Both specific larval food plants and the host ants of Lycaenidae, for example, may be much more widely distributed than the butterfly sometimes associated with them. More generally, local nectar supplies may be boosted enormously during restoration efforts (Chapter 12) and be associated with increased adult richness and abundance – however, as Holl (1995) suggested for butterflies on restored coal mine sites, providing a diversity of larval food plants may be more significant, as many of the nectar-feeding adults may be mobile non-residents.

Several of the examples noted above emphasise that dispersal capability and ecological specialisation are the key parameters in considering impacts of habitat fragmentation on Lepidoptera, with generalists having greater ecological flexibility to persist by changing their resources. Both these parameters, however, are difficult to measure or categorise objectively, due to lack of continuous variables (Dapporto & Dennis 2013), so that the real meanings of 'generalist' and 'specialist' can become confused as these intergrade in various ways. Careful examination of trends amongst the British butterflies led Dapporto and Dennis to conclude that (1) as long as restricted fragmented sites are large, specialists can exploit the restricted resources provided and also benefit from local conservation measures, leading to a primary conservation need to preserve and extend sites for those species; (2) extreme generalists can benefit from any resource and can move easily between suitable patches; whilst (3) intermediate species cannot participate competitively in either of these alternatives and so are those most seriously affected by environmental changes. That predictive analysis also inferred that reversing declines of generalists may in fact be far more difficult to achieve than restoring specialists.

References

Aviron, S., Kindlmann, P. & Burel, F. (2007) Conservation of butterfly populations in dynamic landscapes: the role of farming practices and landscape mosaic. *Ecological Modelling* 205, 135–145.

Baguette, M., Petit, S. & Queva, F. (2000) Population spatial structure and migration of three butterfly species within the same habitat network: consequences for conservation. *Journal of Applied Ecology* 37, 100–108.

Baguette, M. & Stevens, V. (2013) Predicting minimum area requirements of butterflies using life-history traits. *Journal of Insect Conservation* DOI 10.1007/s10841-013-9548-x.

Battin, J. (2004) When good animals love bad habitats: ecological traps and the conservation of animal populations. *Conservation Biology* 18, 1482–1491.

Baz, A. (2002) Nectar plant sources for the threatened Apollo butterfly (*Parnassius apollo* L. 1758) in populations of central Spain. *Biological Conservation* 103, 277–282.

Beck, J., Kitching. I.J. & Linsenmair, K.E. (2006) Effects of habitat disturbances can be subtle yet significant: biodiversity of hawkmoth assemblages (Lepidoptera: Sphingidae) in south-east Asia. *Biodiversity and Conservation* 15, 451–472.

Bila, K., Kuras, T., Sipos, J. & Kindlmann, P. (2012) Lepidopteran species richness of alpine sites in the High Sudetes Mts: effect of area and isolation. *Journal of Insect Conservation* 17, 257–267.

Boggs, C.L. & Murphy, D.D. (1997) Community composition in mountain ecosystems: climatic determinants of montane butterfly distributions. *Global Ecology and Biogeography Letters* 6, 39–48.

Britten, H.B. & Glasford, J.W. (2002) Genetic population structure of the Dakota skipper (Lepidoptera: *Hesperia dacotae*): a North American native prairie obligate. *Conservation Genetics* 3, 363–374.

Bryant, S.R. & Shreeve, T.G. (2002) The use of artificial neural networks in ecological analysis: estimating microhabitat temperature. *Ecological Entomology* 27, 424–432.

Burke, S., Pullin, A.S., Wilson, R.J. & Thomas, C.D. (2005) Selection for discontinuous life history traits along a continuous thermal gradient in the butterfly *Aricia agestis*. *Ecological Entomology* 30, 613–619.

Crozier, L.G. (2003) Winter warming facilitates range expansion: cold tolerance of the butterfly *Atalopedes campestris*. *Oecologia* 135, 648–656.

Crozier, L.G. (2004a) Field transplants reveal summer constraints on a butterfly range expansion. *Oecologia* 141, 148–157.

Crozier, L. (2004b) Warmer winters drive butterfly range expansion by increasing survivorship. *Ecology* 85, 231–241.

Dapporto, L. & Dennis, R.L.H. (2013) The generalist–specialist continuum: testing predictions for distribution and trends in British butterflies. *Biological Conservation* 157, 229–236.

Davies, Z.G., Wilson, R.J., Coles, S. & Thomas, C.D. (2006) Changing habitat associations of a thermally constrained species, the silver-spotted skipper butterfly, in response to climate warming. *Journal of Animal Ecology* 75, 247–256.

Davis, A.J., Jenkinson, I.S., Lawton, J.H., Shorrocks, B. & Wood, S. (1998) Making mistakes when predicting shifts in species range in response to global warming. *Nature* 391, 783–785.

Dennis, R.L.H. (2004) Just how important are structural elements as habitat components? Indications from a declining lycaenid butterfly with priority conservation status. *Journal of Insect Conservation* 8, 37–45.

Dennis, R.L.H. (2009) Changes in butterfly distribution: a simple correction for bias caused by subsampling of atlas records makes no difference to BAP status. *Entomologists' Gazette* 60, 141–149.

Dennis, R.L.H. (2010) A Resource-Based Habitat View for Conservation. Butterflies in the British Landscape. Wiley-Blackwell, Oxford.

Dennis, R.L.H. & Sparks, T.H. (2006) When is a habitat not a habitat? Dramatic resource use changes under differing weather conditions for the butterfly *Plebejus argus*. *Biological Conservation* 129, 291–301.

Dennis, R.L.H., Shreeve, T.G. & Van Dyck, H. (2003) Towards a functional resource-based concept for habitat: a butterfly biology viewpoint. *Oikos* 102, 417–426.

Dennis, R.L.H., Shreeve, T.G. & Van Dyck, H. (2006) Habitats and resources: the need for a resource-based definition to conserve butterflies. *Biodiversity and Conservation* 15, 1943–1968.

Dennis, R.L.H., Hardy, P.B. & Shreeve, T.G. (2008) The importance of resource databanks for conserving insects: a butterfly biology perspective. *Journal of Insect Conservation* 12, 711–719.

Dover, J.W., Sparks, T.H. & Greatorex-Davies, J.N. (1997) The importance of shelter for butterflies in open landscapes. *Journal of Insect Conservation* 1, 89–97.

Early, R., Anderson, B. & Thomas, C.D. (2008) Using habitat distribution models to evaluate large-scale landscape priorities for spatially dynamic species. *Journal of Applied Ecology* 45, 228–238.

Ekroos, J., Heliola, J. & Kuussaari, M. (2010) Homogenisation of lepidopteran communities in intensively cultivated agricultural landscapes. *Journal of Applied Ecology* 47, 459–467.

Franzen, M. & Nilsson, S.G. (2012) Climate-dependent dispersal rates in metapopulations of burnet moths. *Journal of Insect Conservation* 16, 941–947.

Fuentes-Montemayor, E., Goulson, D., Cavin, L., Wallace, J.M. & Park, K.J. (2012) Factors influencing moth assemblages in woodland fragments on farmland; implications for woodland management and creation schemes. *Biological Conservation* 153, 265–275.

Habel, J.C., Zachos, F.E., Finger, A. et al. (2009) Unprecedented long-term genetic monomorphism in an endangered relict butterfly species. *Conservation Letters* 10, 1659–1665.

Habel, J.C., Engler, J.O., Rodder, D. & Schmitt, T. (2012a) Landscape genetics of a recent population extirpation in a burnet moth species. *Conservation Letters* 13, 247–255.

Habel, J.C., Rodder, D., Lens, L. & Schmitt, T. (2012b) The genetic signature of ecologically different grassland Lepidoptera. *Biodiversity and Conservation*. DOI 10.1007/s 10531-012-0407-y.

Haddad, N. (2000) Corridor length and patch colonization by a butterfly, *Junonia coenia. Conservation Biology* **14**, 738–745.

Harper, G.L., Maclean, N. & Goulson, D. (2003) Microsatellite markers to assess the influence of population size, isolation and demographic change on the genetic structure of the UK butterfly *Polyommatus bellargus. Molecular Ecology* **12**, 3349–3357.

Harrison, S. & Bruna, E.M. (1999) Habitat fragmentation and large-scale conservation; what do we know for sure? *Ecography* **22**, 1–8.

Henry, E.H. & Schultz, C.B. (2013) A first step towards successful conservation: understanding local oviposition site selection of an imperiled butterfly, mardon skipper. *Journal of Insect Conservation* **17**, 183–194.

Hill, J.K., Thomas, C.D., Fox, R. et al. (2002) Responses of butterflies to twentieth century climate warming: implications for future ranges. *Proceedings of the Royal Society of London, Series B Biological Sciences* **269**, 2163–2171.

Holl, K.D. (1995) Nectar resources and their influence on butterfly communities on reclaimed coal surface mines. *Restoration Ecology* **3**, 76–85.

Holt, R.D., Lawton, J.H., Polis, G.A. & Martinez, N.D. (1999) Trophic rank and the species-area relationship. *Ecology* **80**, 1495–1504.

Kuussaari, M., Heliola, J., Luoto, M. & Poyry, J. (2007a) Determinants of local species richness of diurnal Lepidoptera in boreal agricultural landscapes. *Agriculture, Ecosystems and Environment* **122**, 366–376.

Louy, D., Habel, J.C., Schmitt, T., Assmann, T., Meyer, M. & Muller, P. (2007) Strongly divergent population genetic patterns of three skipper species; isolation, restricted gene flow and panmixis. *Conservation Genetics* **8**, 671–681.

Maes, D., Vanreusal, W., Talloen, W. & Van Dyck, H. (2004) Functional conservation units for the endangered Alcon Blue butterfly *Maculinea alcon* in Belgium (Lepidoptera: Lycaenidae). *Biological Conservation* **120**, 229–241.

McIntyre, E.J.B., Schultz, C.B. & Crone, E.E. (2007) Designing a network for butterfly habitat restoration: where individuals, populations and landscape interact. *Journal of Applied Ecology* **44**, 725–736.

Merckx, T., Feber, R.E., Mclaughlan, C. et al. (2010) Shelter benefits less mobile moth species: the field-scale effect of hedgerow trees. *Agriculture, Ecosystems and Environment* **138**, 147–151.

Munguira, M.L. & Thomas, J.A. (1992) Use of road verges by butterfly and burnet populations, and the effect of roads on adult dispersal and mortality. *Journal of Applied Ecology* **29**, 316–329.

Parmesan, C., Ryrholm, N., Stefanescu, C. et al. (1999) Poleward shifts in geographical range of butterfly species associated with regional warming. *Nature* **399**, 579–583.

Pavlikova, A. & Konvicka, M. (2012) An ecological classification of Central European macromoths: habitat associations and conservation status returned from life history attributes. *Journal of Insect Conservation* **16**, 187–206.

Pellissier, L., Pradervand, J.-N., Pottier, J., Dubuis, A., Maiorano, L. & Guisan, A. (2012) Climate-based empirical models show biased predictions of butterfly communities along environmental gradients. *Ecography* **35**, 684–692.

Pollard, E. & Yates, T.J. (1993) Monitoring Butterflies for Ecology and Conservation. Chapman & Hall, London.

Pradel, K. & Fischer, K. (2011) Living on the edge: habitat and host-plant selection in the butterfly *Lycaena tityrus* (Lepidoptera: Lycaenidae) close to its northern range limit. *Journal of Research on the Lepidoptera* **44**, 35–41.

Schultz, C.B. (1998) Dispersal behavior and its implications for reserve design in a rare Oregon butterfly. *Conservation Biology* **12**, 284–292.

Schultz, C.H. & Crone, E.E. (2001) Edge-mediated dispersal behavior in a prairie butterfly. *Ecology* **82**, 1879–1892.

Shreeve, T.G. (1995) Butterfly mobility. pp. 37–45 in Pullin, A.S. (ed.) Ecology and Conservation of Butterflies. Chapman & Hall, London.

Shreeve, T.G., Moss, D. & Roy, D. (2001) An ecological classification of British butterflies: ecological attributes and biotope occupancy. *Journal of Insect Conservation* **5**, 145–161.

Steffan-Dewenter, I. & Tscharntke, T. (2000) Butterfly community structure in fragmented landscapes. *Ecology Letters* **3**, 449–456.

Summerville, K.S. (2004) Functional groups and species replacement testing for the effects of habitat loss on moth communities. *Journal of the Lepidopterists' Society* **58**, 129–132.

Summerville, K.S. & Crist, T.O. (2008) Structure and conservation of lepidopteran communities in managed forests of northeastern North America; a review. *Canadian Entomologist* **140**, 475–494.

Thomas, C.D. (2000) Dispersal and extinction in fragmented landscapes. *Proceedings of the Royal Society of London B* **267**, 139–145.

Thomas, J.A. (1983) The ecology and conservation of *Lysandra bellargus* (Lepidoptera: Lycaenidae) in Britain. *Journal of Applied Ecology* **20**, 59–83.

Thomas, J.A. (1993) Holocene climate changes and warm man-made refugia may explain why a sixth of British butterflies possess unnatural early-successional habitats. *Ecography* **16**, 278–284.

Thomas, J.A., Bourn, N.A.D., Clarke, R.T. et al. (2001) The quality and isolation of habitat patches both determine where butterflies persist in fragmented landscapes. *Proceedings of the Royal Society, series B* **268**, 1791–1796.

Turlure, C., Choutt, J. & Baguette, M. (2005) Resource-based analysis of the habitat in two species sharing the same host plant. pp. 29–31 in Kuhn, E., Feldmann, R., Thomas, J. & Settele, J. (eds) Studies on the Ecology and Conservation of Butterflies in Europe. Vol. 1: General concepts and case studies. Pensoft, Sofia and Moscow.

van Langevelde, F. & Wynhoff, I. (2009) What limits the spread of two congeneric butterfly species after their reintroduction: quality of spatial arrangement of habitat? *Animal Conservation* **12**, 540–548.

Webb, N.R. & Thomas, J.A. (1994) Conserving insect habitats in heathland biotopes: a question of scale. pp. 129–152 in Edwards, P.J., May, R.M. & Webb, N.R. (eds). Large-Scale Ecology and Conservation Biology. Blackwell Publishing, Oxford.

Wilcox, B.A., Murphy, D.D., Ehrlich, P.R. & Austin, G.T. (1986) Insular biogeography of the montane butterfly faunas in the Great Basin: comparisons with birds and mammals. *Oecologia* **69**, 188–194.

Zakharov, E.V. & Hellmann, J.J. (2007) Genetic differentiation across a latitudinal gradient in two co-occurring butterfly species: revealing population differences in a context of climate change. *Molecular Ecology* **71**, 289–308.

8

Communities and Assemblages

Introduction: Expanding the context

The conservation needs of any species must clearly include its functional integration in the community within which it exists, as a member of a wider assemblage of more-or-less related and variously interacting species, and for which protection of the wider biotope may necessarily take higher priority than individual species' needs, or a 'management balance' of some sort be established for collective benefit. It is self-evident that different species of Lepidoptera respond in different ways to any specified environmental change, with vulnerability to habitat quality and isolation amongst the numerous differences that can occur. But, for successful conservation of assemblages, the responses of each individual species, other community members, and of the entire community may need to be integrated (Shreeve & Dennis 2011). Some butterflies may respond in opposite ways (such as extending or contracting abundance or distribution), so that the 'best' conservation measures instituted for one species may compromise wellbeing of one or more other, equally significant, co-occurring taxa. The twin needs of greatest importance to counter this appear to be (1) large habitat patches on which mosaic management, progressively informed by and adapted to increased awareness of specific needs, can be undertaken; and (2) diverse vegetation richness and structure, fostering the diversity of resources present both locally and across wider landscapes.

These themes were addressed for a key European butterfly habitat, calcareous grassland in Poland, by Rosin et al. (2012). Correlation of eight environmental variables (Table 8.1) with the butterflies present (n = 36 species, 2685 individuals) along transects on 32 grassland patches revealed patch quality (as number of plant species) to be very important for butterfly richness, with patch size also

Lepidoptera and Conservation, First Edition. T.R. New.
© 2014 John Wiley & Sons, Ltd. Published 2014 by John Wiley & Sons, Ltd.

Table 8.1 Eight environmental variables potentially affecting butterfly populations on patches of calcareous grassland in southern Poland (Rosin et al. 2012. With kind permission from Springer Science+Business Media.), with characteristics across 32 patch sites.

	Mean	*SE*	*Minimum*	*Maximum*
Patch size (ha)	2.73	0.59	0.1	12.3
Distance to nearest calcareous meadow (m)	217	30	43	689
Percentage cover of permanent grassland within 500 m	20	3	2	54
Percentage cover of forest within 500 m	28	4	0	80
Distance to nearest building (m)	345	37	32	860
Plant species richness	14.2	1.3	1	33
Wind protection (%)	40	4	12	78
Mean height of vegetation (cm)	23	3	2	57

influential. Shelter (protection from wind) was also important, leading to recommendations for edge plantings of shrubs and trees to protect sites from strong winds. Some species responded to sward height, with the correlations together supporting some principles of much wider relevance in Lepidoptera conservation management.

In another recently studied environment, mountain meadows in Spain (n = 47), transect samples of 15,046 individuals (n = 75 species) of butterflies were correlated by regression analyses with a variety of environmental variables (Dover et al. 2011). Interpretation was complex but with strong suggestion that, whilst abiotic factors influenced whether a meadow was likely to be colonised or not, population sizes of many taxa were driven by management and other factors that affected the structure and condition of the sward. Individual species' reactions to landscape and management were 'complex and individual' (Dover et al. 2011, 258).

More generally, such studies confirm that declines or increases in butterfly species sharing the same biotopes involve complex and individualistic responses to the spatial and temporal supply of resources within the landscape, with the same broad trends paralleled in all major biotopes.

The entity of 'endangered community' parallels 'endangered species' as a formal category eligible for protection in some conservation legislations, and is most commonly founded by definition of vegetation type or biotope. As pioneered within Victoria's Flora and Fauna Guarantee Act 1988 (Australia), the main purpose is to consistently differentiate some ecologically valid entity from simply a particular site. In practice, this has proved very difficult to achieve, because non-vegetational criteria for delimiting a community flow largely from incidence of nominated species, and the levels of difference needed to separate communities are ambiguous or undecided. In Victoria, the problems are illustrated well through an entity designated formally in 1991 as 'Butterfly Community no. 1', based on two criteria of the Lepidoptera recorded at a single site, Mount Piper, namely the richness of species (in part reflecting sampling intensity) and the presence of several notable threatened Lycaenidae, particularly two species of ant-blues, *Acrodipsas* (New 2011). The designation remains restricted to this site, thwarting the original intention of the legislation, but has been important in stimulating debate on definitions and limitations. Thus, Wainer and Yen (2000) assessed the butterflies of another hill site in central Victoria, The Papps, with the purpose of comparison with Mount Piper. Apparent absence there of *Acrodipsas mymecophila* (Chapter 5, p. 63) and its host ant, *Papyrius* sp., led Wainer and Yen to suggest that the two assemblages could not be included within the same Butterfly Community, despite having many taxa in common. The extent of compositional difference in assemblages needed to designate separate threatened communities is very subjective, and may be increased by uncertainties over resident status of some taxa. Knowledge of any such assemblage may change rapidly with increased study and sampling intensity, and several distinct ecological groups of butterflies occur, as breeding residents, vagrants (occasional visitors, not found every year), more regular migrants (predictably present in most years, but not resident) and aliens. The resident suite of 'core taxa' is clearly the component of greatest interest, and the most informative for conservation need, with the presence of known threatened taxa within this category representing conservation value (New 2011). The considerable variation imposed by non-resident species or common resident taxa constitutes 'noise' that could, should they be included other than in very general terms in community evaluation, necessitate every community being different. Reliance on 'core taxa' may help to clarify focus on this dilemma; solution (or, at least, a broad working consensus) is needed urgently to satisfy the credibility of biologists, managers and politicians alike.

'Vulnerable groups'

As an example of how 'pattern analysis' of Lepidoptera might inform conservation needs, Holloway and Nielsen (1999, following Holloway & Barlow 1992) outlined how analysis of some macromoth groups across different biotopes or ecological regimes might indicate vulnerability in relation to disturbance. For the more diverse higher groups of macromoths in Borneo, they plotted the proportions of species represented in different biogeographical categories (endemic to widespread) against representation in different ecological categories (such as lowland v. montane; forest v. open habitats), and noted that the outcomes might constitute a 'rough guide' to the biodiversity cost of various types of human activity in modifying the environment. Some clear trends in groupings were found. As examples of possible indicator responses, Holloway and Nielsen noted (1) groups with high proportion of endemics restricted to lowland forests might be vulnerable to logging operations; (2) groups with high montane endemism might be vulnerable to hill resort developments (including examples such as golf courses and holiday accommodation complexes); and (3) groups with high proportions of geographically widespread species in open habitats might benefit from human activity. Surveys along spatial habitat gradients, either elevational or more directly across land use regimes, frequently reveal different assemblages of Lepidoptera. These commonly intergrade, but the causes of the

differences in richness and composition are not always straightforward to interpret, notwithstanding their major importance in assessing restoration and reclamation attempts involving reinstatement of successions and their development
to recreate more mature biotopes (Chapter 12), with spatial comparisons a
common surrogate for temporal sequences of change. Studies vary considerably
in their scope and complexity, ranging from 'pairwise comparisons' of single
disturbed and undisturbed sites to more laborious, and often more informative,
appraisals of several sites along a gradient spanning natural to highly disturbed
regimes. In most cases, site individuality precludes full replication. 'Intermediate
disturbance' effects are sometimes evident through increased richness, as implied
for butterflies in Tam Dao, Vietnam, for which richness was highest in disturbed
forests (including forest gaps) and forest edges (Vu 2009: Table 8.2), but with
some nymphalid groups (Satyrinae, Amathusiinae) more restricted to natural
closed forests. Community composition differed clearly between 'within forest'
and 'outside forest' habitats (Fig. 8.1). Elevational and disturbance gradients may
be difficult to distinguish. An earlier elevational comparison at Tam Dao revealed
94 butterfly species at a lower (200–250 m) level and 89 species at the higher
level (950–1000 m) (Vu & Yuan 2003), with substantial overlap (50 species)
between these, but relatively few of these species occurred on all three sample
transects explored at each elevation: many species were more restricted, in
common with many of those found in only one elevational set. Whilst the

Table 8.2 Diversity of butterflies in five different habitat types, from samples over
2002–2004 in Tam Dao National Park, Vietnam (from Vu 2009. Reproduced with
permission of John Wiley & Sons.).

Habitat	Total species	Total individuals
Natural forest	78	1722
Disturbed forest	107	1800
Forest edge	120	3034
Shrub	114	1993
Agricultural land	74	1540

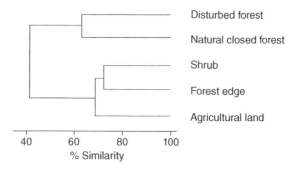

Fig. 8.1 Percentage similarity in composition of butterfly communities of five major
habitat categories in the Tam Dao National Park, Vietnam (Source: Vu 2009.
Reproduced with permission of John Wiley & Sons.).

number of species in closed forests at the lower elevation was the higher, this relativity is clouded by much of the lower forest being disturbed to some extent, in contrast to the more pristine higher level forests. On Grande Comoro (western Indian Ocean) most endemic butterflies are confined to forest habitats, only present above 500 m, with butterfly richness and abundance greatest at the lower elevations of this forest (Lewis et al. 1998).

The relative influences of elevation and disturbance as habitat gradients are often difficult to distinguish and, so, are easily confounded.

More than 500 geometrid moth species were attracted to light traps along two habitat gradients (from primary rainforest to cultivated areas) on Mt Kinabalu, Sabah, over 135 nights of trapping in 1977 (Beck et al. 2002). Old growth forests supported high diversity, whereas samples from farmland were considerably less diverse (Table 8.3). Secondary or regenerating forest sites were more variable – most had diversity almost as low as that on farmland, but one 15-year regenerating site paralleled levels of primary forest. A rich understorey on that site might itself be sufficient to conserve high geometrid diversity even in planted forests, but this might not be so for many other groups. The two gradients surveyed, not true replicates but similar in character, showed strong parallels. As Beck and Chey (2008) noted, elevational gradients in moth diversity can show various patterns, but two are observed commonly in tropical environments – a gradual decline in richness with increased elevation, and, more commonly, some form of mid-elevation peak, for which a number of explanatory hypotheses have

Table 8.3 Geometrid moths along habitat gradients on Mt Kinabalu, Sabah: two gradients are each represented by light trap catches at sites from primary rain forest, through secondary forest to cultivated areas. The gradients (Serinsim: SER; Poring; POR) are ranked from disturbance levels '1' (lowest) to '6' (highest); richness is expressed as number of collected species (individuals), with mean catch/night to reflect different number of trap nights (in parentheses) at the sites (Adapted from Beck et al. 2002).

Disturbance	Site code/characteristics	Richness	Mean catch/night
1	SER1 Primary dipterocarp forest	161 (458)	38.2 (12)
	SER1c Canopy, 30 m, above SER1	122 (441)	36.8 (12)
2	SER2 Old growth forest	134 (362)	32.9 (11)
	SER2c Canopy, 25 m, near SER2	118 (268)	24.4 (11)
3	SER3 Secondary forest, ca 30 years old	75 (167)	11.1 (15)
4	SER4 Secondary forest, ca 15 years old	134 (313)	26.1 (12)
5	SER5 Secondary forest, ca 5 years old	75 (186)	12.4 (15)
6	SER6 Farmland site	99 (575)	47.9 (12)
1	POR1 Primary dipterocarp forest	192 (619)	123.8 (5)
	POR1c Canopy, 45 m, near POR1	135 (427)	85.4 (5)
2	POR2 Disturbed dipterocarp forest	86 (136)	27.2 (5)
3	POR3 Secondary forest with logging gaps	130 (225)	45.0 (5)
4	POR4 Secondary forest, ca 20 m high	62 (118)	23.6 (5)
5	POR5 Secondary forest in cultivated area	65 (118)	23.6 (5)
6	POR6 Cultivated area, plantations	82 (172)	34.4 (5)

been advanced. Comparative investigation of the various possible influences becomes complex, but the pattern transcends taxa and may be found in many herbivorous insect groups. The intensive survey of Borneo Geometridae provided a sound data set for appraisal of five hypotheses, namely the mid-elevation effect, the overlap between lowland and largely endemic montane taxa, water (greater availability at lower elevations) and energy availability (limiting at higher elevations), the relative areas of the elevational bands used for comparison, and influences of host plant specificity. Disturbance and other effects may be superimposed on these patterns, but Beck and Chey suggested that none of the above hypotheses convincingly explained the observed diversity gradients. They also suggested the value of future focus on explaining patterns in the lower parts of elevational gradients, because of possible collector bias toward mid-elevation levels, and with lower levels often not well represented in national parks or other protected areas.

Many studies on tropical Lepidoptera have revealed major losses of richness with habitat changes: for example Miller et al. (2011) compared butterfly richness in rain forest and in plantations of oil palm (*Elaeis guineensis*) in New Britain (Papua New Guinea), to show that 50 of 73 species collected were restricted to rain forest, and only 12 to plantations, as a major impact of habitat conversion. As Miller et al. acknowledged, however, and reflecting a scenario of much wider relevance, their samples were relatively small (total of 312 specimens), seasonally restricted and contained a high proportion of singletons, as common caveats in analysing rapid biodiversity surveys.

Interpreting information such as this is difficult, but the need flows from the reality that habitat-level approaches to studying Lepidoptera for conservation are the most widely adopted approach in the tropics – simply reflecting the impossibility of pursuing species-level or population-level conservation for any but the very few high-profile cases (such as *Ornithoptera alexandrae*, Chapter 5, p. 72). Some basic paradigms (such as understanding metapopulations, Chapter 6, p. 97) implicit in many temperate region-based species studies have scarcely been elucidated amongst tropical taxa, for example (Schultz & Crone 2008). An estimate that about 90% of butterfly species live in the tropics (Bonebrake et al. 2010) is probably at least equalled by moths, and the very limited ecological understanding of the vast majority of taxa in itelf precludes individual species conservation, whilst the logistic support to do this is almost invariably lacking. However, capability to interpret community changes in relation to habitat losses and disturbance is itself very limited, and Bonebrake et al. considered that predictive patterns are still very uncertain, as the examples noted above suggest.

Habitats and landscapes

Loss of suitable habitat is the most pervasive threat facing all native biodiversity, and for Lepidoptera this manifests mainly in changes to natural biotopes resulting from human land use. All major biotopes are at risk, and considerations of butterfly or moth assemblage changes in relation to levels of despoliation are a predominant conservation theme; it is a key theme for almost all the cases noted in Chapter 9, for example, but is introduced here to show some aspects of the impacts that can occur. Large-scale changes that have received particular attention include those associated with deforestation, conversion of land for agriculture and agroforestry (typified by large areas of lowered-diversity monocultures replacing previous high-diversity natural communities), and urbanisation (including industrialisation) and recreational activities. Each, with numerous variations in intensity, extent and distribution, are familiar foci in Lepidoptera species and assemblage conservation, and consequences include (1) needs for planning in land use extent and pattern; to (2) enable protection of key sites and representatives of all major biotopes from further change, ideally in a system of perpetually managed reserves; and (3) facilitate restoration and management of degraded areas through informed land uses less destructive to native biota. Patch size and isolation impose further needs, such as for increased effective sizes and increased connectivity within landscapes, now increasingly urgent as means of securing 'gradient habitats' enabling (or, at least, not actively impeding) movements as conditions change. Studies of distribution and abundance of Lepidoptera along habitat gradients of various kinds have also helped to clarify the scope of various 'specialist' and 'generalist' species allocations (Chapter 5, p. 66) so widely used in setting priorities and designating conservation need. Key parameters for this include larval food plant breadth (monophagous/oligophagous or polyphagous) and voltinism (univoltine/bivoltine or multivoltine/continuously brooded). Their application was demonstrated by Kitahara and Fujii (1994) in a classic study of butterflies along a disturbance gradient in Japan (Fig. 8.2). Of 62 species in total 20 were generalists, 17 were specialists and 25 could not be allocated to either category because of intersecting criteria. Nevertheless, interpretations of nine communities (Table 8.4) showed differences between these groups. Increased diversity and richness linked closely with decreased human disturbance. Mosaics

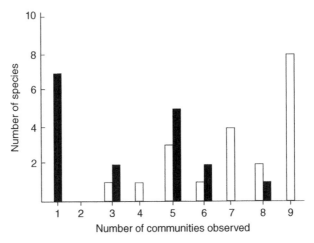

Fig. 8.2 The frequency distributions of generalist (open bars) and specialist (black bars) butterfly species against the number of communities (census routes) in which they were found in Japan (Source: Horner-Devine et al. 2003. Reproduced with permission of John Wiley & Sons.).

Table 8.4 The nine census routes (communities) for butterflies sampled in central Japan as a disturbance gradient by Kitahara and Fujii (1994) (see Fig. 8.2). With kind permission from Springer Science+Business Media.

Group A, Low disturbance
Secondary natural forest, mainly deciduous broad-leaved forest: routes A1, A2; elevation 150–200 m; A2 more disturbed than A1

Group B, Moderate disturbance
Cultivated lands and villages: routes B1, B2, B3-1, B3-2; elevation 50–100 m; disturbance decreased in sequence B1, B3-2, B2, B3-1, with B1 largely including paddy fields/cultivated vegetable fields/villages, to B3-1 with mainly evergreen secondary forest

Group C, High disturbance
Newly designed city parks and connecting pedestrian road with transplanted trees; Routes C1, C2, C3; elevation 25 m in city; intensive human impacts (mowing and related activities), succession essentially prevented

of different successional stages in the study sites suggested that generalist butterflies may be able to colonise and establish in more unstable or unpredictable patches (such as forest gaps), in contrast to the dependence of many specialists on late successional stages, such as mature forest, alone. The general patterns were reinforced by later surveys using a similar approach, but in grasslands with differing levels of disturbance in central Japan (Kitahara et al. 2000). The three butterfly groupings of specialist, generalist and intermediate were evident, with about a third of species in each. Increased levels of disturbance were correlated with decreased richness of specialists, but only weakly with numbers of generalist species.

More generally, Kitahara and Fujii suggested that their interpretation essentially equated 'specialists' to 'K-strategists', and 'generalists' to 'r-strategists', so that clarifying those widespread terms might aid understanding. Application of the concept to noctuid moths in central Europe was also instructive (Spitzer & Leps 1988), with appraisal of 89 species supporting predictions that (1) K-selected species will be more numerous in stable (climax/subclimax) environments – in their study, the earliest sucessional stages supported mainly r-selected taxa, and the most mature stages harboured mostly K-selected species; (2) population fluctuations of r-strategists are generally greater than for K-strategists – endorsed by Spitzer and Leps' light trap surveys over 6–15 years; and (3) lowest variations in abundance occurring in late successional systems.

Parallels between disturbance and successional sequences or gradients are clear, and extend across both space and time, with the latter an important conservation component. Changes in the butterfly fauna of Staten Island (New York) from 1910 (relatively undeveloped) to 1970 (by then highly developed) showed declines of native specialised species and increases of more vagile generalist taxa (Shapiro & Shapiro 1973). One exception was the ecologically specialised

Broad-winged skipper, *Poanes viator*, that succeeded in switching to a weedy larval host plant (*Phragmites*) and then thrived.

It has thus been inferred repeatedly that (1) putative K-selected Lepidoptera species are especially vulnerable to habitat change and disturbance, but also that (2) form and intensity of management of disturbed areas may contribute importantly to that vulnerability. Comparing highly disturbed with relatively undisturbed semi-natural grasslands in central Japan, Kitahara and Sei (2001) found the species of butterfly confined to the latter had lower abundance, lower voltinism, more stable distribution and narrower natural range sizes than other taxa – so that securing as much semi-natural grassland as possible within the pattern of human land use is needed to maintain those species. As in some other contexts, the 'disturbed area species' are largely a subset of those naturally present in the region.

The importance of landscape structure and abrupt changes between adjacent habitats is illustrated well by studies of Lepidoptera across such boundaries, for example between contrasting woodland and abutting open ground – with the 'edge' considered as a distinct third environment, and so incorporating a variety of both anthropogenic (disturbance) and environmental factors in site suitability. One such comparative study used transect samples of butterflies on two sites of each kind around Mt Fuji, Japan (Kitahara 2004). Of 57 species recorded, only five occurred in the forest interior, with only one of these in both such sites; open land yielded 29 species (13 in both sites), and forest edge, 52 species (30 in both sites). These relative numbers indicate the importance of such edge ecotones for butterflies, with many (25) of the species there not found in either of the main habitats. Similar preponderance of forest edge taxa was found in studies based on traditional satoyama systems in Japan (in Kanazawa City: Ohwaki et al. 2007), in which butterfly richness is commonly higher than in mature forests and with secondary forests enhancing the overall richness within the area. High richness in satoyama systems has been attributed to high vegetational diversity, with two study sites that contained forest edges richer than others, with 32 and 36 species of the total 51 recovered. Assemblages differed considerably amongst forest interior (16 species, as the most specialised group), forest edge species (24) and open land (11, as the most generalised taxa). Forest edge species included some that also utilise both forest and open land, and others more closely restricted to the edge areas. Many may depend on secondary forest as a key conservation need in satoyama systems.

Elsewhere, 'forest gaps' created by people provide both opportunity for species 'preferring' such open areas, and also a threat to the more specialized taxa confined to closed forests (Vietnam: Spitzer et al. 1997). Such 'gap dynamics' suggest that even small local disturbances can affect butterfly assemblages in forests.

Correlations of moth abundance and richness with feeding specificity have been debated extensively, with Gaston (1988) unable to confirm this attractive conceptual link. Assessing feeding specificity of caterpillars, particularly of many moths, remains difficult in many places because of (1) incomplete knowledge and (2) local variations in plant preferences or availability. Thus Inkinen (1994) classified the larval feeding habits of British noctuid moths into five categories, as (1) monophages – larvae feeding on one plant species; (2) oligophages, s. str., feeding on several plant species within the same genus; (3) oligophages, s. lat., feeding on several plant genera within the same family; (4) polyphages, feeding on several plant families; and (5) unknown. For this well studied group, categorisation of the 312 species of Lepidoptera in sequence along the above categories was 62, 23, 64, 152 and 11 species respectively. Similar spectra of relative specificity are undoubtedly commonplace amongst Lepidoptera and other insect herbivores. Nectar plant selection can parallel that of larval food plants in complexity, with some Lepidoptera clearly very restricted in the array of sources. Woodland butterflies may be particularly selective in relation to many open country species (Tudor et al. 2004), so that specific nectar sources assume considerable importance in planning site management for these.

Secondary forests may at times reveal more butterfly species than primary or near-primary forests (Cameroon: Bobo et al. 2006). However, many analyses endorse the values and irreplaceability of forests and other stable biotopes as the hosts of specialist species absent from the wider changed landscape. In Costa Rica large forest tracts 'appear key to supporting some regional endemics' (Horner-Devine et al. 2003), within a milieu largely converted for coffee production. Comparison of selected butterfly groups in the large forest remnant studied (Las Cruces Reserve: 227 ha) with coffee plantation sites about 40 years old at different distances from forest and with ('coffee/forest') or without ('coffee') adjacent small forest remnant patches had significantly higher mean species richness and abundance than others (Fig. 8.3), and assemblages on sites more distant from Las Cruces differed in composition. Survey of the moths around Las Cruces (Ricketts et al. 2001) implied that diversity in non-forested sites is correlated with forest cover within 1–1.5 km, a considerably greater distance than for butterflies (Horner-Devine et al. 2003), and possibly reflecting that many moths

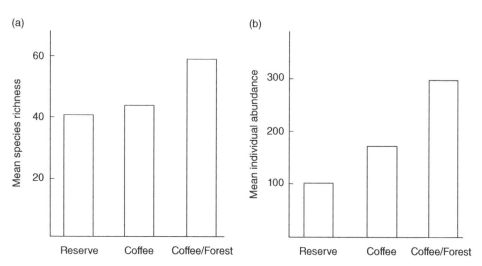

Fig. 8.3 Numbers (mean +/− SE) of (a) species and (b) individuals of butterflies in three main biotopes in coffee-growing areas at Las Cruces, Costa Rica (Source: Kitahara & Fujii. 1994. With permission from Springer Science+Business Media.).

have greater individual dispersal distances. Daily and Ehrlich (1996) had earlier raised the intriguing suggestion that nocturnality may favour use of fragmented landscapes, if hot daytime temperatures in open areas were an impediment for movement by diurnal butterflies.

Much conservation attention has been paid to three such contexts, in particular, and all of which span concerns from localised individual specialised species to major changes in assemblage composition and richness as major vegetation types are lost or modified. Recent wider awareness of 'ecosystem services' has become a major inducement to increase the hospitality of anthropogenic landscapes, particularly agricultural environments, and the values of conserving remnant native vegetation within such landscapes include potential to include the specific resources for individual species of conservation interest as a component of any individual recovery plan (Chapter 9, p. 164). In addition to the more familiar ecosystem services (such as pollination), Gillespie and Wratten (2012) regarded butterfly conservation as an 'aesthetic ecosystem service' in New Zealand viticultural systems, an ethos of very wide relevance in garnering publicity and support, and as a possible encouragement to ecotourism.

Assessing changes

Investigations have considered direct contrasts between 'natural' and 'highly modified' environments or considered various extents of change along 'disturbance gradients' spanning these extremes. Some generalisations emerge from studies on impacts of three of these in particular, namely forest loss, agricultural conversion and urbanisation, and have led to a variety of management practices of wide application. The principles lead to many individual management measures that harmonise or restore natural values – including wellbeing of Lepidoptera – with human uses (Chapter 12).

Forests

Tropical forest butterflies, for example, demonstrate that old-growth forests are important to maintain a large proportion of species, whilst agroforestry systems with smaller plant variety protect only much smaller proportions of species, mostly those that are widespread and can tolerate disturbance. Such widespread taxa thereby tend to dominate assemblages in highly modified land use areas, to leave many of the narrowly endemic taxa occurring only (or mainly) in old-growth forests. The roles of secondary and agroforestry systems as 'refuges' are often very limited. An Indonesian survey (Hill & Hamer 2004) implied strongly that changes in butterfly assemblages following selective logging were associated with changes in vegetation structure, with less variety in both plant richness and vegetation structure. In reviewing this important theme, Schulze et al. (2010) emphasised the methodological and sampling problems of scale that can influence interpretations – such as biases in richness resulting from the vertical stratification of forest butterflies that are difficult to assess from 'ground-restricted methods', but that may still be affected by disturbance. In general, it seems that tropical forest butterflies may be best conserved (other than by protecting and securing undisturbed forests!) through forest management under which a habitat mosaic includes areas of undisturbed primary forest and a network of other patches that vary in level of disturbance and intensity of management – so imposing the greatest variety possible in harmonising human use with conservation need.

The same principle, of reducing intensification of change and incorporating wider values into land use, is widespread. It is sometimes difficult to impose in practice, and numerous studies of Lepidoptera along land use gradients, as above, have helped to demonstrate its worth.

Agriculture

Intensification of agriculture during the middle decades of the twentieth century is associated with losses of key insect habitats, and the replacement of 'traditional' farming methods that were more sympathetic to conserving native biota, through larger-scale and more contrasting approaches under which much cannot persist. Simplifications of landscapes from agriculture involve lower plant diversity that reflects transitions from relatively diverse natural habitats to annual monocultures, and is regarded as a predominant driver of losses of biodiversity. Thus, many of the naturally declining moth species in Britain have become less abundant with increased agricultural intensification (Merckx et al. 2012b). In the United Kingdom, as a related example, changes since the 1940s have involved major losses of grassland which supported 19–28 species of butterflies, in contrast to the 1–3 species now supported by 'improved grassland' (Field et al. 2005). Following Asher et al. (2001), they noted losses of 40% of lowland heath, 97% of lowland flower-rich grassland and 80% of chalk/limestone grassland, as key biotopes together associated with considerable declines in butterfly distribution and abundance. Calcareous grasslands are one of the most species-rich butterfly habitats in Europe, and 274 of the 576 native species have been recorded from them (van Swaay 2002), but few are restricted to these habitats; however, they include 37 of the 71 species then considered threatened in Europe. Threats to grassland were very varied, but agricultural conversion is notable; the first three entries in Table 8.1 are perhaps the most important broad correlations, and it is notable that collecting is only rarely of concern.

Permanent meadows, and careful management of crop edges and surrounds (Chapter 12, p. 223) are important components of management for Lepidoptera (Clausen et al. 1998). Many species become essentially restricted to such areas, or to adjacent grasslands (Oates 1995). The significance of field margins, including hedgerows and conservation headlands, and adjacent uncultivated areas in agricultural environments has been demonstrated repeatedly – with studies such as those on arable field margins in the United Kingdom (Dover 1989) and uncultivated habitats in Denmark (Clausen et al. 2001) helping to elucidate the critical resource factors involved. Both have wide roles in insect conservation as dispersal corridors and/or refuge or reservoir habitats, in which Lepidoptera effectively eliminated from the wider landscape by land use changes may thrive (Dover & Sparks 2000). They and other linear features such as 'green lanes' (unmetalled tracks between fields: Dover et al. 2000), roadsides or powerline easements that are managed regularly in ways that may be regarded as at least 'semi-natural' may all be invaluable compensatory habitats. Unmown roadside verges in Europe can provide a 'nectar refuge', for example (Valtonen 2006), and adjusting mowing regimes elsewhere can enhance the conservation value of many such areas. Thus, mid-summer mowing in Europe may reduce Lepidoptera abundance more than late summer treatment, whilst leaving a mosaic of unmown patches was markedly beneficial (Valtonen 2006). In Finland, both roadsides (Valtonen 2006) and powerline easements (Kuussaari et al. 2003) proffer significant alternative and additional habitats for meadow Lepidoptera – roadside counts yielded 54 butterfly species and 83 moth species, most of them meadow-frequenting taxa.

In linear 'edge habitats', the width of the hedge bank and/or grassy verge, the vegetational diversity, and abundance of nectar sources, shelter and extent of insolation are all positively associated with butterfly richness and abundance. In the wider arena, presence of larval food plants is highly influential. Clausen et al. (1998) found that nectar source species important for butterflies included species with conspicuous purple flower heads, with some yellow-flowering species. Importantly, larval host plants alone are insufficient to foster abundance of many butterflies.

The conservation importance of even very tiny fragments of natural vegetation remaining in agricultural landscapes is high, but often easily overlooked by managers not concerned directly with invertebrate conservation. Kuussaari et al. (2007a) found richness of Lepidoptera in semi-natural grasslands and open forest edges higher than on road and field margins in moderately intensive European agroecosystems. Based on surveys that yielded 60 species of butterflies (19,250 individuals) and 118 species of moths (10,858 individuals of which 10,705 belonged to the 67 species designated as diurnal), grassland patches as small as 0.06–0.3 ha were significant for butterfly richness in typical boreal farmland, with sunny forest edges also valuable – probably reflecting the shelter they confer (Chapter 12, p. 229).

Numerous 'agri-environment schemes', continually increasing in scope and distribution, help to reverse the negative trends from past agricultural intensification. In some of these, Lepidoptera have become useful monitoring tools to evaluate the outcomes from practices such as hedgerow plantings and extent and treatment of field margins. Correlation of these features with macromoths in the United Kingdom demonstrated that increases were associated with increased moth diversity (pool of 311 species across sampled sites), with the two modifications each independently revealing this trend (Merckx et al. 2012b). These authors suggested that in functional terms, hedgerow trees may constitute the only farmland remnants of once-forested systems that can allow persistence of formerly characteristic species. The benefits of agri-environment schemes draw on ecological principles of very wide relevance. For abundance of both Microlepidoptera and macromoths, and richness of the latter, the proportional cover of rough grassland and scrub within 250 m was an important positive predictor, with field margins, hedgerows and multispecies grasslands and riparian margins contributing to this (Fuentes-Montemayor et al. 2011), so that increasing that semi-natural cover over relatively small areas can have marked conservation benefits. Similar inferences were found by Kuussaari et al. (2007a).

Urbanisation

Human settlement represents some of the most intensive changes to landscapes, with symptoms such as roadworks and other communication systems, buildings for residence and business, widespread replacement of natural vegetation with alien introductions, and massive conversion of more natural to highly altered environments inimicable to specialised native biota, and with continuing and expanding despoliation from traffic pollution and general human pressures.

Two main associated themes have been investigated in Lepidoptera conservation: (1) the roles of remnant and other open spaces as refuges or sanctuaries for native taxa, and how their values may be enhanced by management, and (2) the changes along gradients from urban to rural areas. Both inevitably focus on the availability of key resources, and the impacts of urbanisation on those. Thus, the more subtle local changes are illustrated by the highly specialised assemblage of Lepidoptera associated with fungal galls on *Acacia karroo*, through such a gradient from urban to rural Pretoria (South Africa) (McGeoch & Chown 1997). Although that assemblage is small (seven species), richness (three species) was lowest at the inner city sites and increased to suburban sites, and larval density at city sites was 5–10 times lower than at urban reserve sites; McGeoch and Chown regarded the city site assemblages as 'clearly disturbed'. Urban reserves and suburban gardens contributed to the assemblage survival, as a trend acknowledged widely and important in planning urban Lepidoptera conservation. The same assemblage was reviewed by Rosch et al. (2001), who added a parallel set of samples collected 3 years later. Despite differences in gall age structure, habitat-associated differences in assemblages were consistent across years, with urbanisation (distance fom city centre) and influence of roadside disturbance having similar effects. This assemblage was therefore regarded as a 'robust biological indicator of the impacts of urbanisation on an insect assemblage'.

Impacts on individual species are assessed more commonly, and those that are known to be threatened attract particular attention. The Fluminense swallowtail (*Parides ascanius*) was the first insect listed officially as threatened in Brazil (Otero & Brown 1986), because of extensive habitat loss along the southern lowland coastal strip, largely through clearing and draining of coastal swamps for urban expansion in Rio de Janeiro state. A number of populations have been lost to urbanisation and associated recreational expansion. Preliminary modelling to attempt to predict the distribution of *P. ascanius* revealed very low overlap of likely distribution with protected areas (Uehara-Prado & Fonseca 2007), and considerably more field work is needed to explore further for the butterfly.

The values of urban parks for Lepidoptera were explored for eight City of New York parks, mostly modified extensively for human recreation, in which surveys by systematic sweep netting and direct observation yielded 42 species (Giuliano et al. 2004). Individual park richness (7–22 species) and relative abundance of butterflies was related to park size, so that management recommendations included retaining the largest possible land areas, and minimising distances between parks to improve connectivity. Increasing plant species richness was also urged, emphasising the use of native plants for this, together with, where necessary, reducing human impacts by reducing numbers of tracks available. Open areas, such as parks and urban meadows (Clark et al. 2007) within and near cities, tend to support mainly more generalist species of Lepidoptera with the increasing urbanisation pressures affecting mainly the rarer and more ecologically specialized taxa. In Clark et al.'s survey of 20 such meadows in the Boston (Massachusetts) region, the specialist butterfly species – those that were regionally rare, ecologically restricted and had few broods – declined much faster than the more widely distributed generalist species.

The six sites on which butterfly assemblages were compared by Blair and Launer (1997) near Palo Alto, California, were selected to represent a rural–urban gradient from a Biological Preserve to the city Business District. On all sites, richness and abundance of butterflies was assessed visually. Numbers of species (Fig. 8.4a) and individuals (Fig. 8.4b) were both lowest in the most urban areas, and generally increased toward rural environments, with this gradient largely reflecting urban-induced losses from the species pool at the Biological Preserve. However, Blair and Launer regarded 17 species as 'suburban adaptable', as occurring in some well built environments. Most of these species are multivoltine, and none is univoltine; many can also utilise lawn grasses (so can be supported even in managed lawns) or are polyphagous on common weedy plants.

Whilst the maintenance/restoration of native vegetation to foster Lepidoptera in urban areas is urged repeatedly, urban butterflies may come to depend wholly

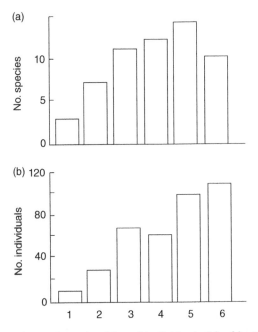

Fig. 8.4 Gradients in numbers of species (a) and individuals (b) of butterflies along a sequence from most urbanised (1) to least urbanised (6) regions (Adapted from Blair & Launer 1997).

or largely on alien plant species (Shapiro 2002), so creating a dilemma for conservation planners to whom such 'weeds' are undesirable and, in many cases, targets for eradication. In Davis (California), 29 of the 32 breeding butterfly species do so at least in part on alien plants, and 13 have no known native host plant in the area. Few of these alien hosts are cultivated deliberately, and their number continues to increase as additional weeds become naturalised (Shapiro 2002). 'Butterfly gardening', habitat enrichment of domestic gardens through establishing larval food plants and nectar sources, provides abundant and diverse opportunities for increasing suitable area: 'Domestic gardens cover millions of hectares in the United States' (Daniels 2012: 155). Whilst most such efforts benefit generalist butterflies, many of the most specialised and threatened taxa do not normally visit urban gardens. However, some surveys (Mauro et al. 2007) have selectively studied these generalist taxa to demonstrate or confirm positive influences of garden size and vegetational diversity on butterfly richness and abundance.

Roadsides are also potential habitats for many Lepidoptera, perhaps particularly species that can adapt to regular mowing regimes (Saarinen et al. 2005). In less urban areas, widened verges and good nectar supplies are important for meadow butterflies, and many moths may gain protection from shelter provided by taller vegetation. In Finland, roadsides are the largest area of regularly mown grasslands in the country, so that conservation of their inhabitants through sympathetic management attracts interest. Verges of highways, urban roads and rural roads were broadly similar in the richness of species found (pool of 99 species of diurnal Lepidoptera: 53 butterflies and 46 moths), with a few species of each relatively abundant. Three roadside species (*Glaucopsyche alexis*, *Euphydryas aurinia* and the sphingid *Hemaris tityus*), although not found commonly, are classified as threatened, and several others are 'declining species'. A high proportion of the species found in previous surveys of local agricultural habitats were found on these roadsides.

However, one counter to fostering Lepidoptera along roadsides is potential losses from vehicle impacts, a topic only rarely documented but occasionally, at least, causing substantial mortality. For Kern's primrose sphinx moth (Chapter 11, p. 203) both direct vehicle strikes and impacts of off-road vehicles on habitat were among threats noted (USFWS 2007). Extrapolating from counts of bodies along roadsides in central Illinois, McKenna et al. (2001) suggested that roadside mortality of Lepidoptera for the state could be more than 20 million individuals in a week of their survey, with more than 500,000 Monarchs (*Danaus plexippus*) included. These impressively high numbers were based on transect counts of road-killed Lepidoptera and daily vehicle counts obtained from the Department of Highways: over 6 weeks, 1824 individuals (dominated by Pieridae, with 1510) were collected with monarchs (99 individuals) the most abundant other taxon. However preliminary these data may be, significant numbers of Lepidoptera are killed by traffic, with some indication that slower traffic is the most harmful, with 'wind currents' from faster vehicles 'lifting' the insects over them. Despite many comments that roads are effective barriers to butterflies, this is clearly not always the case. Whilst many Orange-tips (*Anthocaris cardamines*) turned back at a road edge studied in Britain (Dennis 1986), in other contexts 'numbers of butterflies are often found squashed on roads' (Dennis 2010). Any such estimates may be biased by when counts are attempted – in some Indian national parks, counts were highest on Sundays, for example, with increased traffic flow at weekends (Rao & Girish 2007).

Probably related to their relatively large size and conspicousness, so that corpses are detectable from ground surveys, butterflies and dragonflies are the most studied insect roadkill, but the wide taxonomic array of insects in the 'radiator catch' of vehicles demonstrates much wider impacts. The properties of road verges are major influence on mortality, together with the population sizes of roadside residents. As Skorka et al. (2013) pointed out, this may lead to a conservation management dilemma, in that factors achieving the desired outcomes of increasing population sizes and assemblage richness along roadsides may also contribute to increased mortality. However, their survey (of butterflies in Poland) led to the contrary inference, that road verges most suitable for butterflies may, in fact, suffer the least from road mortality.

References

Asher, J., Warren, M., Fox, R., Harding, P., Jeffcoate, G. & Jeffcoate, S. (2001) The Millennium Atlas of Butterflies in Britain and Ireland. Oxford University Press, Oxford.

Beck, J. & Chey, V.K. (2008) Explaining the elevational diversity pattern of geometrid moths from Borneo: a test of five hypotheses. *Journal of Biogeography* 35, 1452–1464.

Beck, J., Schulze, C.H., Linsenmair, K.E. & Fiedler, K. (2002) From forest to farmland: diversity of geometrid moths along two habitat gradients on Borneo. *Journal of Tropical Ecology* 18, 33–51.

Blair, R.B. & Launer, A.E. (1997) Butterfly diversity and human land use: species assemblages along an urban gradient. *Biological Conservation* 80, 113–125.

Bobo, K.S., Waltert, M., Fermon, H., Njokagbor, J. & Muhlenberg, M. (2006) From forest to farmland; butterfly diversity and habitat associations along a gradient of forest conversion in southwestern Cameroon. *Journal of Insect Conservation* 10, 29–42.

Bonebrake, T.C., Boggs, C.L., McNally, J.M., Ranganathan, J. & Ehrlich, P.R. (2010) Oviposition behavior and offspring performance in herbivorous insects: consequences of climatic and habitat heterogeneity. *Oikos* 119, 927–934.

Clark, P.J., Reed, J.M. & Chew, F.S. (2007) Effects of urbanization on butterfly species richness, guild structure and rarity. *Urban Ecosystems* 10, 321–337.

Clausen, H.D., Holbeck, H.B. & Reddersen, J. (1998) Butterflies on organic farmland: association to uncropped small biotopes and their nectar sources (Papilionoidea and Hesperioidea, Lepidoptera). *Entomologiske Meddeleser* 66, 33–44.

Clausen, H.D., Holbeck, H.B. & Reddersen, J. (2001) Factors influencing abundance of butterflies and burnet moths in the uncultivated habitats of an organic farm in Denmark. *Biological Conservation* 98, 167–178.

Daily, G.C. & Ehrlich, P.R. (1996) Nocturnality and species survival. *Proceedings of the National Academy of Sciences* 93, 11709–11712.

Daniels, J. (2012) Gardening and landscape modification: butterfly gardens. p. 153–168 in Lemelin, R.H. (ed.) The Management of Insects in Recreation and Tourism. Cambridge University Press, Cambridge.

Dennis, R.L.H. (1986) Motorways and cross-movements. An insect's 'mental map' of the M56 in Cheshire. *Bulletin of the Amateur Entomologists' Society* 45, 228–243.

Dennis, R.L.H. (2010) A Resource-Based Habitat View for Conservation. Butterflies in the British Landscape. Wiley-Blackwell, Oxford.

Dover, J.W. (1989) The use of flowers by butterflies foraging in cereal field margins. *Entomologist's Gazette* 40, 283–291.

Dover, J.W. & Sparks, T. (2000) A review of the ecology of butterflies in British hedgerows. *Journal of Environmental Management* **60**, 51–63.

Dover, J., Sparks, T., Clarke, S., Gobbett, H. & Glossop, S. (2000) Linear features and butterflies: the importance of green lanes. *Agriculture, Ecosystems and Environment* **80**, 227–242.

Dover, J., Warren, M. & Shreeve, T. (eds) (2011) Lepidoptera Conservation in a Changing World. Springer, Dordrecht.

Field, R.G., Gardiner, T., Mason, C.F. & Hill, J. (2005) Agri-environment schemes and butterflies: the utilization of 6 m grass margins. *Biodiversity and Conservation* **14**, 1969–1976.

Fuentes-Montemayor, E., Goulson, D. & Park, K.J. (2011) The effectiveness of agri-environment schemes for the conservation of farmland moths: assessing the importance of a landscape-scale management approach. *Journal of Applied Ecology* **48**, 532–542.

Gaston, K.J. (1988) Patterns in the local and regional dynamics of moth populations. *Oikos* **53**, 49–57.

Gillespie, M. & Wratten, S.D. (2012) The importance of viticultural landscape features and ecosystem service enhancement for native butterflies in New Zealand vineyards. *Journal of Insect Conservation* **16**, 13–23.

Giuliano, W.M., Accamando, A.K. & McAdams, E.J. (2004) Lepidoptera-habitat relationships in urban parks. *Urban Ecosystems* **7**, 361–370.

Hill, J.K. & Hamer, K.C. (2004) Determining impacts of habitat modification on diversity of tropical forest fauna: the importance of spatial scale. *Journal of Applied Ecology* **41**, 744–754.

Holloway, J.D. & Barlow, H.S. (1992) Potential for loss of biodiversity in Malaysia, illustrated by the moth fauna. pp. 293–231 in Barlow, H.S. & Kadir, A.A. (eds) Pest Management and the Environment in 2000. CAB International and Agricultural Institute of Malaysia, Kuala Lumpur.

Holloway, J.D. & Nielsen, E.S. (1999) Biogeography of the Lepidoptera. pp. 423–462 in Kristensen, N.P. (ed.) Lepidoptera: Moths and Butterflies. Vol.1. Evolution, systematics and biogeography. Handbuch der Zoologie, Vol. 4, part 35. W. de Gruyter, Berlin and New York.

Horner-Devine, M.C., Daily, G.C., Ehrlich, P.R. & Boggs, C.L. (2003) Countryside biogeography of tropical butterflies. *Conservation Biology* **17**, 168–177.

Inkinen, P. (1994) Distribution and abundance in British noctuid moths revisited. *Annales Zoologicae Fennici* **31**, 235–243.

Kitahara, M. (2004) Butterfly community composition and conservation in and around a primary woodland of Mount Fuji, central Japan. *Biodiversity and Conservation* **13**, 917–942.

Kitahara, M. & Fujii, K. (1994) Biodiversity and community structure of temperate butterfly species within a gradient of human disturbance: an analysis based on the concept of generalist vs. specialist strategies. *Researches on Population Ecology* **36**, 187–199.

Kitahara, M. & Sei, K. (2001) A comparison of the diversity and structure of butterfly communities in semi-natural and human-modified grassland habitats at the foot of Mt. Fujii, central Japan. *Biodiversity and Conservation* **10**, 331–351.

Kitahara, M., Sei, K. & Fujii, K. (2000) Patterns in the structure of grassland butterfly communities along a gradient of human disturbance: further analysis based on the generalist/specialist concept. *Population Ecology* **42**, 135–144.

Kuussaari, M., Ryttari, T., Heikkinen, R. et al. (2003) Significance of power line areas for grassland plants and butterflies. *The Finnish Environment*, no. 638. Helsinki. (in Finnish, English summary)

Kuussaari, M., Heliola, J., Luoto, M. & Poyry, J. (2007a) Determinants of local species richness of diurnal Lepidoptera in boreal agricultural landscapes. *Agriculture, Ecosystems and Environment* **122**, 366–376.

Lewis, O.T., Wilson, R.J. & Harper, M.C. (1998) Endemic butterflies on Grande Comore: habitat preferences and conservation priorities. *Biological Conservation* **85**, 113–121.

Mauro, D.D., Dietz, T. & Rockwood, L. (2007) Determining the effect of urbanization on generalist butterfly species diversity in butterfly gardens. *Urban Ecosystems* **10**, 427–439.

McGeoch, M.A. & Chown, S.L. (1997) Impact of urbanization on a gall-inhabiting Lepidoptera assemblage: the importance of resources in urban areas. *Biodiversity and Conservation* **6**, 979–993.

McKenna, D.D., McKenna, K.M., Malcolm, S.B. & Berenbaum. M.R. (2001) Mortality of Lepidoptera along roadsides in central Illinois. *Journal of the Lepidopterists' Society* **55**, 63–68.

Merckx. T., Marino, L., Feber, R.E. & Macdonald, D.W. (2012b) Hedgerow trees and extended-width field margins enhance macro-moth diversity: implications for management. *Journal of Applied Ecology* **49**, 1396–1404.

Miller, D.G., Lane, J. & Senock, R. (2011) Butterflies as potential bioindicators of primary rainforest and oil palm plantation habitats on New Britain, Papua New Guinea. *Pacific Conservation Biology* 17, 149–159.

New, T.R. (2011) Butterfly Conservation in South-Eastern Australia: Progress and Prospects. Springer, Dordrecht.

Oates, M.R. (1995) Butterfly conservation within the management of grassland habitats. pp. 98–112 in Pullin, A.S. (ed.) Ecology and Conservation of Butterflies. Chapman & Hall, London.

Ohwaki, A., Nakamura, K. & Tanabe, S.-I. (2007) Butterfly assemblages in a traditional agricultural landscape: importance of secondary forests for conserving diversity, life history specialists and endemics. *Biodiversity and Conservation* 16, 1521–1539.

Otero, L.S. & Brown, K.S., Jr (1986) Biology and ecology of *Parides ascanius* (Cramer, 1775) (Lep., Papilionidae), a primitive butterfly threatened with extinction. *Atala* 10, 2–16.

Rao. R.S.P. & Girish, M.K.S. (2007) Road kills: assessing insect casualties using flagship taxon. *Current Science* 92, 830–837.

Ricketts, T.H., Daily, G.C., Ehrlich, P.R. & Fay, J.P. (2001) Countryside biogeography of moths in a fragmented landscape: biodiversity in native and agricultural habitats. *Conservation Biology* 15, 378–388.

Rosch, M., Chown, S.L. & McGeoch, M.A. (2001). Testing a bioindicator assemblage: gall-inhabiting moths and urbanization. *African Entomology* 9, 85–94.

Rosin, Z.M., Myczko, L., Skorka, P. et al. (2012) Butterfly responses to environmental factors in fragmented calcareous grasslands. *Journal of Insect Conservation* 16, 321–329.

Saarinen, K., Valtonen, A., Jantunen, J. & Saarnio, S. (2005) Butterflies and diurnal moths along road verges: does road type affect diversity and abundance? *Biological Conservation* 123, 403–412.

Schulze, C.H., Schneeweihs, S. & Fiedler, K. (2010) The potential of land-use systems for maintaining tropical forest diversity. pp. 73–96 in Tscharntke, T., Leuscher, C., Veldcamp, E. & Faust, H. (eds) Tropical Rainforests and Agroforests under Global Change. Springer, Berlin and Heidelberg.

Schultz, C.B. & Crone, E.E. (2008) Using ecological theory to advance butterfly conservation. *Israel Journal of Ecology and Evolution* 54, 63–68.

Shapiro, A.M. (2002) The Californian urban butterfly fauna is dependent on exotic plants. *Diversity and Distributions* 80, 31–40.

Shapiro, A.M. & Shapiro, A.R. (1973) The ecological associations of the butterflies of Staten Island. *Journal of Research on the Lepidoptera* 12, 65–128.

Shreeve, T.G. & Dennis, R.L.H. (2011) Landscape scale conservation: resources, behaviour, the matrix and opportunities. *Journal of Insect Conservation* 15, 179–188.

Skorka, P., Lenda, M., Moron, D., Kalarus, K. & Tryjanowski, P. (2013) Factors affecting road mortality and the suitability of road verges for butterflies. *Biological Conservation* 159, 148–157.

Spitzer, K., Jaros, J., Havelka, J. & Leps, J. (1997) Effect of small-scale disturbance on butterfly communities of an Indochinese montane rainforest. *Biological Conservation* 80, 9–15.

Spitzer, K. & Leps, J. (1988) Determinants of temporal variation in moth abundance. *Oikos* 53, 31–36.

Tudor, O., Dennis, R.L.H., Greatorex-Davies, J.N. & Sparks, T.H. (2004) Flower preferences of woodland butterflies in the UK: nectaring specialists are species of conservation concern. *Biological Conservation* 119, 397–403.

Uehara-Prado, M. & Fonseca, R.L. (2007) Urbanization and mismatch with protected areas place the conservation of a threatened species at risk. *Biotropica* 39, 264–268.

USFWS (United States Fish and Wildlife Service) (2007) Kern Primrose sphinx moth (*Euproserpinus euterpe*). 5-year review: Summary and evaluation. Sacramento, California.

Valtonen, A. (2006) Roadside environments as habitats for Lepidoptera. Ph.D. Dissertations in Biology, no. 42. University of Joensuu, Joensuu.

van Swaay, C.A.M. (2002) The importance of calcareous grasslands for butterflies in Europe. *Biological Conservation* 104, 315–318.

Vu, V.L. (2009) Diversity and similarity of butterfly communities in five different habitat types at Tam Dao National Park, Vietnam. *Journal of Zoology* 277, 15–22.

Vu, V.L. & Yuan, D. (2003) The differences of butterfly (Lepidoptera, Papilionoidea) communities in habitats with various degrees of disturbance and altitude in tropical forests of Vietnam. *Biodiversity and Conservation* 12, 1099–1111.

Wainer, J.W. & Yen. A.L. (2000) A survey of the butterfly fauna of The Paps Scenic Reserve, Mansfield, Victoria. *Victorian Naturalist* 117, 131–140.

9

Single Species Studies: Benefits and Limitations

Introduction

The origins of interest and concern in Lepidoptera conservation, as for many other animal groups, are in the perceptions and reality of declines and losses for individual taxa, wrought as they progressively disappear, and in the attempts to halt or reverse these losses. Species-level conservation of Lepidoptera, particularly butterflies, since the nineteenth century, has provided the lessons, and founded the practices and principles, that now dominate and drive much insect conservation. Increasingly detailed and sophisticated studies of many taxa have demonstrated the importance of understanding the biological idiosyncrasies of each individual species, and that these can vary considerably across a species' distributional range and between very closely related taxa, so that management may need to be both general in scope and principles and tailored carefully to cater even for individual populations or sites. The accumulated knowledge forms the foundations or templates on which management strategies can be founded and developed. Many such programmes for Lepidoptera are accepted readily as 'worthy'. Some have been highly successful, and many have helped to garner official and public support and sympathy for conservation.

Much of the impetus for species conservation has arisen from concerns over dramatic declines and losses, some of them well publicised, so that many cases are founded in 'crisis management' (linking with threat status; Chapter 3, p. 33), focusing on species (or, in many butterflies, subspecies) that have already become scarce and threatened. Most are very localised in distribution, some occurring on single or few sites by the time concern is evoked, and their decline is commonly linked with ecological specialisations, such as monophagy or dependence on very restricted biotopes or mutualistic associations. Many

Lepidoptera and Conservation, First Edition. T.R. New.
© 2014 John Wiley & Sons, Ltd. Published 2014 by John Wiley & Sons, Ltd.

consumable resource species are themselves of conservation concern, and both the consumer and the consumed species may be highly restricted within altered landscapes. The main foci of species-level conservation are twofold: (1) to conserve, through prevention of further decline and promoting recovery and persistence, those butterflies and moths signalled as 'threatened' and so in danger of extinction; and (2) wider more generalised management to prevent the many other taxa at present not prominent on conservation agendas from becoming threatened. Support and resources available for either activity are scarce, and patently inadequate to deal satisfactorily with all needy cases, so demanding selection of priority taxa by some form of informed triage (Chapter 5), perhaps based on legislative requirement or obligation (Chapter 11). The contexts and needs for insect species conservation (New 2009) can be complex, with background study, planning and coordination, and review based on monitoring, all being central themes and important components of each exercise.

The topic of single species conservation permeates much of this book, but this chapter summarises some of the requirements and lessons provided by individual cases. It focuses initially on the dimensions of a species' conservation strategy, with needs for careful holistic planning and effective coordination and review both being integral components of species management. Selected case histories are then reviewed to demonstrate some of the problems and practicalities of translating plans to practice. These range from classic cases, some historical but well known and persistently influential in Lepidoptera conservation (but perhaps less well known to biologists in other disciplines), to more current exercises. Each exemplifies ways in which ecological knowledge is applied practically in developing management, and the needs for effective organisation and coordination, as well as continuing community interest and support.

The foci range from conserving unique populations, as the only known representatives of particular taxa, to much wider attempts encompassing either multiple populations across a wider distribution, perhaps including a variety of political entities (countries in Europe; States, Provinces or Territories in North America or Australia), in some cases with highly uneven conservation needs across such a range but with varying political obligations from legislative notice. For acknowledged threatened species known from a limited number of sites, setting objective priorities for conservation can be an uncertain process driven by expediency and local interest. More formally, Maes et al. (2004) explored the idea of 'Functional Conservation Units' (FCUs) for *Maculinea alcon* in Belgium, to display the wide range of conservation needs and options that occur across different contexts and scales. They incorporated both occupied sites and those with high occupation potential – for example, historically occupied sites likely to be suitable for re-introductions. *M. alcon* is restricted to wet heathlands, with the single larval food plant the long-lived perennial *Gentiana pneumonanthe*, and has three possible species of *Myrmica* ant hosts. Three scales (categories of FCUs) were recognised by combining data on detailed distribution of *M. alcon*, its host plant and wet heathland biotope, population sizes, and mobility and capacity for colonisation, and applying these to a broad definition of FCU as 'a spatial entity with actual or potential habitat and in which specific management/restoration is recommended'. The categories were (1)

Table 9.1 Management needs and priorities for the three categories of Functional Conservation Unit (FCU) recognized for *Maculinea alcon* in Belgium (Maes et al. 2004: see text).

FCU	Objectives	Management need
1	Increase population size by optimizing habitat conditions on existing patches and restoring potential habitat	Small scale, attention to remaining resources Intensive, relatively expensive Low-intensity grazing; sod-cutting; small-scale burning
2	Larger scale, habitat restoration/ creation to form local networks for metapopulation structure	Larger scale. Emphasis on restoring and creating new habitat (stepping stones). Retain currently unsuitable patches
3	Candidates also for re-introduction after habitat restoration	Larger-scale management (such as large-scale sod-cutting or more intensive grazing) to restore prior to re-introduction

FCU-1, presently occupied sites plus the area within a surrounding range of 500 m, reflecting the maximum local distance movement by butterflies detected by mark–release–recapture studies; (2) FCU-2, areas within a range of 2 km around occupied habitat patches, as the maximum observed colonisation capacity; and (3) FCU-3, potential re-introduction sites on which *M. alcon* had only recently become extinct. Management needs and aims for the three categories differ somewhat (Table 9.1). Modelling studies (Mouquet et al. 2005) on this system incorporated the term 'community module' as the unit for conservation (Chapter 6, p. 108).

The initial task, then, is to determine which primary focal species (or subspecies) to attend, and which resource species must also be considered for specific management; deciding also whether habitats (sites, patches) are secure as the arena for long-term management, or can be rendered secure from future disturbance and loss. Much butterfly or moth species conservation is initially site-focused, but potentially expands to wider landscape considerations. Assuring site security may entail both ownership/proprietorship issues and the need to define as fully as possible the factors that have led to decline and concern, and the ways in which their impacts can be halted (if necessary) and mitigated. The major trajectories in planning include (1) to halt decline and any further losses of the species or its critical resources; (2) to prevent future threats and losses from occurring; and (3) where necessary 'recover' the species to build up its numbers, expand its distribution and increase the number of populations and their security and sustainability. Short-term and long-term perspectives may differ. Crisis management demands rapid and decisive action to reduce risk of imminent extinction, whilst long-term conservation entails providing conditions in which the species can be self-sustaining, continue to thrive and retain its potential for evolutionary development. The underlying rationale has been discussed extensively elsewhere: New (2009) reviewed many of the practical themes in insect species conservation, all relevant here. For each taxon a specific (and, perhaps also site-specific) management plan – a term used broadly to encompass the numerous different epithets: Action Plan, Conservation Plan, Recovery Plan and others, that are in practice often applied synonymously, despite differences in scope, detail and intent – summarises the case for conservation need, and how this is to be addressed, in terms of practical management and the focused research needed to render that management informed and viable. New (2009) listed five desirable components of any such plan, emphasising also that the plan should be understandable clearly by non-experts, and with objectives and actions expressed in SMART (specific, measurable, appropriate, realistic, time-bound) terms and, wherever possible, costed. The components are: (1) express the case for conservation, with relevant biological background on the species; (2) summarise actual and likely threats and causes of concern; (3) develop remedial measures; (4) specify, as comprehensively as possible, the management objectives and details needed, with rationale for these; and (5) assess how these may be undertaken and monitored, and by whom – and funded.

Any management plan should ideally be designed with careful consideration to who should or will read it, and who will bear the responsibility for bringing it into practice. Burbidge (1996) noted three main groups of interested people as (1) those who have a legal responsibility for nature conservation; (2) those who will be funding and implementing the plan; and (3) those who want to know what is being done. Importantly, none of these constituent interest groups is likely to include many entomologists – in many places, entomological expertise (let alone interest in Lepidoptera!) in conservation agencies is sparse, so that construction of a plan that is both comprehensive and comprehensible to non-entomologists is highly desirable.

Whilst a detailed and specific recovery plan is an ideal framework for conservation, other approaches are possible and may encapsulate the needs of multiple species by expressing the common themes and needs. Arnold's (1983a,b) proforma for species-orientated conservation of lycaenid butterflies (Table 9.2) transcends individual species to some extent and, rather, demonstrates the fundamental topics in approaching a species programme, indicating the spectrum of activities that may be needed, and that should be considered in the initial stages of formulating a plan. Many management plans flow from legal obligations, and the variety of these (Chapter 11) necessarily leads to very varied documents, ranging from very short and general summaries to extensive accounts that attempt to cover every individuality. That for the Karner blue butterfly (*Lycaeides melissa samuelis*) in the United States, for example, runs to 239 pages. In contrast, many of the United Kingdom Biodiversity Action Plan statements are only a few pages in length – but are dealing with members of a well known fauna, for which much detailed advice may be somewhat superfluous: the Butterfly Action Plans for the United Kingdom are more comprehensive.

Plans are occasionally couched in terms of 'operating goals' for the programme, as for the goals listed by Sommers and Nye (1994) for the Karner blue butterfly in New York, with each backed by stated operational measures (Table 9.3).

Table 9.2 Points for a proforma approach to species-orientated conservation for lycaenid butterflies (Arnold 1983a,b).

---

1. Preserve, protect and manage known existing habitats to provide conditions needed by the species.
    a. Preserve: prevent further degradation, development or modification
        i.   Cooperative agreements with land owners and/or managers
        ii.   Memoranda or undertakings
        iii. Conservation easements
        iv.   Site acquisition (purchase/donation of private land) or reservation (public land)
    b. Maintain land and adult resources
    c. Minimise threats and external influences
    d. Propose critical habitat
    e. If recovery, clarify taxonomic status of taxon in habitat and other populations
2. Manage and enhance population(s) by habitat maintenance and quality improvement, and reducing effects of limiting factors
    a. Investigate and initiate habitat improvement methods as appropriate
    b. Determine physical and climatic regimes/factors need by species and relate to overall habitat enhancement on site
    c. Investigate ecology of species
        i.   Lifestyle and phenology: dependence on particular plant species or stages
        ii.   Dependence on other animals, and their roles
        iii. Population status
        iv.   Adult behaviour
        v.   Determine natural enemies and other factors causing mortality or limiting population growth
    d. Investigate ecology of tending ant species, if present
    e. Investigate ecology of food plant species
3. Evaluate all the above and incorporate into development of long-term management plan. Computer modeling may assist in making management decisions
4. Monitor population(s) to determine status and evaluate success of management
    a. Determine site(s) to be surveyed, if choice available
    b. Develop methods to estimate population numbers, distribution, and trends in abundance
5. Throughout all of the above, increase public awareness of the species by education/information programmes (such as information signs, interpretative tours, audio and visual programmes, media interviews, etc.)
6. Enforce available regulations and laws to protect species. Determine whether any additional legal steps are needed, and promote these as necessary

---

A further advisory perspective may be even more general, in summarising very broad 'conservation rules' applicable across a species' range and adaptable for any local situation. The European Habitats Directive lists 29 species of butterflies on its Annexes, so giving these taxa high conservation significance and potential protection; most are signalled also in national or other legislations. A recent overview of the needs of these taxa (van Swaay et al. 2012) summarises their status and needs, but includes also a list of 'dos and don'ts' for management of each. These vary across species in numbers and complexity, but collectively

Table 9.3 The early measures taken for Karner blue butterfly conservation in New York since it was declared as Endangered in the State in 1977 (as listed by Sommers & Nye 1994. Reproduced with permission of Minnesota Agricultural Experiment Station.).

1. Extensive and intensive surveys for the Karner blue butterfly and lupine
2. Estimation of population size at selected sites (mark–release–recapture)
3. Annual monitoring of populations and habitats and determination of population trends through index (walk-through) counts
4. Creation and expansion of suitable habitat, especially lupine, through studies of lupine germination and vegetation control techniques
5. Development and implementation of specific habitat management plans for select sites
6. Establishment of cooperative agreements with key landowners for select Karner blue butterfly sites
7. Reviews of project proposals impinging on Karner blue butterfly or lupine habitat, and recommendation of mitigation actions
8. Preparation and dissemination of information and educational materials pertaining to the Karner blue butterfly
9. Direct notification of town governments of locations of Karner blue butterflies, and requests for them to cooperate in site protection and reviews

express many management themes that are noted in this book and elsewhere, emphasising the importance of both habitat protection and focused management. As one example, the Southern swallowtail (*Papilio alexanor*) occurs on warm dry calcareous sites with rich vegetation and, although not of major conservation concern at present, comments (van Swaay et al., p88) are (Do) 'Maintain traditional estate management, for example with light goat grazing and/or controlled burning' and (Don't) 'Abandon the sites'. This brevity contrasts with measures for *Maculinea* (as *Phengaris*) *teleius*, discussed elsewhere in this book (Chapter 7, p. 127), as summarised in Table 9.4. For some species, the only advice listed is under 'Dos', a bias that may reflect lack of knowledge. Thus, for the False comma (*Nymphalis vaualbum*), a highly mobile and migratory woodland/woodland edge butterfly, the single 'Do' (p. 116) is 'Conduct more research on the distribution, ecology and population dynamics', following a statement that it is unclear what factors cause declines in the species' European range. The review thus captures the needs for fundamental research to found the basis of management, and the most detailed recommendations are for those taxa whose biology is best understood. As another example, for the Scarce fritillary (*Euphydryas maturna*) that needs focused woodland management in its conservation 'there is only vague information on . . . ecology, which hinders efficient conservation' (Freese et al. 2006). This species has become extinct in three countries and critically endangered in another five, and the detailed management needs differ across countries in Europe. The necessary open-canopy woods have historically been maintained by coppicing and forest pastures, but these practices have largely become obsolete in the face of 'more efficient forestry'.

Even amongst this best known of all insect faunas, many details of the ecological needs of butterflies in western Europe are not yet clear. Elsewhere, such details are even more fragmentary, or largely non-existent except in the most

Table 9.4 Approaches to conservation of *Maculinea teleius* in Europe: the 'Dos and Don'ts' suggested by van Swaay et al. (2012). Reproduced with permission of Pensoft Publishers.

Dos

In the northern part of the species range, mowing is the best management regime to keep the vegetation open and the soil sunny and warm and to maintain a high *Myrmica* ant nest density

Mow fields once every 1–3 years. In extensive areas, a 3-year rotation of 33% of patches a year is ideal

Mowing should be done either before the second week of June or after mid-September, on sites at lower altitude with a warm microclimate already after the beginning of September. In the first case females can deposit the eggs on the small regrowth of host plants; in the second case the caterpillars have left the host plant before cutting

Maintain 20% of the vegetation per meadow uncut each year on rotation to keep a high level of vegetation structure for a high *Myrmica* ant nest density. Also maintain some patches of scrub cover or hedge to provide shelter and warm conditions

Depending on the productivity of the soil, meadows may be cut once or twice a year. On poor soils best results are achieved by mowing in September or October, except for a cut early in June every 5–6 years to prevent bush encroachment

In the southern part of the species range, grassland habitats may also be managed by low-intensity grazing, preferably by cattle or ponies. Monitor the density of the stock to keep the right level of grazing intensity. In general, *Myrmica scabrinodis* (and *M. rubra*) occur in shorter vegetation on cooler northern sites, resulting in reduced shading of their nests

At landscape scale, create a mosaic of interconnected (within 5 km dispersal potential of species) patches of low-intensity agricultural use with both host plants and host ants for the establishment of a metapopulation. Allow patches of fallow land as refuge for the host ants. Preferably distances between patches are below 500 m and do not exceed 1 km

Monitor populations of the butterfly and its host ants carefully, and adjust management when needed

When the ant nest density is decreasing or at a too low density, apply small-scale management, such as sod-cutting in 3×3 m patches or in narrow long lines, to increase vegetation structure and habitats for the ants

When creating new habitats on former agricultural fields, remove the top soil when the phosphate concentration is too high. Use hay from local origin or local seed mixtures with *Sanguisorba*, or plant it into the new habitat

---

Don'ts

Intensify agricultural use or drain the fields

Graze habitats in the northern part of the species range

Abandon fields with single populations. Abandonment is only acceptable if temporary and the abandoned field is part of a metapopulation

Use manure or biocides

Mow the fields when the butterflies are on the wing and the caterpillars are in the buds of the host plant (between mid-June and the end of August)

---

general terms and by inference from related taxa. Approaches to species conservation must often be far more tentative and 'experimental' than in the most familiar northern temperate region faunas. Extrapolating to other taxa of Lepidoptera, knowledge becomes even more superficial and bland, with even the major food plants of numerous species very incompletely documented (see Robinson et al. 2001, as a regional appraisal). Despite the considerable attractions, and allure from successful cases in temperate regions, parallel examples over much of the tropics and other lesser developed parts of the world, simply do not occur.

Some case histories

The following is not intended as a catalogue of cases in Lepidoptera species conservation – the large number of individual species that have received some attention precludes any such approach here, or the need for it. Rather, I outline a few selected examples from the many candidates available, to illustrate different contexts, principles and approaches to management and setting priority. They demonstrate also the shifting balance between needs for 'research' and 'practical management'. For others, a series of butterfly conservation cases was outlined and discussed by New (1997) and the data in books, such as that edited by Pullin (1995) and many regional compendia, are not repeated here except to illustrate particular, significant points, or where considerable more recent information has accrued and changed previous perspective (as for the Large copper). Lycaenidae are disproportionately represented in species conservation cases – as the largest butterfly family, with numerous narrowly distributed and ecologically intricate and specialised taxa, they have attracted much attention (New 1993) and comparisons across species have helped to emphasise the subtlety of each individual species' requirements. Several studies of moth species conservation were included in New (2004), but in general, these are far fewer than butterfly cases. Examination of lists of 'protected species' and regional synopses, such as Red Lists, helps to reveal the numerous other taxa for which practical management has been individually undertaken, or believed necessary. Many are referred to elsewhere in this book.

Most of the more influential examples refer to localised species on sites scheduled for development, or on which development for industrial or other use has already been initiated, with evidence of undesirable impacts on native biota. The Brenton blue butterfly in South Africa (discussed later in this chapter) is one such example. More notorious, reflecting its geographical position, is the conservation of the El Segundo blue (*Euphilotes battoides allyni*) on the coastal dunes adjacent to Los Angeles International Airport (LAX) in California, that became one of the most significant and influential cases in North America. The small dune area underwent development from the 1880s, and the suitable habitat was divided by establishment of the Chevron oil refinery in 1911, with the southern section subsequently destroyed by residential developments by the 1970s (Mattoni 1992). By about 1990, the butterfly occurred on only three sites, all small and with the largest population on land at the western end of, and belonging to, LAX with pressures for further development to keep pace with the airport's expanding role. A major radar installation was located toward the south end of the dunes, and much of the residential settlement over the LAX dune area was abandoned with increasing disturbance and noise from air traffic, so that between 1965 and 1975, 822 houses were vacated and more than 2000 people relocated (Mattoni 1992). Nevertheless, following the conservation measures proposed by Arnold (1983a) and Mattoni, habitat restoration and active site management was undertaken by the major land owners (LAX and Standard Oil) with focus on removal of alien plants and reinstating native vegetation. This was predominantly the sole host plant, *Eriogonum parvifolium* (seacliff buckwheat), which had been extensively displaced by widespread seeding of the common buckwheat, *E. fasciculatum*, during the previous construction and housing operations, with only a hectare or so of dunes remaining in pristine or near-pristine condition. The progress, guided by the species' recovery plan (USFWS 1998) and the 'LAX Master Plan Final EIS' Habitat Restoration Plan (Sapphos 2005) as impact mitigation, attracted much publicity and appears to have been highly beneficial. Large numbers of butterflies have been reported recently, following decline to about 500 individuals in 1984. Press reports noted about 123,000 individuals present in 2011.

The cases reviewed in this chapter have mostly been documented extensively elsewhere, and many will be familiar to Lepidoptera devotees, if not, perhaps, to other readers. They demonstrate collectively the great variety of individual needs, that knowledge of these often necessitates original (and sometimes risky) field experimentation and that each such innovative example may enhance awareness of the discipline and contribute to the accumulated experience in supporting a stronger foundation for future benefit. Each of these examples has made general contributions to understanding the practical needs of Lepidoptera conservation: collectively, they cover many of the issues and practices used much more widely. Several of the species treated here are noted in more particular contexts in other chapters, so that the accounts serve to consolidate their wider conservation profiles.

The Large blue butterfly, *Maculinea arion*, in England

The extinction of the Large blue in Britain, and its subsequent re-introduction as an outcome of long-term detailed field study is arguably the greatest 'success story' for any species of Lepidoptera, and has stimulated parallel attention to the other threatened European species of the genus. The history of the British exercise has been documented extensively – the account by Thomas et al. (2009) is a self-effacing survey of a remarkable study led by Thomas, and a very readable 'layman's account' by Barkham (2010) helps to emphasise the biological intricacies so critical to understanding reasons for the initial loss and for the ensuing conservation outcome, through which the future of *M. arion* in Britain now seems assured.

At the time of Thomas' initial work on the Large blue in the early 1970s, Lepidoptera conservation in any formal sense scarcely existed, so that this study was truly pioneering in demonstrating the need for biological knowledge and integrating this into practical management, rather than simply basing species conservation on site reservation and preventing disturbance or intrusions. A key finding in this example was the need to provide a very short sward suitable for the host ant, *Myrmica sabuleti*, so that well intentioned abandonment of grazing (or fencing to exclude stock) generated the opposite effect to the protection anticipated, through failing to retain sufficient exposure to warm the ground areas used by the ants.

The initial decline was a long one. Thomas et al. (2009) noted that an estimated 91 colonies over 1795–1840 decreased to about 25 colonies by the 1950s, and to only two colonies in 1972, before extinction in 1979. On the very last site (known only as 'Site X' in the hope of protecting it from collectors), many environmental factors were monitored from 1972–1978, so that far greater understanding of influences on births and mortality of all stages accumulated. The major general outcome, integral to management of *M. arion* and – later – numerous other butterflies, was recognition that resource use is often far more subtle than supposed initially: thus (Thomas et al. 2009: 83) '. . . immature stages . . . typically exploit a narrow subset of their named resources . . . the availability of optimum larval habitat alongside adult metapopulation constraints largely determines population sizes and persistence'. Since its re-introduction from Swedish stock, following extensive searches for suitable donor populations throughout its European range, established British populations have been used to introduce the species to other localities in southern England. It has also spread unaided over relatively short distances by progressive use of 'stepping stone' habitats.

The Large copper butterfly, *Lycaena dispar*, in England

Other than, probably, the parallel of the Large blue in Britain (discussed in the previous section) some time later, no insect species has been as important as *Lycaena dispar* in the development of insect conservation interest and practice, largely because the charismatic endemic British subspecies was one of the earliest recorded butterfly extinctions, and attempts to re-establish the species in England from European stocks of closely related subspecies have taken place since 1909 (Pullin et al. 1995). During the intervening century, work on conservation of the Large copper has contributed experience, example and insight to almost all aspects of species-level conservation – from the importance of habitat security and condition, to resource supply and condition, evaluation and mitigation of threats, captive breeding and re-introduction, and the understanding of complex ecological requirements, all supported over this period by continuing endeavour to redress this 'hugely symbolic loss' (McLean & Key 2012), and facilitated by a major site (Woodwalton Fen) being bequeathed to the Society for the Promotion of Nature by Lord Rothschild in 1910 largely for this purpose, so furnishing a secure environment in which to proceed.

The British endemic *L. d. dispar* was never common in its fenland habitats of eastern England, and was probably already in decline at the time of its discovery in 1749; as Pullin et al. noted, loss was almost certainly due to widespread draining of the fenlands, reducing the butterfly to small isolated populations increasingly vulnerable to over-collecting and stochastic impacts. It became extinct in the mid-1860s. *L. dispar* is distributed over much of Europe, and efforts to reinstate it in the English fens have focused on two continental subspecies, *L. d. batavus* and *L. d. rutilus*, but, despite continuing energetic effort, Pullin et al. (1995) had to admit that 'Sadly, success in the shape of a self-sustaining population still eludes us'. The longest-lasting efforts at re-introduction (of *L. d. batavus* from the Netherlands) were to Woodwalton (from 1927 onward), and depended on caging larvae in spring to protect them from predators and parasitoids: Duffey (1968, 1977) identified larval mortality as the major factor influencing population size. Continued reinforcement (and, at times, re-introduction) from stock reared under greenhouse conditions, was associated with that stock being maintained for more than 20 years and likely to have become weakened and genetically stressed over that period. Survival of wild-caught (Netherlands) caterpillars was superior to that of British captive-reared stock, leading Nicholls and Pullin (2000) to infer genetic divergence between these populations, and to recommend direct translocations rather than reliance on captive breeding (Chapter 10).

The great water-dock, *Rumex hydrolapathum*, is the only food plant for caterpillars of *L. d. batavus* (although captive trials showed that caterpillars can survive on some other docks: Martin and Pullin 2004a,b), as it was previously for *L. d. dispar*. *L. d. rutilus* feeds on several other species of *Rumex*, but seemingly predominantly uses *R. hydrolapathum*, so that maintenance of this plant is a central need.

Caterpillars are also subject to mortality from extreme weather during their obligatory overwintering diapause in the second instar. Impacts of prolonged submergence and flooding led Nicholls and Pullin (2003) to suggest a 'flood tolerance threshold', so that considerations of local hydrology become important in management. Larval survival was significantly higher over the relatively dry winter of 1996–97 than over the wetter winters immediately before and after this, during both of which prolonged flooding occurred (Nicholls & Pullin 2000). 'Flood refugia' for hibernating caterpillars may be an important need.

One important outcome has been the suggestion that the Woodwalton and nearby sites may be too small to support viable populations of *L. dispar*, and considerations have recently extended to include the wider areas of the Norfolk Broads.

The general theme of flood effects and refugia is of concern for some other bog-frequenting Lepidoptera. The Large heath butterfly (*Coenonympha tullia*), for example, needs a mosaic of vegetation in which taller vegetation can be climbed by caterpillars when flood waters rise (Joy & Pullin 1997). The balance is a fine one: total submergence of caterpillars causes increased mortality after 7 days, but there is also the longer-term impact of increased mortality over the next few months. Restoration of the raised mire breeding habitats with the food plant *Eriophorum vaginatum* (cotton-sedge) may need to avoid total submergence of the sedge tussocks through raised water levels until suitable other tussocks have become well established on higher ground. Field trials (Joy & Pullin

1999) confirmed the need for suitable tussocks on higher ground to counter impacts of prolonged winter floods. A similar principle applies to the North American *Lycaena xanthoides*, earlier presumed extinct in western Oregon but rediscovered on some wetland prairie remnants in 2004, and since then a flagship species for preservation of local wetlands (Severns et al. 2006). Parts of its habitat are inundated seasonally, with flooding coinciding with the egg stage. Although numbers of eggs available were relatively small, comparison of outcomes from eggs from flooded and unflooded areas suggested a flooding impact. Only two of 84 'flooded' eggs survived to become third instar caterpillars, whereas seven of 46 'unflooded;' eggs did so (Severns et al. 2006).

The Brenton blue butterfly, *Orachrysops niobe*, in South Africa

As the focus of an extensive and complex conservation exercise to 'Save the Brenton blue', this butterfly has become a leading flagship for insect conservation in South Africa. The campaign involved saving the only known breeding site, a tiny remainder of an apparently always small range along a short stretch of the southern Cape coast. The site, at Brenton-on-Sea, near Knysna, was threatened with loss for housing construction. As with many other Lepidoptera conservation exercises, the initial concerns arose from hobbyists, as the only people becoming aware of the potential loss of the butterfly, and the case was promoted through the Lepidopterists' Society of (then) Southern Africa.

By June 1994, the campaign was formalised as the 'Brenton Blue Project', and complex legal and political arguments arose over the next few years (documented and discussed extensively by Steencamp and Stein 1999). This debate culminated in leading to the first time that a particular legal instrument (section 31A of the Environment Conservation Act 1989: 'Suspension of a development deemed harmful to the environment') was exercised, in April 1997.

O. niobe had not been found elsewhere during extensive targeted surveys by experienced lepidopterists over 17 years (to 1992), and was considered to be both extremely scarce and under high risk of loss. The proposed new housing development at Brenton-on-Sea posed a direct threat to the only known population. The Brenton blue became a major political tool, under media labels such as 'The butterfly that stings like a bee', and during protracted and continued pro-development and contra-development arguments. The major campaign efforts were (1) to protect this last known population and (2) to establish a nature reserve which will ensure that *O. niobe* will be preserved in its fynbos environment, and to be under the control of the (now) Lepidopterists' Society of Africa and the relevant conservation authority.

A number of reasons why the butterfly should **not** be conserved were advanced, to imply that any intervention at Brenton was pointless. These were refuted by original biological studies (Silberbauer & Britton 1999), as follows: (Reason 1) The butterfly could be moved so that development could proceed (counter: it could not be moved safely, and translocation was a high-risk long-term option that could not serve as an effective substitute for in situ conservation); (Reason 2) The development did not actively threaten the population, because the butterfly was on the 'public open spaces' within the development (counter: this was incorrect – most larval food plants [*Indigofera erecta*] and breeding sites were on areas to be developed); (Reason 3) The butterfly was not threatened with extinction, as other populations exist (counter: no other populations are known, and past misidentifications had led to some confusions – Brenton-on-Sea supports the only population); (Reason 4) The butterfly would go extinct in any case as part of its natural progress, or be eliminated by Argentine ant (*Linepithima humile*) infestations (counter: not 'going extinct in any case', and the Argentine ant issue was not proven, with likelihood that the ant could be controlled – but, more substantially, the loss of habitat was the primary driver toward extinction). Management to promote *Indigofera* was clearly feasible.

This case brought the wider issues of urban development forcefully to national attention in South Africa, together with appreciation of the conservation values of urban remnants as, in this example, the only place where a notable species exists.

The Richmond birdwing butterfly, *Ornithoptera richmondia*, in Australia

A notable Australian endemic butterfly and the most intensively studied representative of the most spectacular of all Lepidoptera, the birdwings, *Ornithoptera richmondia* is the most southerly birdwing and occurs within a coastal/subcoastal belt of subtropical rainforest in southern Queensland and northern New South Wales, in which the major larval food plant, the vine *Pararistolochia praevenosa*, grows. Loss of forest, for timber extraction, agricultural conversion and urbanisation, has been the primary cause of range reduction, and extensive fragmentation of the butterfly's distribution. The extent of habitat loss is exemplified by the area known as the 'Big Scrub' of northern New South Wales, of which the original forested area of about 75,000 ha had been reduced to about 700 ha by the early 1990s, with losses mainly much earlier and flowing from timber extraction (mainly for Red cedar, *Toona australis*) and agricultural conversion. The previously common butterfly has disappeared from about two-thirds of its former range.

Impacts of habitat loss have been augmented by the spread from garden cultivation of an alien aristolochiaceous vine. The South American 'Dutchman's pipe vine', *Aristolochia littoralis* (commonly termed *A. elegans*) is very attractive to female birdwings for oviposition, but the foliage is highly toxic to the hatchling caterpillars, which die after they begin to feed.

Continuing conservation interest since the 1990s has led to one of the most enduring campaigns for an insect in Australia. From the beginning, conservation of the strongly flying butterfly, which has become a notable flagship species, has emphasised facilitating its return to the whole of its former range – hence (and in marked contrast to the Lycaenidae discussed earlier) necessitating a wide landscape perspective for restoration, linked with broad habitat rehabilitation and security. Two major groups of activities recommended in the initial recovery plan (1996) have persisted to the present; the systematic removal of Dutchman's pipe vines wherever they occur in the wild, and extensive plantings of nursery-propagated *P. praevenosa* produced in large numbers especially for this purpose (Sands et al. 1997; Sands & New 2013). Rather than haphazardly, much recent planting has been along potential habitat corridors, in the expectation of reducing isolation and enhancing movement of butterflies from existing populations. Captive breeding, initially to investigate inbreeding effects (Orr 1994, Chapter 10, p. 188), in purpose-built flight cages is also a component of the conservation programme.

The entire programme has relied strongly on community support and interest, with innovative participatory schemes involving many constituent groups. Thus, numerous schools (fostered through CSIRO's Double Helix group: Sands et al. 1997) planted vines within their grounds and monitored these for birdwing incidence; some participated in an 'Adopt-a-Caterpillar' exercise, through which caterpillars were supplied (under licence, as a protected species) and which could be nurtured through to the adult stage. The current Richmond Birdwing Conservation Network has around 500 volunteer members, and its activities include a regular newsletter, field days, instruction sessions and plantings and social events supporting the programme and encouraging sustained interest.

Richmond birdwings have reappeared in recent years in many places in which they had not been seen for decades, and there is strong evidence that the conservation measures are working well. Indeed, Sands and New (2002) suggested that the species might by then qualify for downgrading from its formal threatened status to the category they designated as 'rehabilitated species'.

The Golden sun-moth, *Synemon plana*, in south-eastern Australia

All species of the endemic Australian castniid genus *Synemon* are of conservation concern, but *S. plana* has received special attention because it occurs on remnant native grasslands in the south-east that are subjected to severe pressures for development and have already been reduced to a tiny fraction of their formerly considerable extent, so becoming highly fragmented. Much of the remaining grassland is sought energetically for housing or industrial developments around Melbourne and other urban centres, so the land has acquired high commercial value. Although the moth is widely distributed, occurring in New South Wales, Victoria and the Australian Capital Territory (Chapter 6, p. 112), it is listed federally as 'Critically endangered' and has similar importance under the three separate range Acts covering the above regions. This status arose in part because the moth has been difficult to survey, but greater knowledge of its biology has now changed this (Gibson & New 2007, Chapter 5, p. 52), reducing the previously high incidence of 'spurious absences', and it is now known to be widespread, but with many populations being small and on tiny remnant patches that themselves appear vulnerable. The moth has become an important flagship taxon for this highly threatened ecosystem, a status it shares with two threatened reptiles – their generalised common names ('mouthless moth, legless lizard and earless dragon') have led some journalists to refer to these grasslands as a 'refuge for deformity'!

However, the discovery of numerous new populations of *S. plana* has led to debate over its somewhat ambiguous conservation status, with developers urging that human needs should take precedence and that compromises in land use should permit this. The Golden sun-moth has thus become embroiled in controversy over principles of habitat offsets to enable grassland development in periurban areas, and campaigns to minimise its conservation significance, through doubting the credibility of the allocated status. Discovery of the moth on many sites on which it seemed unlikely to occur has led to continuing frustrations from developers (New 2012a).

Many aspects of the moth's biology remain unknown. It is still unclear, for example, whether the life cycle (with the subterranean caterpillars feeding on grasses) takes 1, 2 or even 3 years to complete, and there are thus substantial problems in estimating population size as opposed to simply detecting presence. The cautious level of designating five individual adults detected during a survey to constitute a significant population reflects this uncertainty (DSE 2011), but the numbers found in any given year may simply be one annual cohort of a greater population, so that several continuous years of sampling are needed at this stage to infer population size (and, even, to confirm presence), each year with several visits needed to counter the very short adult lifespan (Chapter 5, p. 52), and continuing emergences over 2–3 months.

Australia's first dedicated sun-moth reserve, in western Victoria, is the only known site on which two species of *Synemon* (*S. plana* and a morph of the Pale sun-moth, *S. selene*) co-occur, and its establishment was an important development (Douglas 2004).

An additional dilemma is the presence of an aggressive alien grass, Chilean needle grass (*Nassella neesiana*) on many grassland patches frequented by *S. plana*. *Nassella* is a declared 'noxious weed' and so targeted for eradication. However, *S. plana* is sometimes common on highly invaded sites and, although not yet confirmed conclusively, there are strong suggestions that its caterpillars can (and do) feed on this alien plant.

The New Forest burnet moth, *Zygaena viciae*, in Scotland

The brightly coloured aposematic diurnal burnet moths, Zygaenidae, are very attractive to collectors, so that over-collecting is a realistic threat to some of the scarcer species. As for the Golden sun-moth in Australia, the New Forest burnet occurs on open ground, and has always had a very circumscribed range in Britain. Despite its common name, from its discovery in the New Forest (southern England) in 1869, it is now known in Britain only from a single population in Scotland. It had become extinct in the New Forest by 1927 (Tremewan 1966), with over-collection probably a contributory factor. The Scottish population was discovered in 1963, and is sometimes referred to as a distinct subspecies, *Z. v. argyllensis*, to differentiate it from the southern *Z. v. anglica*. That population declined severely from 1980 to 1990, and contained – at most – 20 moths in 1990 (Young & Barbour 2004), a level they considered likely to constitute a genetic bottleneck, with consequent low genetic diversity increasing vulnerability. The entire site is only about 1 ha, with the effectively occupied area probably considerably less. The decline was attributed mainly to overgrazing, and also stimulated formation of a Burnet Study Group to help stimulate interest in managing *Z. viciae* and other scarce zygaenids in Scotland.

In 1991, a fence was erected to exclude sheep from the site, but sporadic damage led to intermittent grazing until more effective fencing was constructed, first in 1996 and finally in 2001. Vegetation changed considerably from 1990, with extensive spread of the main larval food plant, *Lathyrus pratensis*, within a taller sward, and a second vetch (*Lotus corniculatus*) also a significant larval food source. Young and Barbour (2004) reported that from 1996 to 2001 a 'dramatic rise' in moth numbers occurred, from about 10–15 to an estimated 7500–9000 individuals. An apparent decline in 2002 (to 5000–6000) was followed by recovery in 2003, when the estimated total population was 8500–10,200 moths. This increase was attributed to exclusion of grazing and the consequent changes in vegetation, and especially to the expansion of food plants. Recent concerns that lack of grazing might induce changes in sward composition so that grasses predominate and *Lathyrus* and nectar plants decline has led to suggestions of reintroducing light grazing to counter this.

Despite a feeling of optimism for the future of *Z. viciae* on this site, Young and Barbour cautioned that the moth remains critically threatened in Britain, reflecting the single site occupancy.

The Essex emerald moth, *Thetidia smaragdaria maritima*, in England

This geometrid is unusual in that its demise has been documented carefully and extensively (Waring 1993, 2005), with a considerable variety of factors involved in this loss, both in the wild and subsequently of captive populations.

The Essex emerald was always regarded as a great rarity in England, with one outcome of this being very thorough recording of specimens, often with numbers captured and many biological notes emanating from collectors, so that is limited distribution was documented in considerable detail (Waring 2005). *T. smaragdaria* was found on salt marshes in Essex and Kent, where the larval food plant (sea wormwood, *Artemisia maritima*) thrived, but this coastal habitat declined substantially to leave only scattered small remnants, mostly associated with sea walls, and patches associated with coastal defences or land reclamation. Waring reported that the last known population 'died out in the absence of any gross change in the habitat', possibly linked with inbreeding depression.

Earlier, some populations may have succumbed to over-collecting, as some reports of collecting caterpillars imply that every individual that could be found was taken, and little restraint was evident.

By 1979 the moth was protected, and collecting of any stage became an offence. The last-known population in Essex was monitored annually from 1978, with the largest number recorded being 90 larvae in 1980, but it then declined, with none seen in 1985–1987. However, more extensive surveys led to discovery of 11 caterpillars in a small salt marsh in Kent. After considerable discussion over chances of survival in the wild, these were removed (November 1987) to constitute the founders of a captive population that was maintained until 1996, when it died out. Over three generations, that small stock was increased to a peak of 600 larvae each generation, and from this adults and larvae were used in five attempts to establish new populations in the wild (1990–1993). All failed within months.

However, in 1988, a Nature Conservancy Council survey discovered a further 56 caterpillars in a site several hundred metres from the other Kentish locality. This population was monitored, but died out in 1991. In 1999, Waring (2005) revisited the sites that had supported the last three populations (above), and found them 'in a completely unsuitable condition' from overgrazing and mowing, management that he believed would have eliminated any remaining populations.

The Fabulous green sphinx of Kaua'i, *Tinostoma smaragditis*, in Hawai'i

This spectacular moth is one of numerous endemic Lepidoptera in the Hawaiian archipelago. Unlike most Lepidoptera of conservation concern, it represents a family (Sphingidae) typified by large strongly flying moths not normally strictly bound by habitat patch other than through larval food plant distribution – so that, despite the scarcity of many hawk moths, they have not often figured highly on conservation agendas.

T. smaragditis is a flagship species, long sought by collectors, for the diverse mesic forest habitats of Hawai'i, but is still very poorly known. It is supposedly endemic to parts of western Kaua'i, a westerly island of the archipelago, and has never been found elsewhere. It is noted here to exemplify this context – of a large, conspicuous, narrowly endemic species, presumed rare and of which little is known, and for which any conservation plan cannot include any details of what is needed, other than to preserve presumed habitat.

The history of *T. smaragditis* was summarisd by Heddle et al. (2000), who commenced their account with 'In 1895, a moth was captured in a mountain home in Makaweli, Kaua'i, that would captivate and elude entomologists for the next century'. It was not reported again until 1961, and since then further specimens attracted to light raised the number of specimens known to Heddle and her colleagues to 18. The only information available on early stages results from eggs laid by a female captured in 1992 – she laid 15 eggs, five of which hatched, but the young caterpillars could not be induced to feed on any of the 130 native plants tested, or on an artificial diet (Cambell & Ishii 1993). The food plant/s remain unknown, and the long list of vascular plants in the region recorded by Heddle et al. confirms the richness of the forest habitat. At present, the only key to survival of the moth (and of many other less-heralded denizens of Hawai'i's Diverse Mesic Forest) would appear to be preservation of the forest itself, echoing the more general comment of Black and Jepsen (2007) that 'Ultimately, to protect any species, one must protect its habitat'.

Blackburn's sphinx moth, *Manduca blackburni*, in Hawai'i

The Hawaiian archipelago also harbours another endemic sphingid, far better documented than *T. smaragditis*, and listed as one of only two moths accorded federal protection as 'Endangered' under the United States Endangered Species Act. A detailed recovery plan has been prepared under that listing obligation

(USFWS 2005). It is noted here both for contrast with the Fabulous green sphinx and to exemplify further problems that arise more generally from an isolated island environment prone to alien invasions.

M. blackburni was formerly widespread and recorded from the seven main islands of Hawai'i, but has disappeared from most of these to leave only small localised populations on Maui, Hawaii and Kahoolawe (Rubinoff & San Jose 2010). It was presumed extinct by the 1970s, but was rediscovered on Maui in 1984. The moth purportedly frequents drier coastal and lowland forests, mostly in areas amenable to easy destruction for ranching, agriculture and settlement, so that more than 90% of these forests has been lost – to the extent that *Nothocestrum* trees (including the known larval food plants of *M. blackburni*, notably *N. breviflorum*) are also federally listed as Endangered. These plants are a major focus in proposed site restorations for the moth, with specific areas for restoration (management units) designated on all the main islands within its historical range – including one on the north west of Kauai coinciding with much of the range of *M. blackburni*.

Rubinoff and San Jose expressed concerns that the moth is becoming dependent on *Nothocestrum glauca* (tree tobacco, an invasive weed), so discouraging removal of this otherwise undesirable plant from forest habitat under restoration (a striking parallel with *Synemon plana* in Australia, above). The 2005 Recovery Plan, evolved from a draft issued for public comment earlier, listed eight major actions needed to accomplish success (Table 9.5). Collectively, the drive is toward protecting, managing and restoring habitat, through plantings of known and presumed larval food plants, accompanied by a captive breeding programme and/or a translocation programme across a number of management units and networks within these. Research to achieve these aims is recognised as an intrinsic need. Downlisting and delisting criteria relate to characteristics of sustainable populations, with the proviso that more specific details can be developed for downlisting once more complete ecological information has accrued (USFWS 2005). Somewhat unusually, and adding enormously to the complexity of the process, all four of the stated requirements must be met for delisting, and the detail is exemplified by simply quoting the first of these: '(1) one moth population, within one management unit, must be naturally reproducing and stable or

Table 9.5 The specific actions needed for conservation of Blackburn's sphinx moth in Hawaii, as listed by USFWS (2005).

1. Protect, manage and restore habitat and control threats
2. Expand existing wild *Nothocestrum* spp. host plant populations
3. Conduct additional research essential to recovery of Blackburn's sphinx moth
4. Develop and implement a detailed monitoring plan for Blackburn's sphinx moth
5. Re-establish and augment, through captive propagation if necessary, wild Blackburn's sphinx moth populations within its historical range
6. Develop and initiate a public information programme for Blackburn's sphinx moth
7. Validate recovery objectives
8. Develop a detailed Post-Recovery Monitoring Plan for Blackburn's sphinx moth

increasing in size, through one to two El Nino events or a minimum of 5 consecutive years of average rainfall within the Kauai-Oahu Recovery Unit'.

Variety of contexts

The representative cases summarised in this chapter could be extended to many other taxa and incorporate a wider taxonomic variety of butterflies and larger moths from many parts of the world. Those cited are sufficient to display a variety of contexts and issues that permeate species-level campaigns for Lepidoptera conservation in places where such campaigns are practicable. With few exceptions, biological knowledge at the start of a programme was inadequate, and for some species their very scarcity has made it impossible to augment this to any substantial extent but also to confirm their parlous state. Other than for sporadic records (captures or sightings) of the most elusive species, we have little idea of true conservation needs, and conservation measures focus largely on attempts to secure the most natural vegetational features of places from where they have been reported, and to protect those sites from despoliation. Such measures often have strong political components, resulting from inferential advocacy of conservation importance competing with vested interests seeking other uses and priorities. In other cases, biological knowledge can guide conservation more effectively, but perhaps only in rather general terms based on information from related taxa or others co-occuring on the same site or similar biotopes elsewhere. And, most satisfactorily, but also most rarely, sound biological understanding may allow firm conservation planning and practice with a high chance of success.

Programmes to conserve (or 'preserve') scarce Lepidoptera on individual small sites become intensive and may demand continuing intensive management – as on small isolated urban sites influenced by succession. These, and many parallels, are contexts in which the focal species is 'conservation dependent' (as for *Paralucia pyrodiscus lucida* near Melbourne, Australia, Chapter 5, p. 62). Conservation of such species inevitably has rather different emphasis from those that are wide-ranging or occur on multiple sites across landscapes, as does *O. richmondia* (Chapter 4, p. 42). Any species may need both field-based and ex situ conservation, the latter having potential to become expensive if multigenerational rearings or captive maintenance is needed. Simply clarifying the approaches, contexts and best integration of the various options available for conservation of any species continues to draw heavily on experiences with Lepidoptera accumulated over many decades.

The cited cases are clearly biased toward temperate regions with a strong tradition of natural history interest and in which conservation concern can be nurtured. As emphasised elsewhere in this book, this is not the case in many parts of the world in which needs are greatest. Whilst examples from other places, such as Japan, China, parts of south-east Asia, India and many others are cited in other chapters, many of those programmes are less detailed and less advanced, and have greater focus on obtaining initial field data for establishing conservation status and basic biological information.

References

Arnold, R.A. (1983a) Ecological studies on six endangered butterflies (Lepidoptera: Lycaenidae): island biogeography, patch dynamics, and the design of habitat preserves. *University of California Publications in Entomology*, 99. Berkeley.

Arnold, R.A. (1983b) Conservation and management of the endangered Smith's blue butterfly *Euphilotes enoptes smithi* (Lepidoptera: Lycaenidae). *Journal of Research on the Lepidoptera* 22, 135–153.

Barkham, P. (2010) The Butterfly Isles. A summer in search of our emperors and admirals. Granta Publications, London.

Black, S.H. & Jepsen, S. (2007) The Xerces Society: 36 years of butterfly conservation. *News of the Lepidopterists' Society* 49, 112–115.

Burbidge, A.A. (1996) Essentials of a good recovery plan. pp. 55–62 in Stephens, S. & Maxwell, S. (eds) Back from the Brink: Refining the Threatened Species Recovery Process. Surrey Beatty & Sons, Chipping Norton, New South Wales.

Cambell, C.L. & Ishii, L.M. (1993) Larval host plant testing of *Tinostoma smaragditis* (Lepidoptera: Sphingidae), the Fabulous Green Sphinx of Kauai. *Proceedings of the Hawaiian Entomological Society* 32, 83–90.

Douglas, F. (2004) A dedicated reserve for conservation of two species of *Synemon* (Lepidoptera: Castniidae) in Australia. *Journal of Insect Conservation* 8, 221–228.

DSE (Department of Sustainability and Environment) (2011) Sub-regional Species Strategy for the Golden Sun moth – draft for public consultation. Melbourne, Victoria.

Duffey, E. (1968) Ecological studies on the large copper butterfly, *Lycaena dispar batavus*, at Woodwalton Fen National Nature Reserve. *Journal of Applied Ecology* 5, 69–96.

Duffey, E. (1977) The reestablishment of the large copper butterfly *Lycaena dispar* (Haw.) *batavus* (Obth.) at Woodwalton Fen National Nature Reserve, Huntingdonshire, England 1969–73. *Biological Conservation* 12, 143–158.

Freese, A., Benes, J., Bolz, R. et al. (2006) Habitat use of the endangered butterfly *Euphydryas maturna* and forestry in Central Europe. *Animal Conservation* 9, 386–397.

Gibson, L & New, T.R. (2007) Problems in studying populations of the Golden sun-moth, Synemon plana (Lepidoptera: Castniidae) in south-eastern Australia. *Journal of Insect Conservation* 11, 309–313.

Heddle, M.L., Wood, K.R., Asquith, A. & Gillespie, R.G. (2000) Conservation status and research on the Fabulous Green Sphinx of Kaua'i, *Tinostoma smaragditis* (Lepidoptera: Sphingidae), including checklists of the vascular plants of the Diverse Mesic Forests of Kaua'i, Hawai'i. *Pacific Science* 54, 1–9.

Joy, J. & Pullin, A.S. (1997) The effects of flooding on the survival and behaviour of the overwintering large heath butterfly *Coenonympha tullia* larvae. *Biological Conservation* 82, 61–66.

Joy, J. & Pullin, A.S. (1999) Field studies on flooding and survival of overwintering large heath butterfly *Coenonympha tullia* larvae on Fenn's and Whixall Mosses in Shrophire and Wrexham, U.K.. *Ecological Entomology* 24, 426–439.

Maes, D., Vanreusal, W., Talloen, W. & Van Dyck, H. (2004) Functional conservation units for the endangered Alcon Blue butterfly *Maculinea alcon* in Belgium (Lepidoptera: Lycaenidae). *Biological Conservation* 120, 229–241.

Martin, L.A. & Pullin, A.S. (2004a) Host-plant specialisation and habitat restriction in an endangered insect, *Lycaena dispar batavus* (Lepidoptera: Lycaenidae): I. Larval feeding and oviposition preferences. *European Journal of Entomology* 101, 51–56.

Martin, L.A. & Pullin, A.S. (2004b) Host-plant specialisation and habitat restriction in an endangered insect, *Lycaena dispar batavus* (Lepidoptera: Lycaenidae): II. Larval survival on alternative hosts in the field. *European Journal of Entomology* 101, 57–62.

Mattoni, R.H.T. (1992) The endangered El Segundo blue butterfly. *Journal of Research on the Lepidoptera* 29 (1990), 277–304.

McLean, I.F.G. & Key, R.S. (2012) A history of invertebrate conservation in the British statutory conservation agencies. pp. 41–74 in New, T.R. (ed.) Insect Conservation: Past, Present and Prospects. Springer, Dordrecht.

Mouquet, N., Belrose, V., Thomas, J.A., Elmes, G.W., Clarke, R.T. & Hochberg, M.E. (2005) Conserving community modules: a case study of the endangered lycaenid butterfly *Maculinea alcon*. *Ecology* **86**, 3160–3173.

New, T.R. (ed.) (1993) Conservation biology of Lycaenidae (Butterflies). Occasional Paper of the IUCN Species Survival Commission, no 8. IUCN, Gland.

New, T.R. (1997a) Butterfly Conservation (2nd edn). Oxford University Press, Melbourne.

New, T.R. (1997b) Are Lepidoptera an effective 'umbrella group' for biodiversity conservation? *Journal of Insect Conservation* **1**, 5–12.

New, T.R. (2004) Moths (Insecta: Lepidoptera) and conservation: background and perspective. *Journal of Insect Conservation* **8**, 79–94.

New, T.R. (2009) Insect Species Conservation. Cambridge University Press, Cambridge.

New, T.R. (2012a) The Golden sun-moth, *Synemon plana* Walker (Castniidae): continuing conservation ambiguity in Victoria. *Victorian Naturalist* **129**, 109–113.

Nicholls, C.N. & Pullin, A.S. (2000) A comparison of larval survivorship in wild and introduced populations of the large copper butterfly (*Lycaena dispar batavus*). *Biological Conservation* **93**, 349–358.

Nicholls, C.N. & Pullin, A.S. (2003) The effects of flooding on survivorship in overwintering larvae of the large copper butterfly *Lycaena dispar batavus* (Lepidoptera: Lycaenidae), and its possible implications for restoration management. *European Journal of Entomology* **100**, 65–72.

Orr, A.G. (1994) Inbreeding depression in Australian butterflies: some implications for conservation. *Memoirs of the Queensland Museum* **36**, 179–184.

Pullin, A.S. (ed.) (1995) Ecology and Conservation of Butterflies. Chapman & Hall, London.

Pullin, A.S., McLean, I.F.G. & Webb, N.R. (1995) Ecology and conservation of *Lycaena dispar* – British and European perspective. pp. 150–164 in Pullin, A.S. (ed.) Ecology and Conservation of Butterflies. Chapman & Hall, London.

Robinson, G.S., Ackery, P.R., Kitching, I.J., Beccaloni, G.W. & Hernandez, L.M. (2001) Hostplants of the Moth and Butterfly Caterpillars of the Oriental Region. Southdene, Kuala Lumpur.

Rubinoff, D. & San Jose, M. (2010) Life history and host range of Hawaii's endangered Blackburn's sphinx moth (*Manduca blackburni* Butler). *Proceedings of the Hawaiian Entomological Society* **42**, 53–59.

Sands, D.P.A. & New, T.R. (2002) The Action Plan for Australian Butterflies. Environment Australia, Canberra.

Sands, D.P.A. & New, T.R. (2013) Conservation of the Richmond Birdwing Butterfly in Australia. Springer, Dordrecht (in press).

Sands, D.P.A., Scott, S.E. & Moffat, R.(1997) The threatened Richmond birdwing butterfly (*Ornithoptera richmondia* [Gray]): a community conservation project. *Memoirs of the Museum of Victoria* **56**, 449–453.

Sapphos (Sapphos Envirohmental Inc.) (2005) Appendix. LAX Master Plan Final EIS. A-3c. Los Angeles/El Segundo Dunes habitat restoration plan. Sapphos Environmental Inc., Pasadena, California.

Severns, P.M., Boldt, L. & Villegas, S. (2006) Conserving a wetland butterfly: quantifying early life-stage survival through seasonal flooding, adult nectar, and habitat preference. *Journal of Insect Conservation* **10**, 361–370.

Silberbauer, L.X. & Britton, D.R. (1999) Holiday houses or habitat: conservation of the Brenton blue butterfly, *Orachrysops niobe* (Trimen) (Lepidoptera: Lycaenidae), in Knysna, South Africa. *Transactions of the Royal Zoological Society of New South Wales* (1999), 394–397.

Sommers, L.A. & Nye, C. (1994) Status, research and management of the Karner blue butterfly in New York. pp. 129–134 in Andow, D.A., Baker, R. & Lane, C. (eds) Karner Blue Butterfly: a Symbol of a Vanishing Landscape. Minnesota Agricultural Experiment Station, University of Minnesota – St Paul. Miscellaneous Publication 84–1994.

Steencamp, C. & Stein, R. (1999) The Brenton Blue Saga. A case study of South African biodiversity conservation. Endangered Wildlife Trust, Parkview.

Thomas, J.A., Simcox, D.J. & Clarke, R.T. (2009) Successful conservation of a threatened *Maculinea* butterfly. *Science* **325**, 80–83.

Tremewan, W.G. (1996) The history of *Zygaena viciae anglica* Reiss (Lep., Zygaenidae) in the New Forest. *Entomologists' Gazette* **17**, 187–211.

USFWS (United States Fish and Wildlife Service) (1998) Recovery plan for the El Segundo Blue butterfly (*Euphilotes battoides allyni*). Portland, Oregon.

USFWS (United States Fish and Wildlife Service) (2005) Recovery plan for Blackburn's sphinx moth (*Manduca blackburni*), Portland, Oregon.

van Swaay C., Collins, S., Dusej, G. et al. (2012) Dos and don'ts for butterflies of the Habitats Directive of the European Union. *Nature Conservation* 1, 73–153.

Waring, P. (1993) Essex Emerald Moth *Thetidia smaragdaria maritima* (Prout 1935) – Species Recovery Plan 1993. Unpublished report to English Nature, Peterborough.

Waring, P. (2005) The history, conservation and presumed extinction of the Essex Emerald moth, *Thetidia smaragdina maritima* (Prout, 1935), in Great Britain. *Entomologists' Gazette* 56, 149–188.

Young, M.R. & Barbour, D.A. (2004) Conserving the New Forest burnet moth (*Zygaena viciae* [Denis and Schiffermueller]) in Scotland; responses to grazing reduction and consequent vegetation changes. *Journal of Insect Conservation* 8, 137–148.

10

Ex Situ Conservation

Introduction: Contexts and needs

The accumulated and extensive hobbyist experiences with rearing Lepidoptera in captive conditions, and the popularity of exhibits such as 'butterfly houses', constitute a unique basis for pursuing ex situ conservation. The information available on husbandry and maintenance techniques is far more detailed for Lepidoptera than the commonly very limited data on most other invertebrates – with the exceptions of some species that are mass-reared as biological control agents or for other 'applied' purposes. Many published accounts describe the minutiae of rearing particular species of butterflies or moths, including rare species of individual conservation concern. For many threatened species, field populations are small and highly vulnerable, and salvage or otherwise informed captive breeding is seen as the only (or most probable) chance of survival. Not all such cases succeed (the Essex emerald moth, Chapter 9, is one notable example) but a number of other invertebrates (such as some species of *Partula* snails from the Pacific region) now persist only in 'captivity', having been lost in the wild. The care of captive populations is then the only way to prevent full extinction, with the long-term aim to re-introduce the species to the wild under secure conditions and from healthy captive-reared or maintained stock.

Two major contexts, other than individual interests, are common:

1 exhibition and education, with large showy butterflies paramount, but used increasingly by zoos and related institutions to incorporate and foster knowledge of species of conservation interest;
2 captive rearing for conservation, as ex situ conservation, has for some taxa become an important, sometimes the predominant, focus for conservation

in (1) protecting insects from field mortality through rearing them in conditions that exclude predators and parasitoids, so increasing numbers, (2) providing stock for release into natural environments, and (3) providing high-quality reared specimens for sale, so reducing impact on wild populations from illicit or undocumented collecting and black market trading.

Either emphasis demands considerable care and specialised knowledge, with needs to sustain healthy stock and prevent losses of species of conservation significance. Some of the lessons relevant to captive breeding of butterflies and moths flow also from the 'intermediate' scenarios of enhancing numbers in the wild through 'butterfly gardening' or 'butterfly ranching', both also important components of conservation management. Both depend on local resource enhancement, but without confining the insects. Commercial value of rare species to collectors (the 'high-value/low-volume' trading component of Collins and Morris 1985), and supply of large long-lived and showy species for the livestock trade have both been influential drivers of ranching or farming developments.

'Butterfly ranching' is, perhaps, most familiar through the pioneering exercises flowing from commercial interests in the spectacular birdwing butterflies of New Guinea, and through which naturally concentrated butterflies could be manipulated as a sustainable harvestable source of high-quality specimens for trade. In Papua New Guinea, and in conjunction with centralised trading through a government agency (the Insect Farming and Trading Agency, through which numbers could be monitored and prices standardised and regulated), the practice was based on plantings of abundant food plant vines for commercially desirable birdwings in gardens and on forest edges to 'decoy' females to lay, and from which progeny could be protected and a proportion of reared adults harvested for sale: a 'farming manual' (Parsons 1982) included full directions for planning such a garden, together with husbandry and specimen preparation measures. Caterpillars could be kept in sleeves on the vines or caged to prevent attack by natural enemies, and the specimens for sale selected carefully for high quality and, thereby, the best financial reward. As Parsons (1999) emphasised, the spectacular species of *Ornithoptera* have long been accorded a special 'mystique' associated with their appearance and distributions remote to northern hemisphere hobbyists, and have long been 'special' within Papua New Guinea, with several species recognised formally from 1966 as 'National Butterflies' – a status that evidently increased their desirability and trading prices dramatically! However, that demand then became the major influence in pursuing their conservation through ranching and associated preservation of rain forest habitats through change in economic balance for contributing participants. Adoption of the practice in less developed countries has a number of benefits, and those listed by Craven (1989) from his experiences in the Arfak Mountains of West Papua (then Irian Jaya) remain pertinent, under the wider guiding principle that community participation and inclusion of local people in decision-making and conservation implementation is important. The reasons for promoting butterfly farming were (1) providing employment in isolated rural areas, allowing participation in a cash economy without disruptive changes to lifestyles and local ecology; (2) culturally appropriate land use, drawing on local knowledge of butterflies and plants; (3) reduced clearing of forests for economic need, by providing an alternative source of income, and so having wider conservation benefit; (4) no expensive equipment or capital needed to start the exercise; (5) with understanding of needs for sustainability, might lead to easier control of exploitation of other flagship taxa such as birds of paradise; (6) easier to control illegal exploitation in the same area; and (7) reduce threat of extinction, as claimed more widely for New Guinea birdwings. Proactive guidelines, involving educational and conservation strategies combined with economic reward have helped to transform the trade from 'extractive' to 'sustainable' sources of supply.

The practice has become more widespread, emulated in parts of Asia, in particular, and leading also to further confined breeding operations in especially designed large flight cages that allow the butterflies to mate and behave normally. The cost of such cages restricts this aspect mainly to the very rarest and most commercially desirable taxa. In contrast, one advantage of the butterfly ranching/ farming procedures that became widespread in Papua New Guinea was simply the very low basic cost in establishing a viable operation (Table 10.1). The outcomes of ranching extend from provision of deadstock from New Guinea for collectors, to that of livestock for butterfly gardens and butterfly house exhibits elsewhere, as an increasingly important facet of rural livelihood in, for example, Tanzania and Cambodia (van der Heyden 2011). Indeed, some ranching/ breeding programmes combine fostering tourism with providing livestock (commonly as pupae that can be transmitted easily) for export to 'Butterfly Houses', so again providing incomes that may be associated also with reduced forest losses (Veltman 2012).

Butterfly gardening or ranching, whereby butterflies are attracted and their local abundance increased through provision of, mainly, nectar sources from deliberate plantings, is an exercise in which people can easily participate, and are encouraged to do so by numerous advice leaflets on suitable plants and plant combinations for any local area. Additional plantings of larval food plants are a valuable conservation adjunct and, whereas most such exercises are by private

Table 10.1 The main benefits of butterfly farming/ranching operations as a rural activity, as summarised for Papua New Guinea by Parsons (1999).

1. Little capital outlay is required to set up the individual operation
2. Forest clearance unnecessary; encouraged creation of new habitats
3. Wild stocks of butterflies are not depleted, and may be enhanced
4. Stock produced by farming can be better sustained, tended, monitored and protected
5. Main husbandry techniques are simple and generally easy to learn and to pass on to other people
6. Perfect specimens produced, reaching top prices and providing high financial returns
7. Deadstock is both low volume and low weight, so handling and shipping easy and economical
8. Self-sustaining stock, with rapid build-up and replenishment

individuals, they can play important roles in wider conservation exercises (Sands et al. 1997); a few have become significant tourist attractions, some in conjunction with more confined 'butterfly house' exhibits. The prolific literature on home garden manipulations to encourage butterflies has much wider implications, and many more public butterfly gardens combine commercial interest through fostering ecotourism and visitors with important conservation and documentation roles. Areas used need not be large, particularly if the site abutts natural habitat and so increases the size and 'effective resources' of local or contiguous areas. Thus, a 0.5 ha degraded forest patch within the Kerala Forest Research Institute (India) was adapted by landscaping and introducing a variety of suitable host plants (Mathew & Anto 2007). This stimulated considerable increase in butterfly richness and abundance: 69 species recorded included a number of species endemic to the Western Ghats and several listed for protection under the Indian Wildlife Act. This exercise in habitat enrichment was viewed as a prototype to help develop methods that could be used elsewhere, in conjunction with captive breeding, in India.

The early history of public exhibitions of living invertebrates, as the primary initial driver of interest and proliferating from the 1900s, was summarised by Collins (1987) and Cooper and Dombrowski (2012). Perhaps the first dedicated such exhibition was London Zoo's 'Insect House', founded in 1881. Many more recent ones have been primarily commercial, some as adjuncts to selling dead-stock, or to nurseries. However, both public education and awareness, in many cases aided by informative signage, and captive breeding and maintenance methods owe much to their influences. Roles of such exhibits in conservation have been debated extensively, and have often been secondary to commercial gain. Nevertheless, conservation awareness is high amongst the proprietor fraternity. The International Association of Butterfly Exhibitors and Suppliers (IABES) has the mission statement of 'Advancing the international butterfly exhibition industry through representation, communication, education, and marketing'. Amongst their notable conservation activities, a symposium on the conservation of the Homerus swallowtail (*Papilio homerus*, a spectacular threatened Jamaican endemic species) was held in 2010, with more than a dozen butterfly farms, from a variety of countries, pledging support for this endeavour. Activities are reported in a continuing web-based newsletter ('International Flutterings'), which provides many insights into running such enterprises. More broadly, the global butterfly house industry has major foci on butterflies imported from south-east Asia and central America, and clear definition of such activities has clear importance in conservation.

Some workers have held that the resources put into such institutions might be better used for other conservation, but such broad generalised opinions across the wide range of butterfly houses and the like, with their variety of standards and primary purpose, and across the wide variety of taxa and protocols involved are, perhaps, too bland. More discussion has occurred on the sources of exhibition material, with importation of pupae for exhibition of adult butterflies formerly widespread and in many cases without overt assurance that their collection was not contributing to over-collecting (Chapter 11) of the wild populations. Most of the more enduring institutions now rear much of their exhibition

Table 10.2 Measures considered an integral part of good stock management in butterfly houses (Cooper & Dombrowski 2012).

Careful assessment of incoming stock, preferably including treatment of eggs or pupae with hypochlorite or formaldehyde, to avoid introduction of sick or contaminated individuals

Quarantine (isolation) of incoming stock for screening before joining the main collections

Clinical, post-mortem and laboratory health monitoring of stock, both in quarantine and afterwards

Comprehensive record-keeping with regular review of data, especially analysis of peaks and troughs of morbidity, mortality and parameters such as level of reproductive success or shortened life span

---

material 'in house', but this is not always possible. The spectrum of taxa might be limited by legislation, for example; Australian institutions cannot import livestock of insects from overseas, so must rely wholly on native taxa for exhibition. Public exhibitions of Lepioptera and other invertebrates for tourist interests are widespread.

As Lees (1989) noted, production of healthy (both disease-free and genetically vigorous, not inbred) stocks is vital, whether for exhibition or release. Legal conditions of health and care arise (Cooper & Dombrowski 2012) for any institution classified as a 'zoo' (that is, for Britain, open to the public for more than 7 days a year) and these authors noted a series of important measures that should be considered essential in health management (Table 10.2). In scope and appearance, butterfly houses have progressed from relatively simple designs, often as modifications to existing structures built for other purposes, to much more elaborate purpose-built accommodation, in which climate controls are incorporated. One set of generalised 'threshold' conditions for desirable climate regimes for butterfly houses recommended maximum temperature of 26 °C, minimum relative humidity of 65% and a photoperiod of 12 hours light – 12 hours dark (Lees 1989), but for conservation, the particular needs of a species may need to be incorporated carefully, perhaps within separate quarantined enclosures. As at the Melbourne Zoological Gardens, separate nursery/glasshouse facilities may be needed to propagate and maintain healthy stocks of food plants and 'behind the scenes' butterfly-rearing operations to provide adults for release into an exhibition hall.

Lepidoptera in captivity

Taking an insect from the wild to captive conditions, even on a temporary holding basis, should be accompanied by careful investigation of the health of the animals. Cooper (2012) recommended four points in a general protocol as (1) quarantining new stock before they are mixed with any other, and where possible treating any eggs or pupae with hypochlorite; (2) clinical, post-mortem and laboratory health monitoring of stock, both whilst in quarantine and later;

(3) isolation and/or culling of individuals or groups that show signs of disease; and (4) comprehensive record keeping, with a variety of applications. In short, health care is needed as an integral component of managing captive stocks of Lepidoptera, and Cooper's essay is a useful introduction to many of the husbandry problems that arise. In an accompanying account on butterfly house maintenance, Cooper and Dombrowski (2012) noted the twin components of concern as infectious diseases and non-infective conditions. The latter includes impacts of suboptimal regimes of temperature and humidity, nutrition and others that reduce 'fitness' and thus may also increase susceptibility to diseases or their impacts. Overcrowding and poor hygiene are often implicated in disease onset.

Effects of captive conditions can be severe, with mortality and loss of 'quality' reported frequently. In view of increasing advocacy for its place in conservation – Schultz et al. (2008) noted that captive breeding is recommended for about half of the threatened or endangered butterflies of the United States, for example – refinement of husbandry approaches for avoiding such declines is clearly important, not least as 'quality control' to assure that stock proposed for subsequent release is of equivalent fitness to normal 'wild' individuals.

Trials with the Puget blue butterfly (*Icaricia icarioides blackmorei*) illustrate some of the practical problems that can arise. Whether derived from eggs collected in the wild or from eggs laid by confined females, survival to adults in captivity was less than 10%, with resulting adults lighter and smaller than wild stock (Schultz et al. 2009), so probably leading to reduced dispersal prowess and lowered fecundity. In cases such as this, uses of captive stock for augmentation of field incidence may be limited, and alternatives (such as direct translocations through which similar trends are not induced) may be preferable.

Causes of captive declines are complex, and range from inadequate nutrition or other resource quality, undetected disease, or overlooking or unawareness of some utility or climatic need, to genetic deterioration.

Inbreeding

In addition to the concerns it generates for small wild populations (Chapter 9, p. 174), inbreeding depression is suggested frequently as a cause of deterioration of captive-reared insect stocks, and may be a serious concern in taxa of conservation interest, simply because the initial stock used to found the captive population is likely to be small – often from a single founder parent – and represent a limited gene pool. Effects manifest over time as either increased mortality or decreased 'performance', with early stages and reproductive processes those most affected. Orr (1994), in noting that inbreeding depression had been reported only infrequently in butterflies (despite extensive breeding programmes, some over many generations), listed indicative traits such as reduced size and fecundity of adults, retarded larval development and failures to pupate or eclose successfully. In general, inbreeding depression has been suggested to be most likely in wide-ranging species that habitually outbreed and need larger areas of habitat to allow this; such species may also be more sensitive to habitat fragmentation. Inbreeding depression is potentially severe in some Australian Papilionidae (Orr 1994), being detected in three species (*Cressida cressida, Ornithoptera rich-*

mondia, Papilio aegeus). The first two, studied in more detail, exhibited reduced rates of egg hatch and larval survival with inbreeding, together with protracted larval development. Orr's suggestion that troidines may be especially susceptible to inbreeding has important implications in view of the emphasis on these spectacular insects in captive-rearing programmes for conservation or display. *O. richmondia* declines in the wild may also reflect inbreeding – eggs of three of 11 wild females assessed by Orr had markedly reduced hatching rate in relation to the other individuals. Inbreeding effects can appear within very few generations, leading to wide suggestion that captive stocks should ideally not be sustained for more than a few generations without some attempt to counter possible deterioration, by outcrossing.

Captive-reared stock for release of the Swallowtail, *Papilio machaon*, at Wicken Fen, England, resulted from hand-pairing to reduce inbreeding (Dempster & Hall 1980). Nevertheless, numbers released were far lower than anticipated (due to losses from frost) but an estimated 20,000 eggs were laid. Following early trends of increase, the population crashed as a result of drought, and failed to recover. This was one of the few early studies in which inbreeding depression was considered. No evidence of this was found, despite low initial release numbers, and stochastic loss following reduction to very low numbers in the field was probably the cause of extinction.

Pathogens

Incidence of diseases or other infections is a persistent concern in any captive maintenance or breeding exercise, and demands high standards of monitoring and hygiene to detect and counter. Not all are immediately obvious; latent viruses may not appear for several generations of putative host 'stress' for example, but might then have serious effects on survival or performance, such as by inducing sterility or loss of fecundity. Many are difficult to diagnose, but lepidopterists are very familiar with incidence of polyhedrosis viruses in breeding stock – and similar pathogens are used deliberately in management of lepidopteran pests on crops, with possibility of 'spill-over' to neighbouring habitats and taxa.

Another North American lycaenid example revealed a less well known context of concern, involving endophytic bacteria, exemplified by *Wolbachia*, that parasitise a wide variety of insect species and are 'an under-appreciated additional threat' to many threatened species (Nice et al. 2009). *Wolbachia* has a number of impacts, such as inducing parthenogenesis, killing the sons of infected mothers and cytoplasmic incompatibility. The last of these has been documented most frequently, with concerns that introduction of such infected individuals into an uninfected population reduces effective population size (through changes in effective sex ratio) and may increase chances of stochastic extinction in already small populations. Few such cases have been studied directly in conservation exercises, and consideration of the Karner blue (*Lycaeides melissa samuelis*) has much wider relevance. Screening populations for *Wolbachia* (Nice et al. 2009) demonstrated that the butterflies throughout the whole range to the west of Lake Michigan (Fig. 10.1) appeared to be infected. These included many of

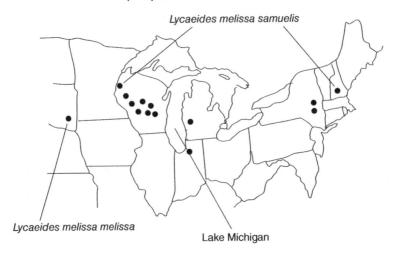

Fig. 10.1 Map of *Lycaeides* sampling localities for surveying presence and impacts of *Wolbachia* (Nice et al. 2009. Reproduced with permission of Elsevier.). Isolated western population is *L. m. melissa*, all others are *L. m. samuelis*. All populations to west of Lake Michigan (n individuals: 26/29 *melissa*,120/120 *samuelis*) infected with *Wolbachia* strain 162; only 1/71 *samuelis* from more easterly populations with *Wolbachia*, a different strain, SS8).

the largest and nominally healthiest populations known. The main conservation relevance was seen, not as a direct threat to those populations, but because those populations are believed to be the most likely candidates from which to initiate captive breeding. Through this, the chances of introducing *Wolbachia* to uninfected populations further east appeared high. Those potentially vulnerable populations are already the most threatened, and Nice et al. recommended screening of any potential introduced stock for *Wolbachia* (and perhaps other endosymbionts) to provide an objective base for deciding whether to proceed in the view of potentially adding further threat.

Translocations and quality control

Movement of species to additional sites, either as re-introductions or augmentations, has been a frequent tactic in butterfly conservation. However, lack of knowledge of the species' requirements has been associated with numerous failures: Pullin (1996) noted that as few as 8% of 323 re-introduction attempts in Britain had succeeded, and reasons for many such failures are complex to ascertain. Much of the rationale of re-introductions and translocations has been summarised by Oates and Warren (1990, Britain) and Schultz et al. (2008, comparing Britain and North America). The latter revealed some practical differences in trends – such as the British tendency to release adults, whilst North American projects often incorporated also pupae and/or larvae. However, the rationale of preferring adults for release, in cases where captive stock was not available and often thus for direct field translocations, was summarised clearly in a Recovery

Plan for the Mission Blue Butterfly (Wayne et al. 2009) in the United States, as 'Eggs, larvae and pupae are difficult to find, collect, and relocate compared with adults'. Reasons for failures in various cases included poor planning, insufficient understanding of the species' biology, inadequate quality of the receptor site/s and very low numbers of individuals used. Many cases were documented inadequately, if at all, and Schultz et al. (2008) emphasised the major practical needs as: (1) increasing systematic recording of efforts made, including number and stages of individuals, source population size and total numbers released over time; (2) enhancing monitoring and experimental design; (3) incorporating ecological modelling; and (4) improving species-specific understanding.

For many re-introduction attempts, simply obtaining donor stock for translocation or release may be a major constraint, and either small numbers from the field or bred from limited parental stocks implying limited genetic variation that could lead to later inbreeding effects. Potential donor populations may become more vulnerable from any unplanned harvest, and their sustainability must be considered carefully. Thus, in planning translocation of caterpillars of the Netted carpet (*Eustroma reticulatum*, Geometridae), reportedly one of the rarest moths in the United Kingdom, to a Lake District site, only a single donor site was identified. This moth has intermittently been regarded as extinct in Britain (Hatcher & Alexander 1993) through undergoing periods of very low abundance, apparently linked to rainfall patterns, and both it and its sole larval food plant (*Impatiens noli-tangere*, touch-me-not balsam) have been lost from many sites. Major threats to sites included increasing shade, choking of balsam by ground vegetation and decreasing soil moisture, with sites becoming too small and too dry to sustain moth populations – so that natural drought is a major threat. For their planned translocation, Hooson and Haw (2008) noted agreement that no more than 15% of larvae found at the donor site would be collected, and that a minimum of 20 and maximum of 30 could be taken; and if fewer larvae were found there than those estimated in the previous season, the attempt would not proceed. Thirty late instar caterpillars were translocated, but no progeny were found at the release site the following year, and the exercise was repeated with the increased total of 40 caterpillars, but to a somewhat more suitable nearby site. Four caterpillars were found there the following year, and monitoring continues. Such 'bottlenecks' of very small numbers are not unusual.

The number of female individuals may be of concern in future bottlenecks and inbreeding depression. In re-introducing *Maculinea nausithous* and *M. teleius* to the Netherlands from Polish stock in 1990, Wynhoff (1998, 2001) was able to obtain reasonably large numbers of adults: *M. nausithous* (22 males and 48 mated females) and *M. teleius* (33 males, 53 mated females), and numbers were monitored in each succeeding year. Loss of genetic variation was very low, and the source populations themselves also showed low genetic variability – a feature that Wynhoff suggested might render them susceptible to stochastic events, and also supporting the creation of metapopulations to compensate for any such local losses. Reflecting the dispersal distances possible, population units for *M. nausithous* should then not be more than 1 km apart, and those for *M. teleius*, within 500 m of each other. The difficulties of achieving this fine-scale patch network within a highly fragmented agricultural landscape were recognised, and would involve cooperation from owners of non-reserved lands.

The timing of relocation attempts may also be influential. Translocations of the Atala (*Eumaeus atala*, Lycaenidae) in Florida have been a key to conserving this butterfly, earlier believed to be extinct (Emmel & Minno 1993) but now out of danger. Early translocations contributing to this success included use of 'trap plants' (of the cycad *Zamia pumila*) on which females laid, after which the plants were transported elsewhere, and releases of adults. Later (Smith 2002) translocations of second instar caterpillars showed a marked seasonal effect on survival to pupation, with considerably higher survival (55.6%) in the wetter summer than (33.2%) in the drier winter. As a further methodological advance, Smith showed survival was also higher if caterpillars were protected by netting the individual *Zamia* plants on which they were released.

This principle of 'soft release', rather than immediate exposure to the wider environment, for protection and monitoring of outcome is recommended widely, particularly when only small numbers are available. In some reported cases, this protection is clearly needed; USFWS (1999) noted that, of 760 pupae of the Schaus swallowtail released in 1995 on seven protected sites, predation by birds was estimated to destroy 85–90%. Later releases of adults, in contrast, were successful.

Assisted colonisation

The reality of climate change inducing range changes in Lepidoptera and numerous other taxa (Chapter 12) links with the equal reality that many taxa may be 'outpaced' by such changes and be unable to naturally colonise new areas, or track resources in the wider landscape, because of poor dispersal prowess. Maintaining mutualisms or close ecological associations poses complex problems in management, and any 'decoupling', such as by differential responses to climate change or spatial separation, increases vulnerability. Long-term site selection and preparation may be needed to conserve conservation modules in expanded ranges. Natural colonisation may be unreliable. Thomas (2011) has argued that the only viable option to conserve wild populations of narrowly endemic species for which natural dispersal is impossible is to translocate them to newly suitable areas. He suggested that artificially increasing the dispersal capacity of endangered species might represent the most effective strategy to counter impacts of climate change. Such 'assisted colonisation' or 'assisted migration' (Hunter 2007) has been investigated for butterflies in a number of contexts, and much depends on the suitability of the receptor sites and the ecological specialisations of the species used. It is one of a portfolio of possible management steps that must be evaluated within the context of the species' ecology and needs (Chapter 12).

Experimental trials to investigate whether populations can survive beyond their current species' climate range are still quite rare (Menendez et al. 2006), and others have probed whether sites broadly within the range of climatic suitability but beyond current distribution can be used. An instructive example involving a skipper butterfly, *Atalopedes campestris*, was noted earlier (Chapter

7, p. 130). However, selection of sites for possible future need is likely to involve the politically challenging milieu of reserving sites that at present are neither inhabited nor habitable by the species intended for conservation. As McLachlan et al. (2007) forewarned, selection of the most deserving candidates for such intensive consideration will be difficult, and any suggestions lead to debate over their relative merits and suitability.

Assisted range expansions for two grassland butterflies, the Marbled white (*Melanargia galathea*) and the Small skipper (*Thymelicus sylvestris*) in northern England (Willis et al. 2009) involved translocating adult butterflies (about 500 Marbled whites and about 400 Small skippers), releasing them the day after capture in sites presumed to be climatically suitable but north of the current range (Fig. 10.2). Numbers of individuals far exceeded those usually available for direct translocations, and populations were monitored annually during the flight season of each species. Both were still present 8 years (generations) after introduction, and had increased in distribution. In this example, costs of the assisted colonisation were minimal, largely because direct transfer obviated need for any captive breeding programme to build up numbers for this purpose (Willis et al. 2009).

More adventurous scenarios have also been advanced – for example, the possibility of re-establishing in Britain two species of butterflies widespread in Europe but which became extinct in Britain early in the twentieth century. Climatic modelling indicated that assisted colonisation to Britain of the Black-veined

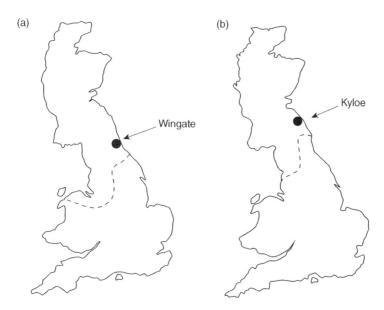

Fig. 10.2 Outline map of the United Kingdom to indicate position (arrowed) of introduction sites during experimental range expansions of two butterflies in relation to recorded northernmost distribution as at 2001 (dashed line): (a) *Melanargia galathea* at Wingate Quarry, Co. Durham; (b) *Thymelicus sylvestris* at Kyloe Quarry, Northumbria (Source: Willis et al. 2009. Reproduced with permission of John Wiley & Sons.).

white (*Aporia crataegi*) and the Mazarine blue (*Polyommatus semiargus*), both of which are declining in Europe, might be viable (Carroll et al. 2009), and helpful also in indicating the suitability of Britain for other European taxa. Thomas (2011) later nominated the Provence chalkhill blue (*Polyommatus hispanus*) and de Pruner's ringlet (*Erebia triaria*), both now threatened by climate changes further south in Europe, amongst a range of other non-British taxa as possible relocation candidates – but also emphasised that all such trans-locations should be within broad geographic regions, and that receptor sites should lack local endemics that might be susceptible to additional species being introduced.

With 'barriers' such as the English Channel and North Sea, most European species of conservation concern are unlikely to undertake such moves to Britain unaided. Nevertheless, such introductions of species that have never been natural residents in the area of introduction will continue to be viewed in very varied ways. Both the ethics and practicality of introducing what are in effect 'alien species' with unknown future impacts in the receptor environments are complex issues that induce strong reactions, negative or positive, from people. Both natural and induced distributional changes of Lepidoptera flowing from climate change are likely to challenge conservationists' attitudes considerably over the next few decades.

Not all human-induced range changes for Lepidoptera are planned for con-servation, and many are undocumented, adding to the confusion of detecting natural trends. Repeated calls for all releases to be documented are difficult, if not impossible, to promote, and outcomes can include possibilities of mixing different genetic stocks and increasing distributional anomalies from specimens that are reported. Contexts include the casual discard of surplus reared material from hobbyists and escapes from captive stock (both outside the natural range). Garnering more attention recently has been the increasing practice of releasing butterflies on ceremonial occasions, predominantly weddings, with particular concerns over the predominant large showy species used in this industry, the Monarch, *Danaus plexippus*. New (2008) and Pyle (2010) demonstrated the concerns. In North America, *D. plexippus* has distinct 'western' and 'eastern' population concentrations using different overwintering sites (in California and Mexico, respectively), and Pyle noted that casual releases at weddings may have conservation implications in that celebrants may be 'helping to undermine our ability to correctly interpret the responses of wild monarchs to all the challenges they face'. The practice will remain controversial, with persistent need also to screen release stocks for disease.

References

Carroll, M.J., Anderson, B.J., Brereton, T.M., Knight, S.J., Kudrna, O. & Thomas, C.D. (2009) Climate change and translocations: the potential to re-establish two regionally-extinct butterfly species in Britain. *Biological Conservation* **142**, 2114–2121.

Collins, N.M. (1987) Butterfly Houses in Britain – Conservation Implications. IUCN, Gland and Cambridge.

Collins, N.M. & Morris, M.G. (1985) Threatened Swallowtail Butterflies of the World. IUCN, Gland and Cambridge.

Cooper, J.E. (2012) Insects. pp. 267–283 in Lewbart, G.A. (ed.) Invertebrate Medicine (2nd edn). John Wiley & Sons, Ames, Iowa.

Cooper, J.E. & Dombrowski, D.S. (2012) Butterfly houses. pp. 323–333 in Lewbart, G.A. (ed.) Invertebrate Medicine (2nd edn). John Wiley & Sons, Ames, Iowa.

Craven, I. (1989) The Arfak Mountains Nature Reserve, Birds Head region, Irian Jaya, Indonesia. *Science in New Guinea* 15, 47–56.

Dempster, J.P. & Hall, M. L. (1980) An attempt at re-establishing the swallowtail butterfly at Wicken Fen. *Ecological Entomology* 5, 327–334.

Emmel, T.C. & Minno, M.C. (1993) The Atala, *Eumaeus atala florida*. pp. 129–130 in New, T.R. (ed.) Conservation Biology of Lycaenidae (Butterflies). Occasional Paper of the IUCN Species Survival Commission, no 8. IUCN, Gland.

Hatcher, P.E. & Alexander, K.N.A. (1993) The status and conservation of the netted carpet *Eustroma reticulatum* (Denis & Schiffermueller, 1775) (Lepidoptera: Geometridae), a threatened moth species in Britain. *Biological Conservation* 67, 41–47.

Hooson, J. & Haw, K. (2008) Reintroduction of the netted carpet moth *Eustroma reticulatum* to Derwentwater, The Lake District, Cumbria, England. *Conservation Evidence* 5, 80–82.

Hunter, M.L. (2007) Climate change and moving species: furthering the debate on assisted colonization. *Conservation Biology* 21, 1356–1358.

Lees, D. (1989) Practical considerations and techniques in the captive breeding of insects for conservation purposes. *Entomologist* 108, 77–96.

Mathew, G. & Anto, M. (2007) *In situ* conservation of butterflies through establishment of butterfly gardens: a case study at Peechi, Kerala, India. *Current Science* 93, 337–347.

McLachlan, J.S., Hellmann, J.J. & Schwarz, M.W. (2007) A framework for debate of assisted migration in an era of climate changes. *Conservation Biology* 21, 297–302.

Menendez, R., Gonzalez-Megias, A., Hill, J.K. et al. (2006) Species richness changes lag behind climate change. *Proceedings of the Royal Society, series B* 273, 1465–1470.

New, T.R. (2008) Are butterfly releases at weddings a conservation concern or opportunity? *Journal of Insect Conservation* 12, 93–95.

Nice, C.C., Gompert, Z., Forister, M.L. & Fordyce, J.A. (2009) An unseen foe in arthropod conservation efforts: the case of *Wolbachia* infections in the Karner blue butterfly. *Biological Conservation* 142, 3137–3146.

Oates, M.R. & Warren, M.S. (1990) A review of butterfly introductions in Britain and Ireland. Joint Committee for the Conservation of British Insects and World Wildlife Fund, Godalming.

Orr, A.G. (1994) Inbreeding depression in Australian butterflies: some implications for conservation. *Memoirs of the Queensland Museum* 36, 179–184.

Parsons, M.J. (1982) Butterfly Farming Manual. Insect Farming and Trading Agency, Bulolo, Papua New Guinea.

Parsons, M.J. (1999) The Butterflies of Papua New Guinea. Their Systematics and Biology. Academic Press, London.

Pullin, A.S. (1996) Restoration of butterfly populations in Britain. *Restoration Ecology* 4, 71–81.

Pyle, R.M. (2010) Under their own steam: the biogeographical case against butterfly releases. *News of the Lepidopterists' Society* 52, 54–57.

Sands, D.P.A., Scott, S.E. & Moffat, R. (1997) The threatened Richmond birdwing butterfly (*Ornithoptera richmondia* [Gray]): a community conservation project. *Memoirs of the Museum of Victoria* 56, 449–453.

Schultz, C.B., Dzurisin, J.D. & Russell, C. (2009) Captive rearing of Puget blue butterflies (*Icaricia icarioides blackmorei*) and implications for conservation. *Journal of Insect Conservation* 13, 309–315.

Schultz, C.B., Russell, C. & Wynn, L. (2008) Restoration, reintroduction and captive propagation for at-risk butterflies: a review of British and American conservation efforts. *Israel Journal of Ecology and Evolution* 54, 41–61.

Smith, E.M. (2002) The effects of season, host plant protection, and ant predators on the survival of *Eumaeus atala* (Lycaenidae) in re-establishments. *Journal of the Lepidopterists' Society* 56, 272-276.

Thomas, C.D. (2011) Translocation of species, climate change, and the end of trying to recreate past ecological communities. *Trends in Ecology and Evolution* 26, 216–221.

USFWS (United States Fish and Wildlife Service) (1999) Schaus swallowtail Butterfly Heraclides aristo-demus ponceanus. pp. 4.743–4.766 in Multi-Species Recovery Plan for South Florida. Atlanta, Georgia.

van der Heyden, T. (2011) Local and effective: two projects of butterfly farming in Cambodia and Tanzania (Insecta: Lepidoptera). *SHILAP Revista Lepidopterologia* **39**, 267–270.

Veltman, K. (2012) Butterfly conservation, butterfly ranches and insectariums: generating income whilst promoting social and environmental justice. pp. 189–197 in Lemelin, R.H. (ed.) The Management of Insects in Recreation and Tourism. Cambridge University Press, Cambridge.

Wayne, L., Weiss, S.B. & Niederer, C. (2009) Recovery Action Plan for the Mission Blue Butterfly (*Icaricia icarioides missionensis*) at Twin Peaks Natural Area. San Franciosco Recreation and Park Department/Creekside Center for Earth Observation (Report prepared for United States Fish and Wildlife Service)

Willis, S.G., Hill, J.K., Thomas, C.D. et al. (2009) Assisted colonization in a changing climate: a test-study using two U.K. butterflies. *Conservation Letters* **2**, 45–51.

Wynhoff, I. (1998) Lessons from the reintroduction of *Maculinea teleius* and M. *nausithous* in the Netherlands. *Journal of Insect Conservation* **2**, 47–57.

Wynhoff, I. (2001) At home on foreign meadows. The reintroduction of two *Maculinea* butterfly species. Ph. D. Thesis. University of Wageningen, the Netherlands.

11

Lepidoptera and Protective Legislation

Introduction

The legal designation of animal or plant species as 'endangered', 'threatened' or 'protected' in some way is widespread, and is seen by many people as a core activity in conservation. Such designations may both impose formal obligations for investigation and action to institute practical conservation measures, and render such 'listed' species eligible or a priority for the limited logistic support available for such activities. In many places, such formal recognition of need is essential, as a 'passport' to government agency attention or support and without which any such support is excluded. Together with fungi and lower plants, invertebrates as a whole are vastly under-represented on such schedules in relation to their very high diversity, with many groups absent or some others tokenly included, notwithstanding their evolutionary and ecological interest and parlous survival potential; most 'protected species' are of the more popular and taxonomically tangible vertebrates and vascular plants. Many differing viewpoints attend the values of such listings – they draw attention to conservation need, and can be powerful and persuasive tools for advocacy in demonstrating the scale and extent of these needs to managers and decision-makers influencing ecosystem use and security. Thus, the inclusion of many moth species in the United Kingdom Biodiversity Action Plan listings had 'enormous benefits for moth conservation, raising awareness of threatened species, stimulating habitat management and generating funds from government for surveys, monitoring, ecological studies and direct conservation action' (Fox et al. 2013). Unfortunately, such reactions and outcomes are not global.

Long lists also demonstrate the impracticality of dealing at an individual species level with all the taxa listed, and so impose further selection to determine

Lepidoptera and Conservation, First Edition. T.R. New.
© 2014 John Wiley & Sons, Ltd. Published 2014 by John Wiley & Sons, Ltd.

the 'most deserving' or 'most meritorious' within schedules of species, whilst inevitably neglecting others, perhaps equally needy, and with selection criteria often uncertain and contentious. Subjectivity (or perceived subjectivity) in status assessment frequently leads to conflicting and strongly held views over whether individual taxa merit formal recognition in this way. 'Listing' is not an end in itself, as sometimes seen, but a step that may facilitate practical conservation and draw attention to the urgent need for this. However, there remains a strong public perception that, in general, 'listing is conservation', and that legal protection is automatically beneficial. Ball (2012) cited the example of 16 South African butterfly taxa listed for protection in Cape Province. Two appear to have become extinct since they were listed, because of habitat destruction, and only five are threatened. Half of the original 16 have no currently perceived threats to their wellbeing, and no practical conservation management or organised data accumulation or monitoring has been undertaken. Many authors have pointed out that legislation tends to highlight prohibitions on collecting (later in this chapter) whilst largely ignoring the far more urgent issues of habitat protection. Ball (2012) summarised the desirable requirements of such legislation in the context of Lepidoptera as that it should be 'appropriate, reasonable, based on good data, enforceable, and should stifle neither research nor interest'. Credibility is critical.

Protective legislation, however well intentioned, can have serious deficiencies in practice – and the intricate biology of many 'protected insects' may mean that those difficulties cannot be appreciated until they are experienced at first hand. Beale (1997) examined the impact of Queensland (Australia) conservation legislation on a protected localised lycaenid, Illidge's ant-blue (*Acrodipsas illidgei*), a myrmecophilous species associated with coastal mangroves threatened by development. He believed that the legislation actually hindered practical conservation, because of deterring amateur interest for fears of penalty, and probably had no effect on the extent of any collecting; in effect the legislation discouraged the study and monitoring of the butterfly and created an atmosphere of distrust between the parties involved. The regulations of permits that could be issued for study appeared 'virtually impossible to enforce'. As other commentators in Australia have also done (Sands & New 2002), Beale suggested a number of species-specific changes to foster constructive progress and discussion (Table 11.1), elements of which are frequently reiterated in 'codes of conduct' and similar documents, such as JCCBI (1987, updated to 2008: Invertebrate Link 2008). Whilst some of the matters raised by Beale may seem excessive in detail, they reflect the disquiet of many lepidopterists in Australia over the restrictions and penalties imposed by legislation and the then widespread feeling of their becoming the victims of a 'witch-hunt' by over-zealous authority. The legislation hampered accumulation and dissemination of information fundamental to conservation assessment and management. However well intentioned, any such regulation may, in Beale's words, raise 'many difficulties which are not fully realized until they have been experienced first hand', and effectively discourage study and monitoring of any listed species.

Lepidoptera are by far the most frequent insect entries on lists of 'protected invertebrate species' from many parts of the world, but the real conservation

Table 11.1 Suggested specific and wider changes needed in conservation legislation in Queensland, Australia to benefit conservation study of *Acrodipsas illidgei* (Adapted from Beale 1997, see text).

General points

Conservation laws that allow new information on *A. illidgei* to be made available (published) without fear of prosecution

Conservation status and form of legal protection of butterflies to be based on recommendations from at least three experts regarded as having extensive knowledge and first-hand experience of the species in question. Recommendations to be reviewed by at least two other recognised experts. A committee that includes scientists, experienced lepidopterists and those associated with commercial insect breeding or trading could be formed to recommend management strategy

Collecting of adults without permits to be allowed at locations which require confirmation or were considered extinct, and at any new locations (excluding National Parks)

Permits not to be required for collecting, keeping or moving adults at identified habitat locations (i.e. species confirmed within the previous 5 years) once every 5 years where 'bag limits' apply (excluding National Parks)

Establishment of register of scientific and amateur lepidopterists who are willing to contribute up-to-date information to species' database, to be supplied without imposition of fees

---

Permit applications

Applications to take, keep or move species with some form of legal protection to be processed using guidelines set out by three recognised experts on the species concerned, and to be granted/denied by the Department according to the guidelines. Guidelines to be made available at all Department offices issuing permits

Reduction of the processing/granting time for permit applications to within 10 working days from receipt of application, through streamlining by (a) inclusion of all types of permit required on a single permit application and (b) requirement for only one government official to grant a permit

Permit extensions to be granted at short notice without requiring resubmission of paperwork accompanying initial application

---

Permit restrictions

Granted permits for *A. illidgei* to include access to State Forest Parks and National Parks, with prior notification to the park ranger. In other case, permits for access to national parks apply as usual, requiring a cooperative approach between relevant government departments

Granted permits not to be restricted to particular geographical regions within Queensland

Collection of immature stages through direct sampling of ant colonies to be prohibited except where collecting is essential for part of a scientific study or for conservation management and, even then, the amount of direct sampling to be limited

(Continued)

Table 11.1 (*Continued*)

Non-destructive sampling techniques (which need to be developed for *A. illidgei*) to be allowed without requirements for a permit

Protocols for collecting, keeping and moving of immature *A. illidgei* for purposes of scientific study to be reviewed by recognised experts and to be permitted according to their recommendations

Preserved specimens

Already preserved specimens of species listed not to be retrospectively covered by law

Preserved specimens not to be restricted by storage location and not to form a source of revenue for a government department

Location of private depositories to be notified and recorded on a government database

needs of many listed taxa remain unknown or uncertain, and criteria for setting priorities (commonly based on the IUCN Red List 'categories of threat') are usually difficult to apply realistically (Cardoso et al. 2011), with difficulties confounded in some instances in which 'advisory' red lists have become the non-critical basis for direct transfer to more formal, legislatively obliging, schedules. In addition, conventions such as CITES target taxa for a specific threat interest, commercial trade, with birdwing butterflies and other Papilionidae the major taxa of current interest. The predominance of butterflies on lists of protected species reflects their attractiveness and popularity (Chapter 1), linked with capability to at least reasonably assess the conservation status and needs of many taxa, throughout the world – based mostly on criteria involving apparent abundance and distribution contractions, rather than the more formal estimations of population sizes and numerical risks of extinction included within the IUCN criteria. However, for credibility, any such list entry should be based on the most accurate and up-to-date evaluation of both absolute and relative conservation needs. In practice, individual zeal has sometimes led to proliferation of numbers of listed species either as (1) a precaution based on incomplete evidence or (2) because the species are rare and/or local, without necessarily any evidence of decline or threat. Whilst expressing sincere and laudable concerns, such confusions can weaken the integrity of such lists to critics, and mask the more real conservation concern needed for some entries. In short, credibility of every entry on such lists rests on it being seen as responsible, with listing criteria transparent and understandable – and applied consistently.

Allocation of conservation status across legislative lists is not always consistent; some recognise the several categories of threat used by IUCN (see Table 5.13), whilst others use broader groupings, such as 'threatened' or 'endangered'. Establishing any such status across different legislations may demand original field surveys, as well as appraisal of all available historical information in which trends of decline may be revealed.

Because of the large accumulations of Lepidoptera in many museum collections, data on specimen labels can be an important source of information on distributions in the past. Whilst such collection information may need to be treated cautiously (in part because common, or formerly common, species are often under-represented as not collected assiduously by many hobbyists), the data on rarer species (some over-represented as sought keenly by collectors) may be particularly useful, and generally incorporate the taxa of greatest current concern. Visits to historical sites may be instructive in revealing the extent of changes and suggesting or confirming threats. Particularly for under-surveyed areas, including much of the tropics, even sporadic collections are likely to augment species inventories based on more recent surveys, simply by increasing sampling effort and serendipity. Examination of a small collection (42 specimens) of hawkmoths made in Honduras in 2006, for example, revealed three new country records and four taxa new to the Cusuco National Park, adding to the 25 species reported earlier from the park (Vanhove et al. 2012). However, and exemplifying a very widespread context, simply because such species are novel in that fauna and not recorded elsewhere in the region, does not necessarily mean that they should be signalled or listed for conservation priority. The ambiguities and possibly misleading biases resulting from being 'data deficient' were noted in Chapter 5.

Prohibition of collecting

A major consequence of formal listing is widespread prohibition of 'take', as implied earlier a theme that has led to widespread controversy within the lepidopterist fraternity, with continuing debates over the real needs for this and its impacts on conservation. More than for any other invertebrate group, the contributions to knowledge of Lepidoptera, and the biological understanding from which most conservation concerns have arisen, have come from hobbyists, rather than from professional scientists, and who have initiated most conservation interest through their informed concerns over declines and losses. Prohibition of collecting specimens imposed by officialdom, often without the practical need for this being justified or evident, is alienating. In some cases, resultant publicity, often well motivated, has led to any collecting being viewed with suspicion. The outcomes include: (1) reduction in collecting/hobbyist interest and of accumulating biological and distributional information on species of significant conservation interest, in part through (2) any such interests becoming clandestine for fears of prosecution; and (3) the role of collecting as a threatening process often being accorded higher importance than it merits in relation to more pervasive threats associated with habitat changes (Chapter 7). 'Overcollecting' of Lepidoptera is an emotive topic, with legacy of suspicion flowing from the nineteenth century, when collectors were in part blamed for loss of *Lycaena dispar* from the fens of eastern England, to the present day, when even carrying a butterfly net in some places can be a formal offence; and more widely lead to challenges from a well intending public. Collecting is now met with official and social disapproval in Britain and some other countries. As Barkham (2010) elegantly put it, during the period between the 1950s and the 1980s in Britain 'Collecting went from being an expression of a love for nature to being judged not merely environmentally damaging but socially deviant'.

Whilst cases for prohibiting collection can be justified – for example from tiny isolated populations where effective population size is already marginal, or from the best known faunas, such as the butterflies of the United Kingdom, where taxonomic confusion is minimal – the converse, of needs for further responsible collecting, are not uncommon elsewhere, even for butterflies and far more so for poorly documented moth assemblages. Thus, for Australian butterflies (Sands & New 2002) taxonomic uncertainties amongst complex endemic radiations and the incomplete distributional knowledge of many taxa render well documented voucher specimens, particularly from beyond the recorded distributional range, or from additional sites within it, important contributions to advancing documentation and assessing variety. Rather than full prohibition, enlightened permit systems to allow (even, encourage) responsible capture of voucher specimens of 'protected species' (with the numbers restricted to short series, and recognition that some vulnerable populations or sites may be excluded from any such permit system) may help to conserve the vital contributory resource of hobbyist interest. The occasional 'rogue' collectors seeking long series for commercial gain, and attracted by rarity or protected status are by far outnumbered by responsible concerned lepidopterists sympathetic to reasoned conservation. Voluntary 'codes for collectors' (such as JCCBI 1987, the first version of which was produced as 100,000 copies in Britain: Morris & Cheesman 2012) emphasise needs for responsible constraint, and have been emulated widely. At one extreme, even a number of the most responsible hobbyists restricting take to just a single pair of adults might collectively jeopardise the future of tiny populations. In Japan, collecting is listed widely as a cause of butterfly decline (Nakamura 2011), with impacts exacerbated by populations of many desirable species being now small and isolated from habitat loss, but with the conservation committee of the Lepidopterological Society of Japan recommending that collecting of all endangered species be discouraged. Nakamura also noted that 'conflicts between collectors and local people have recently become a serious problem'.

Concerns over collecting of the rarer British butterflies for trade, in an environment in which 'chequebook collecting' had long been established, were initiated formally through the Large blue (Chapter 9, p. 169), for which any form of take or trade became illegal (unless under a specific licence) when it was listed under the Conservation of Wild Creatures and Wild Plants Act 1975. The later Wildlife and Countryside Act 1981 added three other butterflies as fully protected – one of which (the Chequered skipper, *Carterocephalus palaemon*) was later removed from this level. However, it and others for a total of 22 species were later (June 1989) listed as available for trading only under licence, with the purpose of monitoring trade effectively whilst still permiting individuals to capture specimens. The conditions and changes caused some confusions, and were clarified by Stubbs (1991), who noted that sale of any individuals of the 22 fully protected species was illegal unless (1) the specimens were bred in captivity, with the onus on the trader to demonstrate this in order to avoid prosecution, or (2) a sale licence issued by the Department of the Environment was held. It remained legal to capture individuals of any of these species, but this condition emphasised the responsibility of entomologists in recognising the need not to harm wild populations.

Nevertheless, it must be recognised that prohibition of take also creates value, and may increase the activities of unscrupulous dealer–collectors. Collecting was regarded as a threat to Kern's primrose sphinx moth (*Euproserpinus euterpe*, Sphingidae), in part because the slower-flying females of this listed species are the easier sex to capture and they often land to oviposit (USFWS 2007). This moth was reportedly subject to increased collecting pressure when it became known that collecting would become illegal by listing under the Endangered Species Act. Most such activities are extraordinarily difficult to police or deter, and are major justification for legal authority to help conservation. Many Lepidoptera are traded commercial commodities, with rare or difficult-to-access taxa demanding increasingly high prices from collectors as they become more difficult to obtain; one influence of formal protection is to induce 'black market' activities that are extremely difficult to regulate or even detect. Entering the name of a protected butterfly, or to a lesser extent, moth, into an internet search engine will often reveal specimens offered for sale far from their country of origin. Collectors may go to extraordinary lengths to obtain desirable taxa – the early tradition of importing rare European species for local release in Britain (Preface, p. xi) was one forerunner of much wider deceptive trading, in which rarity and perceived provenance led to increased value based on supply and demand. The development of centralised 'butterfly farming' operations (Chapter 10) , pioneered through Parsons' work in Papua New Guinea (Parsons 1992a, 1995) was stimulated through the commercial prospects of supplying birdwings and other butterflies desired by collectors in a manner that avoided direct take from small field populations and continued degradation of tropical forest habitats, and so has much wider conservation importance. A thoughtful critique (Parsons 1995) of the outcomes of butterfly collecting in the tropics, and conservation ramifications (including education through increased ecotourism) of the farming/ranching operations alleviating pressures on wild populations merits considerable attention. Occasional other problems can arise. Proposals to ranch *Ornithoptera alexandrae* in Papua New Guinea (Chapter 5, p. 72) to satisfy collector demand from control-reared stock depend on those specimens being available to an international market. Because of current CITES Appendix 1 listing, this is currently not the case, with export and import of specimens illegal. Thus, in order to fulfill the basic need of income provision for local people through ranching for trade, it would be necessary to downgrade the birdwing to Appendix 2. This move has not been supported by the Papua New Guinea Department of Environment and Conservation, pending more thorough exploration of the butterfly's ecology and status, and preparation of an up-to-date conservation plan.

The enforcement of any legislative listing of Lepidoptera depends on ability to recognise and differentiate the individual taxa listed, as the legal entities designated. This is not always straightforward. Many of the numerous butterfly taxa listed, both of species and subspecies, are difficult to diagnose; groups of 'look-alike taxa' are not infrequent, and members of some can be differentiated only by careful specialist examination. In some, very similar related taxa may be either common and secure or extremely scarce and threatened. One precautionary position, not without controversy, is to list all members of particular 'look-alike taxa' as protected, whereby even the most common species are legally protected, for the benefit of those of genuine conservation concern. This reasoning underpins, for example, the listing of all birdwing butterflies traded under CITES: border protection officers can easily recognise the general appearance of 'a birdwing', but species identification is often much more difficult.

The functional purpose of listing taxa is to facilitate practical conservation and accumulation of information. Ideally, it is not a permanent or irrevocable condition, with, desirably, (1) allocated conservation status open to periodic review at intervals, as dynamic, and (2) formal capability to de-list taxa should they be no longer deemed threatened. Such provisions recognise the two most common outcomes of listing a species for conservation interest and priority as (1) practical conservation measures rendering it secure and with markedly reduced threats, or (2) investigations prompted by initial listing revealing that the taxon is more widespread or abundant, so less threatened, than known earlier and does not fulfill the primary listing criteria. Evidence of these conditions might create a case for de-listing of the species, as no longer threatened, so enabling resources to be deployed to then more needy taxa. One concern over this, particularly for those taxa that have been actively recovered after (perhaps considerable) conservation investment is that once 'dropped' they may again become threatened. Sands and New (2002) discussed the need to protect that investment, by continuing to provide monitoring of such de-listed recovered taxa, as 'rehabilitated species', simply to enable early detection of any new or recurrent threats, and act as necessary should these be found.

Ideally, every listing should encourage further study and the needed conservation measures. Often this has not occurred, and no further actions have eventuated. Many species of Lepidoptera have been on protected species lists for many years with little practical progress to enhance their security; they contrast markedly with those species that have indeed been actively conserved (Chapter 9), but that contrast indicates also the enormity of the tasks needed. There is no doubt that responsible listing can be an important conservation tool and adjunct, but that this step is only an initial one in the practical work needed.

However, in principle, every species of butterfly or moth formally listed should be sufficiently well understood for its status to be justified objectively, and its needs defined. Those needs, and the background perspective, are the basis for preparation of a subsequent 'recovery plan' (Chapters 9, 12) and, whilst some such documents are commonly obliged under legislation, they have often not been prepared, or are well beyond their projected review dates. For many listed species, little or no additional information has accrued. In practice, these lacunae and failings are likely to continue: the dual outcomes of listing numerous deserving Lepidoptera for protection and conservation need are that (1) long lists help to express a realistic scale of need to managers and decision-makers but also (2) increase the burden of choice and achievement through necessary selection within these for optimal deployment of very limited resources. In many regions, long lists of taxa that are difficult to recognise or identify, however well intentioned, are essentially impracticable to enforce for practical conservation.

References

Ball, J.B. (2012) Lepidopterology in southern Africa: past, present and future. pp. 279–300 in New, T.R. (ed.) Insect Conservation: Past, Present and Prospects. Springer, Dordrecht.

Barkham, P. (2010) The Butterfly Isles. A summer in search of our emperors and admirals. Granta Publications, London.

Beale, J.P. (1997) Comment on the efficacy of Queensland nature conservation legislation in relation to *Acrodipsas illidgei* (Waterhouse and Lyell) (Lepidoptera: Lycaenidae: Theclinae). *Pacific Conservation Biology* 3, 392–396.

Cardoso, P., Erwin, T.L., Borges, P.A.V. & New, T.R. (2011) The seven impediments in invertebrate conservation and how to overcome them. *Biological Conservation* **144**, 2647–2655.

Fox, R., Parsons, M.S., Chapman, J.W., Woiwod, I.P., Warren, M.S. & Brooks, D.R. (2013) The State of Britain's Larger Moths 2013. Butterfly Conservation and Rothamsted Research, Wareham, Dorset.

Invertebrate Link (2008) Statement on the appropriate role of legislation in controlling activities likely to harm specified taxa of terrestrial and freshwater invertebrates, with particular reference to taking and killing. *British Journal of Entomology and Natural History* **21**, 202–204.

JCCBI (Joint Committee for the Conservation of British Insects) (1987) A code for Insect Collecting. Forestry Commission, Farnham.

Morris, M.G. & Cheesman, O.D. (2012) Insect conservation in the United Kingdom – the role of the Joint Committee for the Conservation of British Insects and Invertebrate Link (JCCBI). pp. 21–40 in New, T.R. (ed.) Insect Conservation: Past, Present and Prospects. Springer, Dordrecht.

Nakamura, Y. (2011) Conservation of butterflies in Japan: status, actions and strategy. *Journal of Insect Conservation* **15**, 5–22.

Parsons, M.J. (1992a) The butterfly farming and trading industry in the Indo-Australian region and its role in tropical forest conservation. *Tropical Lepidoptera* 3, Supplement 1, 1–31.

Parsons, M.J. (1995) Butterfly farming and trading in the Indo-Australian region and its benefits in the conservation of swallowtails and their tropical forest habitats. pp. 371–391 in Scriber, J.M., Tsubaki, Y. & Lederhouse, R.C. (eds) Swallowtail Butterflies. Their Ecology and Evolutionary Biology. Scientific Publishers, Gainsville, Florida.

Sands, D.P.A. & New T.R. (2002) The Action Plan for Australian Butterflies. Environment Australia, Canberra.

Stubbs, A.E. (1991). Protected British butterflies: interpretation of Section 9 and Schedule 5 of the Wildlife and Countryside Act 1981. *Entomologist* **110**, 100–102.

USFWS (United States Fish and Wildlife Service) (2007) Kern Primrose sphinx moth (*Euproserpinus euterpe*). 5-year review: Summary and evaluation. Sacramento, California.

Vanhove, M.P.M., Jocque, M., Mann, D.J. et al. (2012) Small sample, substantial contribution: additions to the Honduran hawkmoth (Lepidoptera: Sphingidae) fauna based on collections from a mountainous protected area (Cusuco National Park). *Journal of Insect Conservation* **16**, 629–633.

12

Defining and Alleviating Threats: Recovery Planning

Introduction: The variety of threats to Lepidoptera

The practical conservation of Lepidoptera, as for other taxa, has two major groups of need: the measures that can be taken as informed positive management, and the things that should no longer be done, as threatening. Much of the text of this book illustrates these by specific examples of species or biotopes, in which a considerable variety of threats and remedial and preventative actions are illustrated. Defining and alleviating threats, and preventing further threats from arising, comprise the major bases for conservation assessment, action and management, with the practicalities incorporating any suitable or available legislative support for both field-based and ex situ management components. The tools available, and the practical needs, differ for any individual case but the relatively high number of practical exercises for individual species of Lepidoptera have been instrumental in demonstrating both the individual subtleties and the more general principles available for insect species conservation, and which may be extended to other taxa and to wider units such as assemblages with little conceptual change. Securing and restoring habitat, both as space and resources, is the most universally pervasive practical theme. Whilst other threats must be appraised carefully, and are likely to differ in balance across different populations or sites supporting the same focal taxon/a, the core conservation needs flowing from understanding of critical resource needs are almost invariably paramount in planning conservation measures, with these extending from site to landscape scales. Initial 'crisis-management', with a keen sense of urgency, is likely to be followed by need for a considered, long-term conservation plan. The latter may need to be instigated with very little background knowledge or awareness of the consequences, with the ideal of being able to monitor carefully and conduct parallel more basic studies as grounds for adaptive management changes. Such

Lepidoptera and Conservation, First Edition. T.R. New.
© 2014 John Wiley & Sons, Ltd. Published 2014 by John Wiley & Sons, Ltd.

ideals are often not assured. The best possible advice at this stage is itself a vital resource.

This chapter includes a brief overview and summary of major and other purported threats other than direct habitat loss to Lepidoptera, and some of the major practical measures used to counter these, and in habitat restoration and management.

Alien species

Implications of threats to Lepidoptera from introduced or accidentally arriving alien species have devolved mainly on the non-target impacts of parasitoids and predators introduced as classical biological control agents, most of them undertaken in an era preceding the current concerns for formal screening protocols to investigate specificity and likely non-target effects in areas where they are released. Adverse effects on Lepidoptera have been claimed in both continental and the more isolated island faunas, with the latter especially susceptible as environments in which long periods of isolation and intricate coevolutionary developments can be upset by introduced species. Howarth's (1983, 1991) concerns over spatial and ecological spread of classical biological control agents prompted considerable debate around this theme, but concerns arose much earlier. Zimmerman (1958), for example, implicated introduced parasitoids in 'the wholesale slaughter and near or complete extinction of countless species' within the extensive moth fauna of Hawai'i, a theme developed by Gagné and Howarth (1985). On Guam (Nafus 1993), introduced parasitoids attacked early stages of native *Hypolimnas* butterflies and might have contributed also to declines of several other taxa. Examples can be multiplied across different environments (New 2012b), but many suggestions of harm to native taxa flowing from the spread of introduced natural enemies from cropping systems into natural environments remain to be investigated in detail. That introduced pupal parasitoids introduced to New Zealand to control the Small cabbage white butterfly, *Pieris rapae*, on crops contribute to the decline of the endemic Admiral, *Bassaris gonerilla* (Barron et al. 2003) is one such example.

Purported impacts of any such species may need to be investigated very carefully. One of the same pupal parasitoids of *P. rapae* introduced into North America as a control agent has received such attention recently (Benson et al 2003, van Driesche et al. 2004). The native *Pieris napi oleracea* (a member of the *P. napi* 'superspecies' which has numerous named forms across the Holarctic region) has declined substantially in range since the mid-nineteenth century, and several hypotheses have been ventured to explain this. They include the implication of losses from the introduced braconid *Cotesia glomerata*. The scenario was complex, but it seemed that *C. glomerata* preferred *P. napi* to the target *P. rapae* and that parasitoid populations were sufficiently high in populations of the latter in agricultural habitats to affect neighbouring meadow populations of *P. napi*. This circumstance linked with a north – south cline in *P. napi* pupal diapause regimes – the more northerly populations were functionally a univoltine taxon in parasitoid-free woodland habitats, whilst southerly populations were bivoltine and went extinct due to high exposure to *C. glomerata* in open meadows colonised by the woodland first-generation

adults. Non-target effects of introduced parasitoids may thereby vary geographically and in relation to host habitat and life history features across its range.

Any parasitoid or predator with a tendency to polyphagy has potential to
expand host range in a novel environment, but clarifying impacts is difficult. The
polyphagous tachinid fly *Compsilura concinnata* has been reported from about
180 non-target host species in North America, with experimental exposures of
early stages of native Saturniidae revealing their susceptibility (Boettner et al.
2000). The principle of deploying 'sentinel targets' was pursued further in this
context by exposing second to fifth instar caterpillars of *Actias luna* (Saturniidae)
within and beyond the current range of *Compsilura* (Kellogg et al. 2003). The
outcomes supported the notion of *Compsilura* becoming a threat to such native
species and their native parasitoids, and Kellogg et al. urged the importance of
collecting data on the fly's impact and on population dynamics of native forest
Lepidoptera and their parasitoids beyond the current ranges of introduced agents
likely to spread. Such impacts will continue to be debated, and additional management of introduced agents may be advised if they co-occur with known threatened species that are potentially vulnerable. The introduced predatory ladybird
Coccinella septempunctata occurs within the range of the endangered Karner blue
(Chapter 10, p. 189) in North America (Minnesota, Wisconsin), with its seasonal
development suggesting vulnerability of the butterfly's early stages (Fig. 12.1).
An adult beetle was observed eating second instar caterpillars, and Schellhorn
et al. (2005) suggested maintaining an (at present undefined) minimum isolation
distance between crops where the beetle is needed to control aphids, and extant
Lycaeides populations.

Not all such aliens have been introduced deliberately, and the mode of arrival
of many of them in expanded range areas is simply unknown. The northward

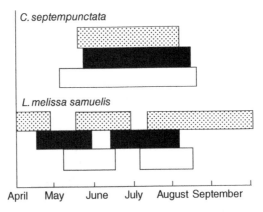

Fig. 12.1 Seasonal development patterns of the introduced ladybird *Coccinella septempunctata* (upper group of bars) and the endangered Karner blue butterfly *Lycaeides melissa samuelis* (lower bars) where they co-occur in Minnesota and Wisconsin, United States. Seasonal presence of eggs (top line: dotted), larvae (middle line: black) and pupae (bottom line: open) of each are shown against months of the year (Source: Schellhorn et al. 2005. With kind permission from Springer Science+Business Media.).

spread of the tachinid *Sturmia bella* in Britain since it was reported first in southern England in 1998 has been linked with declines of the Small tortoise-shell butterfly, *Aglais urticae* (Gripenberg et al. 2011). How the fly arrived in Britain is not certain, but it has been suggested to have been from released imported caterpillars; it was reared first in Britain from larvae of the Peacock (*Inachis io*), a suitable but somewhat secondary nymphalid host for *Sturmia*. Whilst the fly, on evidence accumulated by Gripenberg et al., is unlikely to be the sole cause of recent severe declines of *A. urticae*, it may well have contributed to this through adding to the host mortality already occurring from other parasitoids.

Alien insect predators, notably amongst social Hymenoptera, are acknowledged widely as likely to cause severe changes in the receiving fauna, both directly and in affecting ecological interactions amongst native species.

Vertebrate predators are less often noted specifically in relation to impacts on Lepidoptera, although general presence of introduced rats on islands, for example, has widespread and severe impacts on native invertebrates. The North American green anole lizard (*Anolis carolinensis*) arrived on the Ogasawara (Bonin) Islands of southern Japan following the Second World War and by 2010 had an estimated population there of more than six million. Predation by the anole is considered the main cause of decline of the endemic Ogasawara blue butterfly (*Celastrina ogasawaraensis*) that declined rapidly after the 1970s and is now found only on one island (Nakamura 2010), where it is regarded as 'in extreme peril'.

Host-switching of Lepidoptera to newly present alien food plants can occur rapidly (Singer et al. 1993, Chapter 1, p. 5), and newly evolved 'preferences' for these can introduce complex implications for future developments as such associations become entrenched. Alien plants may prove to be threats (*Ornithoptera richmondia*, Chapter 9, p. 173) or valued additional resources (*Synemon plana*, Chapter 9, p. 175; *Poanes viator*, Chapter 8, p. 150; *Manduca blackburni*, Chapter 9, p. 177), with introduced plants commonly important in urban and other garden environments as nectar sources and, less commonly, as additional larval food plants. In general terms, alien species are integral to many habitat changes affecting Lepidoptera. Replacement of diverse native vegetation with introduced monoculture crops, with attendant alien weeds and the various alien pests and control agents that may flow from needs to sustain crop production, can lead to well founded conservation concerns. A further, but rarer, or more rarely appraised, concern is that of hybridisation between alien and native taxa that are closely related. Gibbs (1980) expressed concerns that declines of the New Zealand endemic *Zizina oxleyi* (Lycaenidae) were in part due to hybridisation with the introduced Australian *Z. labradus*.

Diseases

As with the more conventional natural enemies noted in the preceding section, both the impacts and places of origin of many pathogens of Lepidoptera are not understood, and remain difficult to assess. In part this reflects that many impacts are sublethal, and may be dose-dependent or differ across sexes.

The neogregarine protist *Ophryocystis elektroscirrha* was first recorded from Monarch butterflies (*Danaus plexippus*) in 1966, and its incidence varies considerably amongst populations (Altizer et al. 2004). However, experimental trials demonstrated that the life span, mating success and weight loss of adult butterflies were affected by infections, with some sexual differences. The male life span was shortened in the highest *Ophryocystis* dose tested, whilst females were not affected. Larval survival and adult size were also less at the highest parasite density (Altizer & Oberhauser 1999). Many individuals with that dose had difficulty in emerging from the pupal case, and died – possibly resulting from the large numbers of gregarines in hypodermal cells disrupting development of the host integument. High parasite numbers were also associated with wing condition in overwintering monarchs, a phase of great significance in conservation (Chapter 4, p. 45); greater scale loss and levels of 'tattiness' were related to increased likelihood of parasite spores being present.

The roles of many such parasites are very subtle, and in most their impacts and importance in host conservation are simply unknown. *Ophryocystis* has received attention as common in, especially, non-migratory populations of a host of massive conservation interest, but this is relatively unusual. It also infects the related Queen butterfly (*Danaus gilippus*), but that relationship has not been investigated in comparable detail.

Climate change

Future climate warming was considered likely to 'become a major force in shifting species' distributions' (Parmesan et al. 1999), but these authors also suggested that northward range expansions in Europe could 'prove difficult for all but the most efficient colonizers' because of the highly fragmented nature of the landscapes encountered. Although commonly considered as a threat, climate change is also a major opportunity for many taxa in facilitating range expansions, but those species then become, in effect, aliens in the receptor environments they colonise. Changing climatic conditions can affect insects in many ways, not all of which are yet clearly understood, but actual and potential changes in abundance and distribution of individual species have attracted most attention, predominantly with range changes to encompass higher elevations and latitudes as warming occurs. Such changes are particularly well documented amongst temperate region butterflies, reflecting the propensity of hobbyists to report specimens observed outside their 'normal' expected distribution areas. Most such observations are based on data from northern temperate regions, but with additional evidence from Australia (Beaumont & Hughes 2002, Peters et al. 2010). Only more rarely are range changes linked with documented phenological changes. The southward extension of the Large banded awl (*Hasora khodia haslia*, Hesperiidae) of some 400 km in eastern Australia (Fig 12.2, with breeding confined at the southern edge) since the 1970s is supported by indication that the adult flight period now commences earlier in the season and is longer, aspects that led Peters et al. (2010) to suggest that climate may be driving this change.

In Borneo, as a rarer tropical region example, the elevational range of 102 species of montane geometrid moths shifted upward by an average of 67 m over

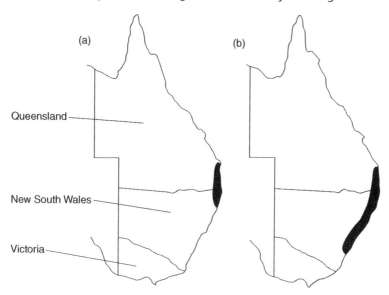

Fig. 12.2 The southward range extension of the skipper *Hasora khoda haslia* in eastern coastal Australia: (a) distribution (black) prior to 1964; (b) distribution as at 2010 (Adapted from Peters et al. 2010).

42 years (Chen et al. 2009), so that whole assemblages may be influenced strongly. That inference arose from repeating (in 2007) an earlier elevational transect survey for moths made in 1965, using the original trap design, protocols and sampling effort, with comparable data obtained from the six sites (spanning 1885–3675 m elevation) for which vegetation had remained largely unaltered. The upward shifts were viewed as consistent with either physiological responses to climate change or as the outcomes of changed interactions with other species.

Following from Dennis (1993), Woiwod (1997) and other earlier essays, Kocsis and Hufnagel (2011) summarised many aspects of climate effects and emphasised the difficulty of unambiguous attribution of observed changes to climate impacts alone, with community context being influential. Woiwod's (1997) three main impact classes in Lepidoptera, as (1) changes in abundance; (2) changes in range distribution or area; and (3) changes in phenology, have been the most frequently cited conservation concerns. The last is reflected in developmental rate and the earlier appearance of adults each flight season, and was considered by Woiwod to be the most readily observable early effect that can be attributed directly to climate change. In contrast, the disruption of interactions between species, almost inevitably reflecting differing rates and extents of response to climate change may disrupt or 'de-couple' highly specific patterns that have co-evolved in both space and time. Associations with consumable resources and natural enemies are, perhaps, the most obvious of these, but any component of a conservation module may become unbalanced. A butterfly's main nectar plant source may not keep pace with an earlier flight season, for example, or suitable caterpillar food not be available at a changed time of need.

Composition of local Lepidoptera assemblages may change with advent of species from elsewhere, essentially as 'community aliens', some possibly with competitive impacts akin to those of more commonly considered alien species. Many such initially successful colonists are likely to be relatively generalist species, and resident specialists may be increasingly susceptible to impacts of any competing novel taxon. The balance between 'gain' from a species increasing or changing its distributional range may be countered by increased threat to species already present in the receiving area. Studies of species at range margins can help to reveal both ecological and evolutionary processes that affect probability of range expansions (Thomas et al. 2001), with some evidence that northward expansion of European butterfly communities lags substantially behind climate changes. Thus, van Swaay and Warren (2012) noted that over 20 years, those butterfly communities have shifted northward by an equivalent of 75 km, whilst the temperature regime has shifted north by 246 km. Regional warming at range margins is likely to both facilitate and provoke range expansions on 'proximal' grounds of population dynamics and ecological capacity. Once such expansions occur, selection may ensue for the 'most able' expanders, as discussed by Thomas et al.

Faunal changes are exemplified well by the increased number of migratory Lepidoptera species moving northward in Europe. At Portland in southern England, the number of migratory species detected has risen steadily from 1982–2005 (Fig. 12.3), a trend linked strongly with increasing temperatures in south-western Europe (Sparks et al. 2007). Over that period, 75 species were recorded, but only 10 of these were found in 20 or more years; 31 occurred in only 1 or 2 years. Distributional changes have, perhaps, received more attention than other climatic impacts, numerous though these may be. The burgeoning evidence of climate change impacts on European butterflies, in particular, continues to reveal subtle interplays between distributions, resources and seasonality, and also that species differ considerably in their individual responses. Even when

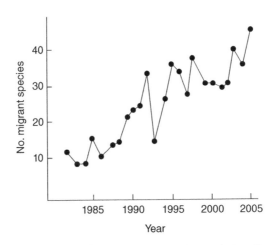

Fig. 12.3 Number of species of migrant Lepidoptera recorded at Portland Bird Observatory, southern England, over 1982–2005 (Source: Sparks et al. 2007. Reproduced with permission of Institute of Entomology.).

sharing very similar ecological traits and distributions, the responses of closely related species may be very different, and affect their vulnerability in different ways. Need for species-level data emphasises the difficulties of proposing any general mitigation processes, but the very general framework given in Arribas et al. (2012) (Fig. 12.4) summarises much current perspective, advanced to complement other aspects of considering species' vulnerability. The 'butterfly trends' of northward movement are paralleled by numerous moths (for example in Estonia: Kruus 2003). Implications for the Netherlands Microlepidoptera were summarised by Kuchlein and Ellis (1997).

As Altermatt (2012) emphasised, climate change and habitat structure (Chapter 7) and change can have combined impacts on phenology, with the individual impacts of each factor often hard to distinguish. For 28 butterfly species

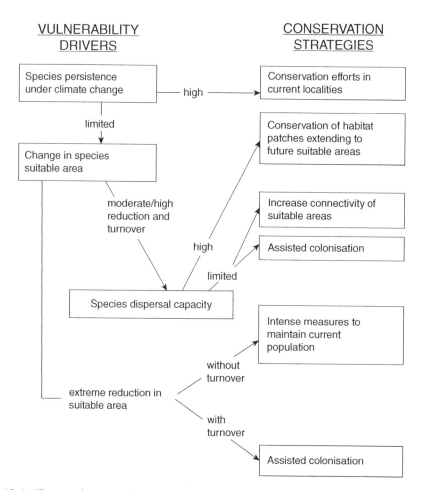

Fig. 12.4 Climate change and conservation strategies: a decision framework based on interpretation of vulnerability to climate change, and strategies that might counter this (Source: Arribas et al. 2012. Reproduced with permission of John Wiley & Sons.).

in Switzerland, seasonal appearance was correlated significantly with (1) mean ambient seasonal temperature; (2) habitat type – whether forest, agricultural systems or human settlement areas; and (3) an interaction between these. Despite the higher temperatures associated with settlement areas, some species commenced the flight season there significantly later than in agriculture or forest habitats. Effects of dispersal, however, could not be eliminated in this interpretation, and complexities continue to thwart any 'simple' analysis of such patterns. Altermatt's (2012) warning that habitat type should be considered carefully in interpreting phenological change is important. Likewise, simple attributions of range changes to climate alone, however attractive, are most obvious at the larger scales of latitude or elevation, whilst small-scale local changes are both more subtle and, often, more central to conservation strategies that facilitate local sustainability or expansion of the species (Oliver et al. 2012).

Many of the British butterfly species appear to have been unable to exploit the opportunity provided by climate change and (over the last three decades) have decreased their habitat breadth as populations have declined. Despite favourable warm climate conditions for range expansions (as for *Aricia agestis*, Chapter 6, p. 108) reduced habitat breadth may hamper or prevent this, and thwart the species' persistence. Oliver et al. suggested that such species may be able to exploit the new opportunities presented only when the other, non-climate, constraints to the populations are reduced or removed. Individual butterfly species showed 'idiosyncratic responses' (Mair et al. 2012) likely to reflect the balance between climatic and habitat changes influencing their distribution and abundance. Mair and her colleagues compared responses of 37 British butterflies over two periods (1970–82 to 1995–99, and 1995–99 to 2005–09) by measuring changes in range margin, distribution area and abundance, as possible responses to climate change. Only five species showed qualitatively similar responses in all three variables over time: three generalists (*Aphantopus hyperantus, Pararge aegeria, Polygonia c-album*) had consistently positive responses, whereas two specialists (*Leptidea sinapis, Pyrgus malvae*) had consistently negative responses; all other species were more variable. The trends differed somewhat across the two periods (Fig. 12.5), implying that species' responses differed over time. Whilst generalists tended to expand northward at greater rates than specialist species during the second recording period, changes in distribution areas were similar in the two groups.

Deliberate recommendations of areas to be protected to cater for future range changes of Lepidoptera from climate impacts are rare. *Melitaea cinxia* occurs at the lower levels of the Tianshan Mountains, China, whilst meadows at higher elevations (above 2050 m) have suitable host plants, but no butterflies. Zhou et al. (2012) emphasised the need to maintain those meadows as investment for the future, as potential habitats for *M. cinxia* in response to global warming.

Exploitation for human need

Exploitation of Lepidoptera is not always deliberate, and may even be unintended and unrecognised. The myrmecophagous 'moth butterfly', *Liphyra brassolis* (Lycaenidae) occurs widely, but perhaps sporadically, from south-east Asia to northern Australia, and is rarely collected. The flattened testudinate larvae

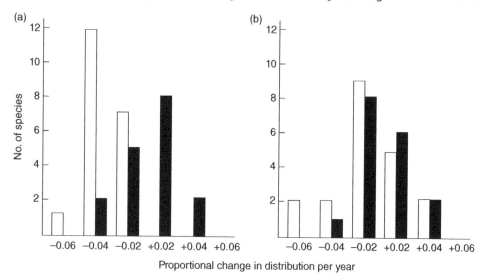

Proportional change in distribution per year

Fig. 12.5 Changes in species distribution of British butterflies, depicted as the proportional changes in the number of 10 km Ordnance Survey squares occupied per year for generalist (black bars) and specialist (open bars) from (a) 1970/82–1995/99 and (b) 1995/99–2005/09. Categorisation of species after Asher et al. 2001 (Source: Mair et al. 2012. Reproduced with permission of John Wiley & Sons.).

live within nests of weaver ants (*Oecophylla smaragdina*), and feed on the ant brood. Those ant brood are also sought and collected widely for medicinal and other uses, including human food. Surveys of people in Thailand (Eastwood et al. 2010) revealed that many people had indeed eaten these characteristic caterpillars when found in *Oecophylla* nests, and regarded than simply as 'big' or 'queen' ants. There is no evidence that they are sought deliberately as individual delicacies, but this survey remains an intriguing approach to monitoring an elusive butterfly.

However, this is not the case for some other Lepidoptera, participant in organised entomophagy and long familiar to many ethnic groups as major sources of protein and contributors to food security (Gahukar 2011). Caterpillars are amongst many groups of insects targeted for human consumption, in some cases extending to 'industry-level' exploitation of wild populations with conservation implications through excessive take, with non-target take, as described in the previous paragraph, also possible.

In southern Africa, caterpillars of the emperor moth *Imbrasia belina* (Saturniidae) ('mopane worms') are harvested as a traditional protein-rich food, and marketing is the basis for lucrative livelihoods of many people. The moth undergoes population outbreaks, during which enormous numbers of caterpillars can be collected easily (McGeoch 2002), and which have been associated with regional losses of the moth. Over-exploitation is thus a real concern, with mopane worms described as 'a vital trading commodity' (Akpalu et al. 2007). Although accurate figures for levels of take are elusive, one older published estimate, for 1982, suggested that about 16,000 tonnes of mopane worms entered commerce in that year, with a canning factory closing after a few years as supply declined. Harvesting embargos have been suggested as a conservation measure, but wide variation in population dynamics renders setting such thresholds difficult. Some form of 'restriction of take', such as by licensing or site protection, may be a key aspect of assuring sustainability.

Medicinal uses are exemplified by the rather complex case of the Chinese 'caterpillar fungus' (*Ophiocordyceps sinensis*, endemic to the Tibetan plateau and the Himalayan region) parasitising subterranean caterpillars of Hepialidae (mainly of species of *Thitarodes*, previously referred to as *Hepialus* in the region) and harvested as units of fungus plus caterpillar for a wide variety of traditional medical uses. The fungus is listed as an endangered species, and investigations of its possible host range within the swift moths were a major stimulus for taxonomic work on the endemic members of the family Hepialidae (Wang & Yao 2011). Market demand for the fungus has increased, with hand-harvesting by individually digging up caterpillars, with the protruding fungus stroma ('summer grass') revealing individual presence, on high-elevation grassland areas a major component of providing livelihoods for local people, and strong economic incentives to increase collection of this luxury product (Guo et al. 2012). Many people gain at least half their annual income from the practice. Retail prices for *O. sinensis* reported by Guo et al. can reach up to more than US\$53,700/kg; and a caterpillar-steeped liquor was quoted to me at about US\$170/litre in Tianjin in June 2012. Harvest quantity, although very hard to confirm, has been estimated at 80–100 tonnes/year (Guo et al. 2012). Extent of harvesting, with diggings of grassland has led to conservation concerns for the most exploited fragile alpine tundra systems, to the extent that a small levy is used to restore local grassland vegetation each year. Whilst needs to sustain harvesting capability are appreciated clearly, nothing is known of the impact of the fungus on the host moth populations. Continuing removal of diseased individuals (as dead caterpillars) might, possibly, be a valuable conservation measure for the moths, but any such suggestion must be extremely tentative.

Sustainable harvesting levels of any utilised species of Lepidoptera are difficult to determine. Whilst silk production is best known for the domestic silk-worm (*Bombyx mori*), now cultivated for, perhaps, up to 5000 years for this purpose, silk from other species of bombycoid moths is also in demand for high-quality 'niche marketing' (Peigler 1994). Interest in silk moths and their commercial uses is very widespread, as manifest in a specialist Japanese Society and its publication, the 'Journal of Wild Silk Moth and Silk'. Much of this trade flows from field-harvesting of cocoons. The variety of taxa used is considerable, with growth of cottage industry based on non-mulberry-feeding silk moths in India now substantial (Ahmed & Rajan 2011) and encouraged under the Forest (Conservation) Act 1980. The activity is viewed as an *in situ* forest activity that strengthens interest in forest conservation. Although still a small component of the overall silk industry in India, from which more than six million peple reportedly gain livelihoods, this sector is growing rapidly and involves a variety of Saturniidae of the genera *Antheraea*, *Attacus*, *Cricula* and *Samia*.

Two indigenous species of Lasiocampidae (*Gonometa postica*, *G. rufobrunnea*) produce high-quality, commercially desirable silk in southern Africa, where the harvesting of wild cocoons from populations of unknown viability and sizes is viewed with some concern over assuring sustainability (Veldtman et al. 2002). Both species and sex can be determined from cocoon features, as the first practical step toward effective field monitoring of these moths. Silk moth cocoons are also used widely as cultural artefacts. Peigler (1994) cited many examples, some historical, but with ornaments such as anklet rattles of *Gonometa* and *Argema* cocoons current in parts of Africa.

Light pollution

Recent discussions of 'light pollution' (Longcore & Rich 2004) raise concerns for many nocturnal Lepidoptera (Frank 1988, Kolligs 2000), such as the impacts of street lighting in concentrating moths (Eisenbeis & Hassel 2000). Precise consequences are not always clear, but matters raised include (1) concentration of moth numbers to form population sinks attracting individuals from surrounding habitats; (2) those concentrations forming easily available massed prey for insectivores such as bats, amphibians and some birds, increasing predation in both intensity and time – with moths attracted by night remaining near the lights over the following daytime, so accessible also to diurnal predators: (3) disruption of normal behaviour mechanisms such as mate-finding and diurnal activity patterns; and (4) disruptions of intraspecific interactions and those within wider communities. A related topic is the increased urban and domestic concentration and use of 'insectocutors', based on ultraviolet light attraction and aimed mainly at pest or nuisance taxa, but without selectivity, and impacts little documented. Substantial increases in direct mortality and behavioural disruption can be inferred from lighting, perhaps particularly within declines of urban Lepidoptera, with increased electric lighting having 'transformed the nighttime environment over substantial portions of the Earth's surface' (Longcore & Rich 2004: 191).

Frank (1988) reviewed many aspects of the dilemma, and emphasised that the outcomes of reactions of moths to light are very difficult to evaluate, with opinions and evidence very varied. That hundreds to thousands of individual moths can be attracted to a single light source is the behavioural basis of light traps used widely in sampling and collecting (Chapter 5, p. 55), and has understandably led to concerns over excessive take of rare and local species – often those most desired by collectors in the past – but with emphasis on direct take rather than disruption, perhaps significant, of complex behaviour patterns. 'Codes of collecting' emphasise needs for responsible limits on take, with heed to prohibitions on captures of 'listed species', of persistent trapping in small or specialised habitats, and for care in releasing individuals retained overnight in order to prevent predation.

Moth behaviour at light is complex, but global proliferation of artificial night lighting is regarded as one of the major threats to moth populations, with high ultra-violet wavelengths especially attractive. Spectral composition of artificial light may be influential, with lamps dominated by smaller wavelengths attracting more moth species and individuals (van Langevelde et al. 2011), suggesting that use of lamps with larger wavelengths might help to reduce negative impacts of light pollution on moth assemblages and populations. Fox (2013) commented on the lack of firm evidence of impacts of light pollution on moths, particularly the need to assess these separately from usually accompanying impacts of urbanisation and habitat loss. His claim that 'light pollution remains uninvestigated as a possible cause of population level changes in moths' remains challenging. More broadly, it reflects that the considerable variety of 'drivers of change' to Lepidoptera populations (Fox 2013) are commonly difficult to interpret because of the complex interactions and synergies that arise. Few such threats act permanently in isolation, and natural and anthropogenic changes to environments can interact strongly. Much 'knowledge' of butterfly or moth declines may be open to severe reappraisal as understanding of the various interactions increases, and conservation management should be adapted accordingly.

Pesticides

Insecticide and herbicide drift from crop treatments into nearby habitats can be harmful, with direct mortality to Lepidoptera (Davis et al. 1991) and loss of larval food plants. Increasing diversity and use of 'biopesticides', many of them based on viruses or bacteria, can also pose concerns. An earlier review of impacts of pesticides on farmland butterflies (Longley & Sotherton 1997) remains highly relevant, and emphasises the importance of protecting surrounds such as arable field margins from both direct applications and drift. The principle extends easily to considering pesticide use on any areas bordering sites occupied by species or assemblages of interest or concern. Russell and Schultz (2010) considered that many species could be placed at risk from herbicides applied primarily to crops, and urged careful application techniques to reduce non-target impacts. However, in some situations it is impracticable to avoid areas where non-target taxa are likely to occur. Formulations of *Bacillus thuringiensis* ('Bt') have frequently been a major component of controlling forest pest Lepidoptera, such as Gypsy moth (*Lymantria dispar*) and Spruce budworm (*Choristoneura fumiferana*) in North America and *Tortrix viridana* (Green oak tortrix) in Europe. A review by Miller (1990) suggested that this practice was associated with declines of some of the numerous co-occurring species in forests, but the precise causes for reductions of non-target taxa are usually not wholly understood. Concerns over the continued spread of the Gypsy moth in North America are exemplified by studies in hardwood forests in Virginia and West Virginia (Rastall et al. 2003), in which impacts of aerial applications of Bt var. *kurstaki* and of a target-specific formulation of a polyhedrosis virus were assessed through surveys of caterpillars and adults of 19 species of representative native forest Lepidoptera over five families (Arctiidae 2 species, Geometridae 8, Lasiocampidae 1, Noctuidae 6, Notodontidae 2) using trap bands and foliage surveys for caterpillars and light trap catches for adult moths. Surveys extended over 2 pretreatment years and the following 2 treatment years, over six plots of each treatment and six control plots. Outcomes varied considerably; some species showed increased numbers in treatment over pretreatment years, whilst some others (notably in Arctiidae, Geometridae and Notodontidae) showed significant declines. Inferences that early-season Bt applications may potentially harm some common non-target taxa was countered for other taxa by their being late-season developers, so that caterpillars were not exposed directly to Bt applications. The higher numbers of such species found in treatment years may reflect patterns of seasonal development, but interpretation of any such impacts is difficult. Concerns over non-target effects of Bt have extended to crops containing transgene material, with the debate over possible impacts on *Danaus plexippus* in North America stimulating wider interests. The concern devolved largely on possible impacts on caterpillars from exposure to toxic amounts of Bt-affected pollen from corn plants on the milkweed host plants nearby. A recent 2-year study and risk assessment (Sears et al. 2001) concluded that risk was negligible. However, debate has continued, with accounts by Losey et al. (2003), Shirai and Takahashi (2005) and others suggesting possible negative effects on non-target Lepidoptera. Thus, in Japan Shirai and Takahashi found that larval survival of the Pale grass blue butterfly (*Pseudozizeeria maha*) was significantly affected at pollen densities of more than 20 grains/cm^2 on exposed test leaf discs, with equivalent natural risks possible within 5 m of corn field edges.

Very careful appraisal is needed when pesticides are used to suppress pest taxa on habitats of known threatened species, perhaps employed as a component of the species' conservation strategy. Invasive alien plant weeds threatened the endangered Lange's metalmark (*Apodemia mormo langei*) on the Antioch Dunes National Wildlife Refuge of California (Arnold 1983a), so that control of these to secure the larval food plant supply became a key conservation need. Options available included hand removal, burning and herbicide use. As a surrogate trial, Stark et al. (2012) investigated likely herbicide impacts on the butterfly by trials of three candidate formulations on the closely related Behr's metalmark (*Apodemia virgulti*) by exposing first instar caterpillars and monitoring their development. All three herbicides significantly reduced adult butterfly emergences from untreated control levels, with implication that similar impacts might occur on the threatened metalmark.

Use of commercial pesticides to control mosquitoes has contributed substantially to decline of the Schaus swallowtail (*Heraclides* (or *Papilio*) *aristodemus ponceanus*) in Florida. The revised recovery plan (USFWS 1999) included a component on eliminating impacts of this practice on the swallowtail, as well as interpreting the effects of the insecticides on other butterflies of the Florida Keys.

Habitat manipulation and management

Whilst it is commonly possible to single out an individual threat as the primary cause of concern for any given species, a combination of threats may cause greater concern once habitat loss or degradation has occurred as a major precursor of increased vulnerability. Distribution of the Schaus swallowtail in Florida is restricted to subtropical hardwood hammocks, a biotope that has been cleared extensively – but the butterfly occurs in regrown areas after they have been cleared and farmed, so that habitat planting and restoration is clearly a major and rewarding conservation measure. The hammock clearing was largely for residential and agricultural purposes, and has extirpated the butterfly from parts of its historical range. A report by the USFWS (United States Fish and Wildlife Service) (1999) noted wryly that lesser disturbances such as formation of dirt roads and trails seemed to be harmful only by facilitating access by collectors! Acquisition and restoration of habitat, in conjunction with restrictions on capture and reducing impacts of mosquito-control measures form a three-pronged approach to conservation, aided by a captive breeding programme to furnish butterflies for release. The nine measures proposed for its habitat protection illustrate the variety of cooperations that may be needed in any such case, with details necessarily varying to accommodate idiosyncrasies of the species' biology.

'Sympathetic management' of anthropogenic landscapes and biotopes has potential to reverse many of the trends of Lepidoptera declines and losses, but scales of consideration are universally important. Lepidoptera broadly characteristic of 'forests' or 'grasslands' may have greater ecological amplitude (and resilience) than many found only in very restricted biotopes, such as boreal mires (Vaisanen 1992). The model for optimal restoration needed for a given species is itself often not complete, and is derived from the site or few sites on which the species occurs, not necessarily representing optimal conditions. Kern's primrose sphinx (Chapter 11, p. 203) was for long known from only one site, so that the 1984 recovery plan necessarily focused on this. Following discovery of a second site in California (Jump et al. 2006), a review of that plan (USFWS 2007) could refine that outlook somewhat, and suggest characteristics of potentially suitable habitat to be recommended for surveys.

Setting priorities for habitat restoration for Lepidoptera is complex. However, following Schtickzelle et al.'s (2005) example of *Euphydryas aurinia* in Belgium, the most constructive and feasible approach to improve carrying capacity is to improve currently occupied patches through targeted low-risk management based on sound biological understanding. In this particular example, this involved restoring habitat suitability by 'rejuvenation' and halting overgrowth of the desired open vegetation through maintaining a short sward by light grazing or mowing to encourage growth of (and butterfly access to) the food plant *Succisa pratensis*. More generally, a high proportion of Lepidoptera conservation exercises deal with successional control and restoration of early successional stages of vegetation. The many species and assemblages of concern that depend largely or entirely on particular 'open habitats' or early successional stages and are susceptible to the inevitable natural changes that occur, pose recurring problems in management in both time and space. Successional trajectories may come to depend on unusual and unpredictable events. The endangered St Francis' satyr butterfly (*Neonympha mitchellii francisci*, Chapter 5, p. 63) is restricted to North Carolina (USA), where it occurs on wetland created by beavers that build (and subsequently abandon) their dams, after which *Carex* sedge wetlands regenerate, providing the likely food plant for the butterfly (Kuefler et al. 2008). The decline of the butterfly may have followed regional extinction of beavers (in 1897) but their reintroduction in 1939 may have again provided suitable habitat, particularly within the Fort Bragg military installation. Bombing activity may also have helped to establish habitat, with fires aiding succession restoration, and preventing development of tree cover leading to unsuitable conditions. At present the butterfly is known only from Fort Bragg (Haddad et al. 2008).

Restoration has many components, from enhancing resources on individual sites, through expanding hospitality of the matrix and edges abutting sites of interest, to creating networks of resources at a landscape level to reduce isolation and promote connectivity between otherwise isolated sites and populations. Much enhancement of critical resources involves simply direct plantings of consumable plant species to increase numbers, density or distribution, either within and around secure sites or across the wider landscape – such as the 'corridor plantings' of *Pararistolochia praevenosa* undertaken for the Richmond birdwing (Chapter 9, p. 173). Plantings are necessarily often accompanied by encouragement of naturally occurring food plants, through measures such as successional regeneration and weed control. Major responses sought thus include increases in abundance, richness and distribution from changes to vegetation structure and condition, and improvements in landscape architecture to foster connectivity, together with heterogeneity and assurance of suitable microclimatic regimes. Many grasslands and other 'open habitats' are in some way dependent on disturbance, whether through factors such as natural fires or, as in Europe, a long history of traditional low-intensity human intervention in agricultural landscapes. Persistence or restoration of such disturbance then becomes a management need, so that some regime of cutting or grazing (with stocking rates and times controlled) is a core need for conserving Lepidoptera and other taxa occupying, and dependent on, such biotopes. Maintaining or regenerating early successional stages or 'open habitats' is perhaps the most frequent activity in Lepidoptera conservation, with preservation of woodland or forest as important but not, in general, demanding equivalently intensive ecological manipulations. Activities needed for open habitat conservation include weed (including woody weed) removal and suppression, coppicing of woodlands, control of sward height and composition (by grazing, mowing, haying, burning) and restorative planting of larval food plants, nectar plants and associated native vegetation. On a landscape scale, the activities may need to be coordinated effectively to enhance practical connectivity or otherwise influence spatial arrangement. Terms such as 'restoration' are used in various senses (as is 're-introduction', Chapter 11) but is broadly the introduction, replacement or enhancement of resources (most notably food plants, nectar plants and accompanying native flora, reflecting the bias toward 'consumables' noted earlier), and the associated engineering and site preparation needed.

Whilst restoration is frequently needed in Lepidoptera conservation (as for 47 of the 50 butterfly cases reviewed by Schultz et al. 2008), success is predicated on removal or mitigation of other threats, if these measures are not associated directly with the restoration process. Very few species examined by Schultz et al. did not need active management once the major threat was removed: their North American example was for the Uncompahgre fritillary (*Boloria acrocnema*), previously threatened by overgrazing of its moist alpine habitats above the treeline. Restoration is sometimes regarded as a 'quick fix' approach for single species whose requirements are well known. Often, it is not, and wider scale reclamations such as by planting trees as recovery from surface mining and to restore hardwood forests on mined sites are inevitably much longer-term exercises – at least 25 years of recovery may be needed for the 'new communities' to mature sufficiently to parallel those on natural sites (Holl 1996), with the normally used monitoring point for assessing success or failure, of only 5

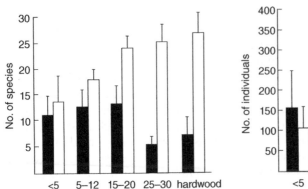
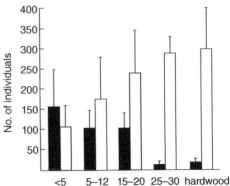

Fig. 12.6 Numbers of species (a) and individuals (b) of Lepidopterara recorded on reclaimed coal mine sites (periods of reclamation given in years) and the reference hardwood biotope in Virginia, United States, in 1993 (black bars: butterflies; open bars; moths) (Source: Holl 1996. Reproduced with permission of John Wiley & Sons.).

years, manifestly insufficient. Holl's cross-site survey, of up to 23 quarter-hectare sites representing four stages (periods) of mine reclamation and the parental hardwood biotope, yielded 52 species of butterflies and 99 moth species. Butterfly species richness and abundance decreased significantly with site age, whereas moths displayed the reverse trend (Fig. 12.6). Even after 30 years, moth communites in the restored forests lacked the kinds of beta-diversity found in undisturbed forests.

Summerville and Crist (2008) suggested that broad-scale forest restoration may be capable of providing the habitat needs of most common lepidopteran species, but that such efforts are limited by the large numbers of localised and ecologically specialised taxa. The general principles for forest management and long-term Lepidoptera conservation advanced by Summerville and Crist merit wide consideration, and are encompassed as (1) managers should attempt to allow natural disturbances to function, because disturbances such as fire maintain habitat heterogeneity and enable a mosaic of tree age classes to occur on single sites; (2) approaches to timber management and conservation planning should consider importance of spatial scale as a predictor of moth diversity, and planned at the landscape or regional scale; and (3) maintaining forest linkages among unlogged concessions may provide for refugia following cutting and facilitate recolonisation after timber extraction. Further, regional conservation planning should involve consideration of regionally unique Lepidoptera and, if knowledge is insufficent to nominate these, strategies that preserve large tracts of forests with high plant diversity may protect significant Lepidoptera diversity. Wider appreciation of local variations in moth assemblages are needed in many land use contexts.

The re-creation of forests (by extensive plantings of shrubs and trees) is a very different process from re-creation of grasslands (typically involving removal of scrub, imposition of grazing and cutting, often accompanied by extensive re-seeding). Either trajectory can be a protracted and uncertain process. Even for grasslands, a period of **at least** 10 years may be needed from commencement of the exercise in order to establish representative typical butterfly communities (Woodcock et al. 2012).

Changes in land-use regimes to reduce intensity and extent of disturbance and change may be fundamental. Thus, the basic principles of organic farming are largely compatible with many aims of conservation management, with Lepidoptera amongst the numerous beneficiaries of practices such as increasing amounts of unsprayed areas, crop rotations, reduced tillage, absence of synthetic chemicals and generally increased plant diversity. For butterflies in Britain (Feber et al. 2007) pair-wise comparisons of up to 10 each of organic and conventional farms included assessments of richness and abundance in both margins and cropping areas. In all 3 years, these were higher in margins than crops, and consistently greater in organic farms than conventional ones. Differences were not always significant – but no species was significantly more abundant on the conventional farms in any year. Conservation headlands (Dover et al. 1990) and unsprayed margins (Feber et al. 2007) are both valued as butterfly habitats, and the above inference of the influences of wider farming practice broadens the conservation perspective and endorses the considerable relevance of unsprayed edge areas.

Habitat edges or linear habitats – including often-overlooked areas such as roadsides (Chapter 8, p. 157) – have two, not always independent, roles in Lepidoptera conservation, as residential habitats supporting permanent populations, and as corridors facilitating dispersal, and reducing isolation in fragmented landscapes. Samples of insects are thus likely to include residents and transient 'tourist' species, categories that are not always easy to distinguish. Differences in edge uses between species are common, and are sometimes difficult to interpret, as based on presence and estimates of abundance. Thus the preference for forest edges alongside meadows by *Maculinea nausithous* and lack of it by *M. teleius* detected by Korosi et al. (2012) and related to microclimate conditions was interpreted rather differently by Nowicki et al. (2013), who showed that **both** species tended to avoid interior habitat in favour of edge zones, with all available natural edges (adjacent to reed and grassland as well as forest) used to a similar extent. This pattern was linked with nesting needs of the host *Myrmica* ants, for which edges constituted refuges (see also Nowicki et al. 2007). This edge effect was evident in three large patches, each exceeding 1 ha in area, but not on two smaller patches, which in functional terms probably lacked any interior habitat. The practicalities of conserving and restoring edge habitats are substantial, not least because they are often not specifically included in management regimes for the sites they border or enclose (discussed for *Maculinea* by Nowicki et al. 2013).

Wider landscape restorations within agricultural areas may involve very extensive revegetation exercises. The New Zealand 'Greening Waipara' project reported by Gillespie and Wratten (2012), sought to restore previously common native plants in and around lowland agricultural areas formerly covered by forests but cleared extensively from the mid-nineteenth century on, with the target of fostering native butterflies through providing resources. Some endemic

Lycaenidae were almost wholly restricted to native vegetation remnants, and, particularly for the Common copper, *Lycaena salustius*, provision of additional nectar sources elsewhere may aid functional connectivity between such remnants.

However, much restoration or other habitat manipulation has targeted individual species or populations. But, even when the critical resources are well known and few, satisfactory restoration can be a very complex process. The endangered North American Fender's blue butterfly (*Icaricia icarioides fenderi*) requires only a single larval food plant (Kincaid's lupin, *Lupinus sulphureus kincaidii*) and native wildflower nectar for foods, in a short-grass upland prairie environment with grasses and forbs. Threshold amounts of food resources needed were calculated by Schultz and Dlugosch (1999), as around 20 mg of sugar/m^2 of nectar from native forbs, and larval food plants at around 40 leaves/m^2 of the lupin. Such knowledge is an invaluable guide to minimum restoration targets and in helping to overcome one of the major reasons for failure of re-introductions and translocation attempts (Chapter 10, p. 190). Habitat restoration for Fender's blue also benefited from availability of suitable sites distributed across the landscape and within natural dispersal ranges, to serve as 'stepping stones'. An experimental trial to compare several possible restoration options demonstrated the importance of linking focused restoration to the individual needs of the focal species (Schultz 2001: Fig. 12.7; Schultz & Crone 1998), as well as the values of practical trials on the various options that might

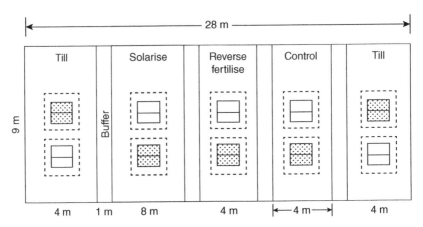

Fig. 12.7 Experimental design of plots to evaluate restoration measures for habitats of Fender's blue butterfly. A sample experimental plot is shown, with five soil treatments: two till treatments (identical); solarisation (laying sheets of clear plastic over freshly tilled soil as a 'heat shield; note treatment area larger than others to accommodate need for a 2–3 m buffer around edge of sheets); reverse fertilisation (restoring historically nutrient-poor soil condition); and a control. In each block positions of treatments are randomised, in each treatment, dark boxes are high forb: low grass plantings and open boxes are low forb: high grass plantings; each is divided into a weeded and an unweeded half and surrounded (dashed line) by an area clipped to assess weed biomass (Source: Schultz 2001. Reproduced with permission of John Wiley & Sons.).

be available, and of careful monitoring to evaluate these. Thus, attempted establishment of the lupin by seeding (using >12,000 seeds) and transplanted seedlings (600) from 1995 on led to fewer than 60 plants present in 1999, a number insufficient to support a butterfly population. Establishment of nectar plants was also complex, and no single nectar plant species was highly successful at both sites used. Planting of several species provided greater opportunity than relying on any single species. For both lupins and nectar plants, site treatments were an important influence on success.

Perhaps the major environments in which practical conservation management is needed frequently for Lepidoptera are those most influenced by people, namely agricultural ecosystems, forest biotopes and the wider landscape subject to urban and recreational pressures. Across these, a variety of general management tactics have been adopted, with varying outcomes. Examples are outlined in the following sections. Whilst these are noted separately, opting for any single management approach – for example to grasslands for butterfly conservation – may be very uncertain. Comparing responses to prairie burning, grazing or both, Vogel et al. (2007) concluded that no single management practice would benefit all species, or even all species that were either habitat specialists or habitat generalists. Responses of individual species were highly variable. Rotational management of grassland and other open habitats in agricultural landscapes can have the dual roles of maintaining variability by mosaics ensuring successional regenerations, and providing refuges from any single disturbance process. Treatments of field margins, for example by burning or disking (Dollar et al. 2012), had rather different effects on butterflies, but also different levels of risk – so that disking, disturbing the ground surface, could be applied more precisely than fire.

Grazing

Livestock grazing is used widely for direct conservation management of grasslands, and has considerable value in changing composition and structure of vegetation and influencing the local microclimates of the areas treated. Extensive investigations in Europe have demonstrated its roles in management of Lepidoptera and other invertebrates. Erhardt (1985) emphasised its importance in maintaining early successional stages with, in general, high plant diversity, and sward heights maintained by mosaic grazing treatments leading to environments that are more hospitable than ungrazed, heavily grazed, or mown areas. As demonstrated clearly for British chalk grassland (BUTT 1986), grazing regimes can be tailored for particular seasonal impacts and to maintain particular sward heights needed by individual species; height and structure of sward in grasslands are perhaps the most influential features on habitat quality for many insects living there. High grazing intensity is common in grassland management, to create and maintain high floristic diversity, with variety in grazing regimes introduced through rotation, stocking intensity, seasonality and animal type – with many such treatments designed to meet the requirements of individual species of conservation significance.

Grazing impacts on grasslands are far more complex to interpret than mowing treatments (below), because the latter is essentially non-selective whilst grazing may induce subtle changes in vegetation composition and ground exposure. Following other authors, Saarinen and Jantunen (2005) noted that grazing creates a mosaic of habitat conditions likely to foster assemblages of species living there, with rotational grazing increasing further the successional mosaics that may be present. Both treatments, however, participate in traditional management practices in Europe. Comparisons of 37 butterfly species across mown (meadows) and grazed (pastures) revealed rather similar richness in assemblages, with impacts of the treatments rather similar in a region of Russia and Finland where traditional management still occurs. Slightly higher diversity and richness in mown meadows was attributed in part to grazing reducing the most important nectar plants in pastures, leading to reduced butterfly populations, in a trend increased with more intensive grazing.

However, as Poyry et al. (2004, 2005) noted, the roles of grazing management in restoring wider insect communities have been relatively neglected. They compared diurnal Lepidoptera on three categories of semi-natural grasslands in Finland, namely (1) 'old pastures' (grazed annually for at least several decades), (2) 'abandoned pastures' (on which grazing had ceased more than 10 years previously), and (3) 'restored pastures' (on which grazing had been re-instated for at least 3 years after >10 years of abandonment). Transect counts over 2 years revealed 96 species (64 in 1999, 88 in 2000), with simple measures of species richness and abundance highest in abandoned pastures (Table 12.1), and no major differences between the two grazed regimes. More species (15) were associated with abandoned than with old pastures (3), with some being sufficiently characteristic to be considered indicative of those biotopes. However, diversity and evenness were highest in the old pastures, and lowest in either restored or abandoned pastures. These increases were attributed in part to disappearance of abundant species typical of abandoned pastures, so that the number of potential host plants is probably highest in old pastures. After 5 years of

Table 12.1 Species richness and abundance (mean ± 1SD) of butterflies and diurnal moths on mesic semi-natural grasslands in Finland under different management regimes (after Poyry et al. 2004).

	Pasture regime*		
	1	2	3
Number of patches 1999/2000	6/5	6/6	6/4
All species			
Species richness	25.9 ± 6.5	22.4 ± 4.7	33.3 ± 7.8
No. of individuals	126.2 ± 57.9	126.0 ± 73.8	306.3 ± 141.8
Grassland species			
Species richness	20.8 ± 3.3	18.1 ± 2.9	24.3 ± 3.6
No. of individuals	118 ± 54.1	120 ± 71.0	282.8 ± 131.2

*Regimes are (1) old pastures, grazed annually for at least several decades; (2) abandoned patches, where grazing ceased >10 years ago; (3) restored patches where grazing reinstated 3–8 years ago after >10 years abandonment

restoration management, some species characteristic of old pastures had not recovered in restored sites, so that a considerable period may be needed for host plants and vegetational structure to render such areas suitable. Poyry et al. supported earlier suggestions of increasing management flexibility to allow for different grazing intensity or rotational grazing on individual patches, but also assuring continuity of coordinated management on a wider regional level. Used thus, grazing has considerable potential in restoration of abandoned or degraded semi-natural grasslands, perhaps in concert with wider restoration measures. The extensive losses of grazed or mown calcareous grasslands in central Europe have been associated clearly with declines of many Lepidoptera, including Red-Listed butterflies (Balmer & Erhardt 2000). Later successional stages, such as old fallow land may, however, continue to support more species, including Red List taxa in Switzerland, than earlier stage (more heavily managed) grasslands. This land was considered to be 'at least as important for butterflies as extensively cultivated grassland'.

Grazing and other management impacts can be substantial, and also very difficult to interpret – notwithstanding apparently clear responses. Moth diversity is influenced strongly by grazing management: both in Scottish grasslands (Littlewood 2008) and coastal saltmarshes in Germany (Rickert et al. 2012), the lowest species richness was associated with the most intensive grazing regime applied, supporting the principle that severe disturbance is highly detrimental. The Scottish moth samples (taken by light trapping and comprising an overall 6230 individuals across 152 species) were categorised by feeding habit, overwintering stage and conservation status (Table 12.2) across the grazing regimes compared (Fig. 12.8). The trends mirrored those found by Poyry et al. (2006)

Table 12.2 Numbers of individual moths recorded in light traps (38 nights) in four grazing regimes (T1–T4) in Scotland, categorised according to conservation status ('BAP', Biodiversity Action Plan listed), food plant preferences and overwintering stage (Source: Littlewood 2008. Reproduced with permission of John Wiley & Sons.).

	Grazing regime*			
	T1	T2	T3	T4
Category				
Graminivorous	434	599	461	637
Non-graminivorous	749	1297	1014	1039
Overwintering stage (127 species):				
Egg	299	706	440	289
Larva	696	1008	865	1161
Pupa	121	117	100	152
BAP species (16 species)	149	163	136	187
Non-BAP species	882	1317	1121	1294

*T1, high-intensity sheep grazing at commercial stocking rate of 2.7 sheep/hectare; T2, low-intensity sheep grazing at 0.9 sheep/hectare; T3, low-intensity mixed grazing with 2 sheep/plot for 4 weeks in winter, plus 2 cows each with 1 calf to represent equivalence to T2; T4, ungrazed control

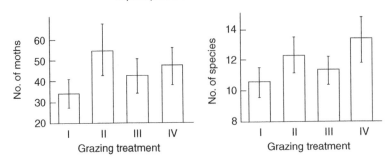

Fig. 12.8 Relationships of grazing treatment and moth species richness (a) and abundance (b) in Scotland. Grazing treatments; I, high-intensity sheep grazing; II, low-intensity sheep grazing; III, low-intensity mixed grazing; IV, ungrazed control (see text) (Source: Littlewood 2008. Reproduced with permission of John Wiley & Sons.).

in suggesting that Lepidoptera in grassland may be less tolerant to grazing disturbance than the plants themselves. Grazing might also select against graminivorous species, whilst being less detrimental to herbivorous species in the same environments. Similar inferences arose for butterflies in Oregon, in response to cattle grazing in which the role of grazing itself appeared secondary to influences of plant species richness and diversity (Runquist 2011). Whilst butterflies with grass host plants declined at high grazing intensity, this did not predict total butterfly diversity. More local and site-specific impacts of grazing may result from trampling and manuring.

Mowing

Mowing (including 'haying') and grazing are both used extensively in management in which reduction of sward height is needed, and mowing can be tailored for greater uniformity and control in treatment. Mowing regimes can thus also be integrated with ecology of particular focal species, as demonstrated for conservation outcomes for wetland meadows inhabited by *Maculinea teleius* and *M. nausithous* in Europe (Johst et al. 2006). Mowing had two major impacts – on the suitability of the larval food plant, with mowing at 'the wrong time' causing mortality of eggs and larvae, and affecting the microclimate needed by the host *Myrmica* ants. Without mowing, succession may displace both *Sanguisorba* and the host ants. Regimes suitable for wider landscape use were one cut every year, or every second, or every third year, with mowing to take place in the weeks before or after the flight period; it was considered essential that no more than one cut occur in any year, so departing from the farmers' traditional mowing regime of two cuts/year.

Modelling the outcomes to incorporate ecological differences between these two species (Johst et al. 2006) suggested that *M. teleius* could persist only at a regional scale, with dispersal between meadows, whilst *M. nausithous* could survive at both local and regional levels.

An issue of much wide relevance is that a disturbance from a practice such as mowing may have unforeseen indirect impacts. The possibility of mowing roadside verges leading to increased roadkills of butterflies (Chapter 8, p. 157) from heightened activity (Skorka et al. 2013), exemplifies the wider needs for careful planning in seasonal treatments to reduce such effects.

Coppicing

An important traditional component of temperate region forest management, coppicing has considerable significance in promoting and retaining open spaces in forest environments, and as a manipulable tool for adjusting succession. Thus, the only management choice to conserve the last remaining population of the Scarce fritillary butterfly (*Euphydryas maturna*) in the Czech Republic is considered to be re-establishment of traditional short-rotation coppicing (Konvicka et al. 2005), both at the sole occupied site and in surrounding woods to facilitate re-colonisation, perhaps aided by re-introductions. As in some other cases, the proposal was considered risky as not wholly compatible with current forestry practices and standards – but Konvicka et al. emphasised that no procrastination could be tolerated in view of the vulnerability of the population, which comprised fewer than 200 individuals. The butterfly is not a strong disperser, and is restricted to clearings within deciduous forests, so that restoration of such early successional environments is critical.

Coppice management can strongly influence the Lepidoptera assemblages present, and promoted clear changes in assemblage composition in Britain (Broome et al. 2011). In that study, based on light trap comparisons over 3 years in coppiced sweet chestnut (*Castanea sativa*) woodland, the time since coppicing influenced moth assemblages through food plant accessibility and exposure. Many species largely depending on ground flora benefited from newly cut areas, whilst others were restricted to more mature systems of around 20 years since coppicing. Many of the latter were regarded as threatened or scarce, and cannot be supported in more frequently coppiced systems, leading Broome et al. to endorse earlier opinion that very old coppice regrowth should not be felled if such species are to be safeguarded. Coppicing is thus an important conservation consideration for many macromoths (Merckx et al. 2012a), but the outcomes may be complex to interpret. Overall, nocturnal moth richness and abundance (from the pool of about 265 species in that study) was **lower** at coppiced sites and on wide woodland rides than on 'standard' woodland and rides in contrast to diurnal taxa, possibly reflecting the need of the latter to gain heat by exposure in open areas, so accounting for the frequent observations of heightened butterfly abundance in open areas within woodlands. Many nocturnal moths, in contrast, need dark areas (for shade and moisture) and this group may be susceptible to woodland clearing or increased open areas. Merckx et al. stressed the importance of these 'core' habitats for many declining moths, and the accompanying need to buffer these by forms of zoning that include coppicing and widened rides near their edges, so also providing accessible woodlands for the numerous species that inhabit mixed and more open areas, through diversifiying conditions for these relatively generalist taxa.

Burning

Deliberate 'control' burns may be variously a severe threat or valued management tool in Lepidoptera conservation. Used increasingly for 'fuel reduction', often in conjunction with human asset protection, controlled fires can be important agents for regenerating successions, and also have roles in restoration through, for example, effecting weed control. Thus, the recent absence of fire (long used traditionally by native Americans) on natural grasslands in western North America has led to succession of these through to shrublands and Sitka spruce forests, essentially eliminating the habitats needed by many species. The uses of planned burns and other management to reduce risks from future wildfires are complex. Comparison of three approaches to fire suppression (total clearance of non-herbaceous vegetation to provide refuges for people, bush removal but retention of trees, livestock overgrazing) revealed the first two of these reduced numbers of specialist butterfly species but favoured some generalists through 'opening up' areas for grass invasions, whilst grazing alone retained the trees and shrubs and favoured endemic or specialised butterflies (Ricouart et al. 2013).

Even without vegetational succession, Hammond and McCorkle (1984) suggested that the thick smothering mat of grass thatch acumulated progressively in long-unburned grasslands crowds out the specific *Viola* food plants of rare *Speyeria* fritillaries, and that a regime of burning every 5–10 years would render the areas suitable again.

The ambiguities are illustrated well from this notable nymphalid butterfly genus. Prescribed burning was regarded as a serious concern for *Speyeria idalia* in Kansas (Powell et al. 2007). It was more abundant on unburned refuge areas, but on some burned sites recovered several years after burning (Swengel & Swengel 2007), whilst some other species may never recover. The Swengels recommended that non-fired refugia should be established for Lepidoptera in prairies, with these sited where the greatest numbers of specialist species may benefit, and that they should be managed unintensively, such as by mowing when needed. They thus endorsed widespread advocacy for mosaic burning to ensure habitat heterogeneity, by adding permanent unburned refuges within the traditional treatments. As for *Paralucia pyrodiscus*, discussed later in this section, refuges to be protected during any planned burn may focus on the densest parts of the focal population. However, the influences of fires are not always clearly definable for (1) mortality of early stages; (2) impacts on food plants; (3) impacts on nectar resources; and (4) changes in microclimate. The first two of these are perhaps those most often acknowledged, together with wider community changes wrought through changes in vegetation composition and structure, and spatial aspects of connectivity. Prescribed fires were a tool in restoring a pine-bluestem community associated with abundance of two *Speyeria* species, through provision of high-quality nectar resources, with the abundance of both species greatest in the second season post-fire and both essentially absent from control (unrestored) sites (Rudolph et al. 2006b). Also in Arkansas, Rudolph et al. (2006a) suggested that widespread suppression of fires throughout the twentieth century had substantially reduced the nectar available to migrating Monarchs (*Danaus plexippus*).

Use of deliberate burns in Lepidoptera resource management is relatively rare, but has been undertaken or appraised in prairies of North America and in fire-prone environments of Australia, with the practical considerations including many components of a fire regime, such as area of burns (from broad areas to fine 'micro-mosaic' local burns), intensity of fire, season, 'time-since-fire' and interval between burns (New et al. 2010). Very localised burns may be feasible with adequate control and care – to the level of individual sedge tussocks in stands to regenerate older plants and induce new foliage suitable for oviposition and con- sumption. Thus individual senescing tussocks of *Gahnia filum* in Victoria, Australia, can rapidly regenerate fresh foliage attractive for ovipositing females of the Yellowish sedge-skipper (*Hesperilla flavescens flavescens*) if burned 2–3 months previously (Chapter 2, p. 27) (Relf & New 2009) – and in this case individual tussocks can be selected as not harbouring the skipper by simple inspection to confirm the absence of the conspicuous larval foliage shelters. Whilst these appli- cations are clearly conservation-orientated, rotational burning may also be linked with harvest levels of exploitable species. High caterpillar yields of mopane worms sought for human food in Africa (Munthali & Mughogho 1992) can be improved by rotational burns, with the putative impacts of reduction of caterpillar preda- tors, regeneration of nutritious foliage and restricting heights of occupied trees, so facilitating hand harvest.

Fire can cause direct mortality to exposed insects, especially non-vagile early stages, and loss of host plants, with visually severe – and often long-lasting – struc- tural changes to sites. It is one of the two main management measures (with haying) used in the extensive tall-grass prairies of the central United States (Swengel 1996). Many prairie patches are managed primarily by cool season fires across about a quarter of the patch area each year. Of 90 species of butterflies surveyed by Swengel in comparing prairies subject to burning or haying, 42% of the total 80,906 individuals occurred in recently burned (since the last growing season) plots, with the proportion of individuals of the 34 more abundant species ranging from 0.8–84% of the totals. Analyses demonstrated that specialists declined strongly and significantly after fires, with effects persisting for 3–5 years or more, whilst many generalists with the broadest habitat tolerances were most abundant in the most recently burned areas, and scarcest in those longest unburned. Haying was more beneficial to specialist butterflies and also for total butterflies. Swengel (1996, also Swengel and Swengel, 1996, for the Karner blue butterfly as an example) recognised five major factors affecting butterfly responses to fire management, as (1) extent of specialisation; (2) vagility (low vagility increases susceptibility, whilst high vagility facilitates recolonisation); (3) above ground stages at time of fire (vulnerable, compared with under ground as refuge); (4) voltinism (more than one generation increases chances of recolonisation); and (5) responses of key food plants to burning. Each may be important in relation to mosaic burning regimes. For the Karner blue, early successional oak barrens habitat can only be conserved by some form of disturbance to prevent succession to oak woodland and forest. Autumn burns are a common approach to this, in what Shuey (1997) referred to elegantly as 'dancing with fire' in a milieu in which isolated butterfly populations are vulnerable both to succession and ecosystem disturbance regimes, and in which 'metapopulations swirled with patch dynamics to the music of fire'.

Understanding impacts on individual species depends on understanding responses of key food plants to burning, with fires regarded as 'the most significant factor that influences the composition of tallgrass communities' (Vogel et al. 2010), and fires are regarded in North America, as in Australia and South Africa, as compo- nents of the historical moulding of many biological communities across the landscape.

Vogel et al. (2010) emphasised the parameter 'time since burn' in relation to butterfly abundance, floral resources, cover of bare ground and cover of warm season grasses: both total butterfly richness and abundance and those of habitat specialists were correlated positively with time since burn, and did not level off or decline during the 5-year study, so that recovery times for some species may be longer than documented previously. Fire intervals on the managed prairies are often less than 5 years, so that recovery rates and fire dispersion are important considerations for specialist species. Potential for recolonisation of burned areas depends on persistence of the species in refuge areas nearby, but small numbers may be sufficient for this to occur. A single egg mass from a female *Euphydryas gillettii*, for example, can lead to population establishment on burned sites, but there is also a strong element of chance in this succeeding (Williams 1995). Spatial and temporal fire mosaics are highly relevant considerations (New et al. 2010) – but in some instances the use of fire in management for individual species may be adjusted for compatibility with any 'refuges' in space or time that the focal taxon may have. For example, these adjustments can include that fires should not coincide with active feeding seasons (time) and avoiding local hot spots of abundance (space). Direct exposure of immobile early stages is clearly undesirable – impacts of fire on the diurnal saturniid *Hemileuca eglanterina* in its wet prairie biotopes in western North America can be assessed only by examining these separately on various life stages, in contrast to the more usual approach of using only the adult stage in monitoring changes (Severns 2003). Thus, surveys of burned and unburned patches suggested that examination of egg distribution implied adult bias toward burnt areas for oviposition (with eggs destroyed if laid before a burn occurs). This bias was not evident from adult surveys, and if caterpillar abundance was surveyed the fire appeared catastrophic. When planning use of fire in rejuvenating the small isolated urban sites occupied by the Eltham copper (*Paralucia pyrodiscus lucida*) in Australia, considerations in designing a suitable site burn included (1) that it should take place as late in the summer as possible, at a time when no adults or eggs are present, and caterpillars are relatively well grown before their overwintering phase, during which little feeding occurs; (2) that it should take place during the day, when the nocturnally active caterpillars are shielded from direct exposure to flames, by being in the underground nest chambers of their host ant; and (3) that small patches with high caterpillar numbers, determined by surveys in previous weeks, should be damped down and protected from fire. The exercise was discussed in more detail by New et al. (2000) and New (2011), but emphasised the relevance of examining whether possible refuges exist, and whether their roles can be tailored within management schemes to reduce chances of the possibly catastrophic outcomes that could result from less informed burning. A close parallel has been suggested for the Karkloof blue, *Orachrysops ariadne*, endemic to part of the mistbelt grassland of KwaZulu-Natal, South Africa, and there known from four sites. Fire is both an important ecological factor and a tool in habitat management for this butterfly and, as for the Eltham copper in Australia, management recommendations include that burning should take place during the larval refuge stage, and be undertaken only after caterpillars have moved underground to nests of their host ant, *Camponotus natalensis* (Lu & Samways 2001, 2002), with prescribed 'rotational burning' between June and

October. This timing avoids any damage to eggs or adults, and the populations at two sites survived burning of the entire sites – although Lu and Samways considered it prudent not to do this. And, as for the Eltham copper food plant (*Bursaria spinosa*), mature larval food plants of *O. ariadne* (*Indigofera woodii* var. *laxa*) survive and respond well to fire. The host ants of both these taxa are least active in winter and are mainly nocturnal.

A further parallel, in North America, is for the Dakota skipper (*Hesperia dakotae*), for which management prescriptions include burning, grazing and mowing (but with burning also implicated in historical loss of populations), and for which very specific prescribed burn conditions were summarised in detailed guidelines (Table 12.3) (Cochrane & Delphey 2002). Likewise, short-term burning

Table 12.3 Suggestions for use of prescribed fire for habitat management of the Dakota skipper butterfly, as recommended by Cochrane & Delphey (2002) (slightly abbreviated).

Divide the Dakota skipper habitat at the site into as many burn units as feasible.

Use the maximum length fire return interval that is adequate to maintain or restore high-quality native prairie habitat on each burn unit. Allow at least 3 years to elapse without fire (i.e. minimum 4-year rotation) before re-burning any area. Longer intervals may be necessary where Dakota skipper populations are small and/or isolated

Never attempt to burn an entire Dakota skipper habitat patch in any single year

Allow fires to burn in a patchy ('fingering') pattern within units – i.e. do not make concerted effort to burn 'every square inch', and leave fire 'skips' unburned. Burning under cool or damp conditions may increase survival of insects present in the litter layer within the burned unit

Consider the use of proactive techniques to increase the patchiness of fires, especially if habitats that would serve as sources of recolonising adults are small or greater than 0.5 km from the burn area

Conduct pre-burn surveys and evaluate other applicable information to understand the distribution and relative abundance of Dakota skippers within and among burn units. Poor weather or other conditions (e.g. persistent high winds) may reduce the likelihood of adequate survey conditions during the flight period in any given year. Therefore, it may be prudent to plan surveys for at least 2 consecutive years before a planned burn

Spring burns should be conducted as early as possible to limit larval mortality – in southwest Minnesota, for example, burns on or before May 1 may be early enough to ensure that Dakota skipper larvae have not yet emerged from their buried shelters. Dakota skipper larvae are less vulnerable to fire before they have resumed activity in the spring and after they have ceased activity in the fall (i.e. when they are in shelters at or below the ground surface). Moreover, late spring burns may delay flowering of early and midsummer blooming forbs, thereby limiting nectar sources for Dakota skippers during their flight period. Fall burns, however, may result in higher soil temperatures than early spring burns and greater mortality of larvae, even after they have retreated for the season to shelters at or below the ground surface. In addition, the removal of plant material by fall burns may expose larvae to greater temperature extremes during winter, and which may reduce their survival

(Continued)

Table 12.3 (*Continued*)

If fires may need to be conducted in late spring to address a particular management need, other precautionary measures will be especially important. These include the division of occupied Dakota skipper habitat into multiple burn units, ensuring that fires stay within planned burn areas, maximising the number of years between fires and reducing fuel loads (such as by haying) in Dakota skipper habitat in units where frequent or intense fire is not necessary

If a site is managed with prescribed fire, subdivide Dakota skipper habitat into rotational burn units even if all burning will likely be done when Dakota skippers are in sub-surface shelters. Other species of butterfly that rely on native prairie may still be vulnerable to fire mortality even during early spring fires

If you plan to change the configuration of burn units or make other changes to your prescribed fire plan, review the location and timing of recent burns to understand the potential effects of these previous fires on the current abundance and distribution of Dakota skippers on the management area

Be sure to consider any other rare, prairie-dependent species present on sites when designing burn plans

Plan for escape of fires out of burn units if that is a reasonable possibility. That is, plan for the contingency that a prescribed fire will escape a burn unit and burn one or more additional units that contain Dakota skipper habitat. If this is reasonably likely, determine how the Dakota skipper population would persist despite such a scenario

High fuel levels increase the likelihood that fires will kill Dakota skipper, even during early spring burns when larvae are still in their sub-surface shelters. Therefore consider reducing fuel levels (e.g. by haying the previous fall) before conducting burns where fuel levels seem to be high

of sites supporting the Marsh fritillary, *Eurodryas aurinia*, in Glamorgan, Wales, appeared to be beneficial, with high numbers of larval webs on a site wholly burned about 6 months previously supporting the notion that 'flash fires' may not necessarily be harmful (Lewis & Hurford 1997). Nevertheless, risks occur, and these authors endorsed earlier remarks by Warren (1994) that grazing management is preferable and 'must always be the management regime of choice on any *E. aurinia* site'.

Burning has been considered essential to maintain the grassland habitat of *Shijimiaeoides divinus asonis*, one of the rarest lycaenids in the Aso region of Japan, and the grasslands are burned routinely by farmers in spring (Murata et al. 1998). Butterfly numbers declined markedly in areas where burning was neglected, and appeared to increase once burning was resumed. Burning also benefits the larval food plant (*Sophora flavescens*) and a range of nectar plants, in part by reducing tall grasses that compete with, and overgrow, these.

Any such pioneering case is high risk, with potential for loss of resident populations a prime consideration. In many cases, preservation of resident populations may not be practicable, so that landscape issues have increased relevance in considering potential for site/patch recolonisation from elsewhere. However any generalities are difficult to endorse, because every case is different and responses of the same species under different conditions may also differ. Separating the effects of historical treatment on sites and the impacts of currently imposed management such as fire or grazing remains uncertain. Thus, despite the advocacy of planning for refuges, as above, two specialist prairie butterflies with caterpillars that overwinter above ground (*Speyeria idalia, Cercyonis pegala*) were expected to be less abundant during the summer immediately after spring fires. This was not the case, and *C. pegala* actually increased in density over control unburned plots (Moranz et al. 2012): this was attributed to mosaic effects, allowing for rapid colonisation from nearby pastures and that the fires themselves left a pattern of less disturbed habitat. In addition, fire impacts may well change in the future, as Mollenbeck et al. (2009) suggested for *Hipparchia fagi* in Europe. At present, winter burning of vineyards has little effect on over-wintering caterpillars of this species, because the habitats are sparsely vegetated and, in many cases, legally protected. However, with any expansion into more densely vegetated areas with climate warming, more combustible habitats may become important and fire impacts change accordingly.

References

Ahmed, S.A. & Rajan, R.K. (2011) Exploration of vanya silk biodiversity in north eastern region of India: sustainable livelihood and poverty alleviation. International Conference on Management, Economics and Social Sciences (ICMESS'2011), Bangkok, 485–489.

Akpalu, W., Muchapondwa, E. & Zikhali, P. (2007) Can the restrictive harvest period policy conserve mopane worms in Southern Africa? A bioeconomic modeling approach. *University of Pretoria Working Paper*, no. 65, 1–11.

Altermatt, F. (2012) Temperature-related shifts in butterfly phenology depend on the habitat. *Global Change Biology* 18, 2429–2438.

Altizer, S.M. & Oberhauser, K.S. (1999) Effects of the protozoan parasite *Ophryocystis elektroscirrha* on the fitness of monarch butterflies (*Danaus plexippus*). *Journal of Invertebrate Pathology* 74, 76–88.

Altizer, S.M., Oberhauser, K.S. & Geurtz. K.A. (2004) Transmission of the protozoan parasite *Ophryocystis elektroscirrha* in monarch butterfly populations: implications for prevalence and population-level impacts. pp. 203–215 in Oberhauser, K.S. & Solensky, M.S. (eds) The Monarch Butterfly. Biology and Conservation. Cornell University Press, Ithaca and London.

Arnold, R.A. (1983a) Ecological studies on six endangered butterflies (Lepidoptera: Lycaenidae): island biogeography, patch dynamics, and the design of habitat preserves. *University of California Publications in Entomology*, 99. Berkeley.

Arribas, P., Abellan, P., Velasco, J., Bilton, D.T., Millan, A. & Sanchez-Fernandez, D. (2012) Evaluating drivers of vulnerability to climate change: a guide for insect conservation strategies. *Global Change Biology* 18, 2135–2146.

Asher, J., Warren, M., Fox, R., Harding, P., Jeffcoate, G. & Jeffcoate, S. (2001) The Millennium Atlas of Butterflies in Britain and Ireland. Oxford University Press, Oxford.

Balmer, O. & Erhardt, A. (2000) Consequence of succession on extensively grazed grasslands for Central European butterfly communities: rethinking conservation priorities. *Conservation Biology* 14, 746–757.

Barron, M.C., Barlow, N.D. & Wratten, S.D. (2003) Non-target parasitism of the endemic New Zealand red admiral butterfly (*Bassaris gonerilla*) by the introduced biocontrol agent *Pteromalus puparum*. *Biological Control* 27, 329–335.

Beaumont, L.J. & Hughes, L. (2002) Potential changes in the distributions of latitudinally restricted Australian butterflies in response to climate change. *Global Change Biology* 8, 954–971.

Benson, J., Pasquale, A, Van Driesche, R. & Elkinton, J. (2003) Assessment of risk posed by introduced braconid wasps to *Pieris virginiensis*, a native woodland butterfly in New England. *Biological Control* 26, 83–93.

Boettner, G.H., Elkinton, J.S. & Boettner, C.J. (2000) Effects of a biological control introduction on three non-target native species of saturniid moths. *Conservation Biology* 14, 1798–1806.

Broome, A., Clarke, S., Peace, A. & Parsons, M. (2011) The effect of coppice management on moth assemblages in an English woodland. *Biodiversity and Conservation* 20, 729–749.

BUTT (Butterflies Under Threat Team) (1986) The management of chalk grassland for butterflies. Focus on Nature Conservation no 17. Nature Conservancy Council, Peterborough.

Chen, I.-C., Shiu, H.-J., Benedick, S. et al. (2009) Elevation increases in moth assemblages over 42 years on a tropical mountain. *Proceedings of the National Academy of Sciences* 106, 1479–1483.

Cochrane, J.F. & Delphey, P. (2002) Status assessment and conservation guidelines: Dakota Skipper, *Hesperia dacotae* (Skinner) (Lepidoptera: Hesperiidae), Iowa, Minnesota, North Dakota, South Dakota, Manitoba and Saskatchewan. United States Fish and Wildlife Service, Department of the Interior, Minneapolis, Minnesota.

Davis, B.N.K., Lakhani, K.H. & Yates, T.J. (1991) The hazards of insecticides to butterflies of field margins. *Agriculture, Ecosystems and Environment* 36, 151–161.

Dennis, R.L.H. (1993) Butterflies and Climate Change. Manchester University Press, Manchester.

Dollar, J.G., Riffell, S.K. & Burger, L.W. (2012) Effects of managing semi-natural grassland buffers on butterflies. *Journal of Insect Conservation* DOI 10.1007/s 10841-012-9543-7.

Dover, J.W., Sotherton, N. & Gobbett, K. (1990) Reduced pesticide inputs on cereal field margins – the effects on butterfly abundance. *Ecological Entomology* 15, 17–24.

Eastwood, R., Kongnoo, P. & Reinkaw, M. (2010) Collecting and eating *Liphyra brassolis* (Lepidoptera: Lycaenidae) in southern Thailand. *Journal of Research on the Lepidoptera* 43, 19–22.

Erhardt, A. (1985) Diurnal Lepidoptera: sensitive indicators of cultivated and abandoned grassland. *Journal of Applied Ecology* 22, 849–862.

Eisenbeis, G. & Hassel, F. (2000) Zur Anziehung nachtaktiver insekten durch Strassenlaternen: eine Studie kommunaler Beleuchtungseinruchtungen in der Agrarlandschaft Rheinhessens. [Attraction of nocturnal insects to street lighting – a study of municipal lighting systems in a rural area of Rheinhessen (Germany)]. *Natur und Landschaft* 75, 145–156 (in German).

Feber, R.E., Johnson, P.J., Firbank, L.G., Hopkins, A. & Macdonald, D.W. (2007) A comparison of butterfly populations on organically and conventionally managed farmland. *Journal of Zoology* 273, 30–38.

Fox, R. (2013) The decline of moths in Great Britain: a review of possible causes. *Insect Conservation and Diversity* 6, 5–19.

Frank, K.D. (1988) Impact of outdoor lighting on moths: an assessment. *Journal of the Lepidopterists' Society* 42, 63–93.

Gagné, W.C. & Howarth, F.G. (1985) Conservation status of endemic Hawaiian Lepidoptera. Proceedings of the 3rd Congress of European Lepidoptera, 74–84.

Gahukar, R.T. (2011) Entomophagy and human food security. *International Journal of Tropical Insect Science* 31, 129–144.

Gibbs, G.W. (1980) New Zealand Butterflies. Collins, Auckland.

Gillespie, M. & Wratten, S.D. (2012) The importance of viticultural landscape features and ecosystem service enhancement for native butterflies in New Zealand vineyards. *Journal of Insect Conservation* 16, 13–23.

Gripenberg, S., Hamer, S., Brereton, T., Roy, D.B. & Lewis, O.T. (2011) A novel parasitoid and a declining butterfly: cause or coincidence? *Ecological Entomology* 36, 271–281.

Guo, J., Lin, X. & Kanari, K. (2012) Towards sustainable livelihoods from sustainable medicinal products: economic aspects of harvesting and trading the Chinese Caterpillar Fungus *Ophiocordyceps sinensis* and Southern Schisandra *Schisandra sphenanthera* in China's Upper Yangtze ecoregion. *Traffic Bulletin* 24, 16–24.

Haddad, N.M., Hudgens, B., Damiani, C., Gross, K., Kuefler, D. & Pollock, K. (2008) Determining optimal population monitoring for rare butterflies. *Conservation Biology* 22, 929–940.

Hammond, P.C. & McCorkle, D.V. (1984) The decline and extinction of *Speyeria* populations resulting from human environmental disturbances (Nymphalidae: Argynninae). *Journal of Research on the Lepidoptera* 22, 217–224.

Holl, K.D. (1996) The effect of coal surface mine reclamation on diurnal lepidopteran conservation. *Journal of Applied Ecology* 33, 225–236.

Howarth, F.G. (1983) Biological control: panacea or Pandora's box? *Proceedings of the Hawaiian Entomological Society* **24**, 239–244.

Howarth, F.G. (1991) Environmental impacts of classical biological control. *Annual Review of Entomology* **36**, 485–509.

Johst, K., Drechsler, M., Thomas, J.A. & Settele, J. (2006) Influence of mowing on the persistence of two endangered large blue butterfly species. *Journal of Applied Ecology* **43**, 333–342.

Jump, P.M., Longcore, T. & Rich, C. (2006) Ecology and distribution of a newly-discovered population of the federally threatened *Euproserpinus euterpe* (Sphingidae). *Journal of the Lepidopterists' Society* **60**, 41–50.

Kellogg, S.K., Fink, L.S. & Brower, L.P. (2003) Parasitism of native luna moths, *Actias luna* (L.) (Lepidoptera: Saturniidae) by the introduced *Compsilura concinnata* (Meigen) (Diptera: Tachinidae) in Central Virginia, and their hyperparasitism by trigonalid waps (Hymenoptera: Trigonalidae). *Environmental Entomology* **32**, 1019–1027.

Kocsis, M. & Hufnagel, L. (2011) Impacts of climate change on Lepidoptera species and communities. *Applied Ecology and Environmental Research* **9**, 43–72.

Kolligs, D. (2000) Okologische Auswirkungen Kunstlicher Lichtquellen auf nachtaktive insekten, inbesondere Schmetterlinge (Lepidoptera). [Ecological effects of artificial light sources on nocturnally active insects, in particular moths (Lepidoptera)] Faunistisch-Okologische Mitteilungen, Supplement 28, 1–136.

Konvicka, M., Cizek, O., Filipova, L. et al. (2005) For whom the bells toll: demography of the last population of the butterfly *Euphydryas maturna* in the Czech Republic. *Biologia, Bratislava* **60**, 551–557.

Korosi, A., Orvossy, N., Batary, P., Harnos, A. & Peregovits, L. (2012) Different habitat selection by two sympatric *Maculinea* butterflies at small spatial scale. *Insect Conservation and Diversity* **5**, 118–126.

Kruus, M. (2003) The greenhouse effect and moths' responses to it. 1. How to compare climatic and insect phenology databases. *Agronomy Research* **1**, 49–62.

Kuchlein, J.H. & Ellis, W.N. (1997) Climate-induced changes in the microlepidoptera fauna of the Netherlands and the implications for nature conservation. *Journal of Insect Conservation* **1**, 73–80.

Kuefler, D., Haddad, N.M., Hall, S., Hudgens, B., Bartel, B. & Hoffman, E. (2008) Distribution, population structure, and habitat of the endangered St Francis' satyr butterfly, *Neonympha mitchellii francisci*. *American Midland Naturalist* **159**, 298–320.

Lewis, O.T. & Hurford, C. (1997) Assessing the status of the marsh fritillary butterfly (*Eurodryas aurinia*): an example from Glamorgan, UK. *Journal of Insect Conservation* **1**, 159–166.

Littlewood, N.A. (2008) Grazing impacts on moth diversity and abundance on a Scottish upland estate. *Insect Conservation and Diversity* **1**, 151–160.

Longcore, T. & Rich, C. (2004) Ecological light pollution. *Frontiers in Ecology and the Environment* **2**, 191–192.

Longley, M. & Sotherton, N.W. (1997) Factors determining the effects of pesticide upon butterflies inhabiting arable farmland. *Agriculture, Ecosystems and Environment* **61**, 1–12.

Losey, J.E., Hufbauer, R.A. & Hartzler, R.G. (2003) Enumerating lepidopteran species associated with maize as a first step in risk assessment in the USA. *Environmental Biosafety Research* **2**, 247–261.

Lu, S.S. & Samways, M.J. (2001) Life history of the threatened Karkloof blue butterfly, *Orachrysops ariadne* (Lepidoptera: Lycaenidae). *African Entomology* **9**, 137–151.

Lu. S.S. & Samways, M.J. (2002) Conservation management recommendations for the Karkloof blue butterfly, *Orachrysops ariadne* (Lepidoptera: Lycaenidae). *African Entomology* **10**, 149–159.

Mair, L., Thomas, C.D., Anderson, B.J., Fox, R., Botham, M. & Hill, J.K. (2012) Temporal variation in responses of species to four decades of climate warming. *Global Change Biology* **18**, 2439–2447.

McGeoch, M.A. (2002) Insect conservation in South Africa: an overview. *African Entomology* **10**, 1–10.

Merckx, T., Feber, R.E., Hoare, D.J. et al. (2012a) Conserving threatened Lepidoptera: towards an effective woodland management policy in landscapes under intense human land-use. *Biological Conservation* **149**, 32–39.

Miller, J.C. (1990) Field assessment of the effects of a microbial pest control agent on nontarget Lepidoptera. *American Entomologist* (summer), 135–139.

Mollenbeck, V., Hermann, G. & Fartmann, T. (2009) Does prescribed burning mean a threat to the rare satyrine butterfly *Hipparchia fagi*? Larval-habitat preferences give the answer. *Journal of Insect Conservation* 13, 77–87.

Moranz, R.A., Debinski, D.M., McGranahan, D.A., Engle, D.M. & Miller, J.R. (2012) Untangling the effects of fire, grazing and land-use legacies on grassland butterfly communities. *Biodiversity and Conservation* 21, 2719–2746.

Munthali, S.M. & Mughogho, D.E.C. (1992) Economic incentives for conservation: bee-keeping and Saturniidae caterpillar utilization by rural communities. *Biodiversity and Conservation* 1, 143–154.

Murata, K., Nohara, K. & Abe, M. (1998) Effect of routine fire-burning of the habitat on emergence of the butterfly, *Shijimaeoides divinus asonis* (Matsumura). *Japanese Journal of Entomology* (N.S.) 1, 21–33.

Nafus, D.M. (1993) Movement of introduced biological control agents onto nontarget butterflies, *Hypolimnas* spp. (Lepidoptera: Nymphalidae). *Environmental Entomology* 22, 265–272.

Nakamura, Y. (2010) Conservation of butterflies in Japan: changing environments. *Wings* (Fall 2010), 13–17.

New, T.R. (2011) Butterfly Conservation in South-Eastern Australia: Progress and Prospects. Springer, Dordrecht.

New, T.R. (2012b) Hymenoptera and Conservation. Wiley-Blackwell, Oxford.

New, T.R., Van Praagh, B.D. & Yen, A.L. (2000) Fire and the management of habitat quality in an Australian lycaenid butterfly, *Paralucia pyrodiscus lucida*, the Eltham copper. *Metamorphosis* 11, 154–163.

New, T.R., Yen, A.L., Sands, D.P.A. et al. (2010) Planned fires and invertebrate conservation in south east Australia. *Journal of Insect Conservation* 14, 567–574.

Nowicki, P., Pepkowska, A., Kudlek, J. et al. (2007) From metapopulation theory to conservation recommendations: lessons from spatial occurrence and abundance patterns of *Maculinea* butterflies. *Biological Conservation* 140, 119–129.

Nowicki, P., Halecki, W. & Kalarus, K. (2013) All natural habitat edges matter equally for endangered *Maculinea* butterflies. *Journal of Insect Conservation* 17, 139–146.

Oliver, T.H., Thomas, C.D., Hill, J.K., Brereton, T. & Roy, D.B. (2012) Habitat associations of thermophilous butterflies are reduced despite global warming. *Global Change Biology* 18, 2720–2729.

Parmesan, C., Ryrholm, N., Stefanescu, C. et al. (1999) Poleward shifts in geographical range of butterfly species associated with regional warming. *Nature* 399, 579–583.

Peigler, R.S. (1994) Non-sericultural uses of moth cocoons in diverse cultures. *Proceedings of the Denver Museum of Natural History, series* 5 (3), 1–20.

Peters, J.V., Smithers, C.N. & Rushworth, G.D. (2010) Changes in the range of *Hasora khodia haslia* (Swinhoe) (Insecta: Lepidoptera: Hesperiidae) in eastern Australia – a response to climate change? pp. 219–226 in Lunney, D., Hutchings, P. & Hochuli, D. (eds) The Natural History of Sydney. Royal Zoological Society of New South Wales, Mosman.

Powell, A.F.L.A., Busby, W.H. & Kindscher, K. (2007) Status of the regal fritillary (*Speyeria idalia*) and effects of fire management on its abundance in northeastern Kansas, USA. *Journal of Insect Conservation* 11, 299–308.

Poyry, J., Lindgren, S.W., Salminen, J. & Kuussaari, M. (2004) Restoration of butterfly and moth communities in semi-natural grasslands by cattle grazing. *Ecological Applications* 14, 1656–1670.

Poyry, J., Lindgren, S., Salminen, J. & Kuussaari, M. (2005) Responses of butterfly and moth species to restored cattle grazing in semi-natural grasslands. *Biological Conservation* 12, 465–478.

Poyry, J., Luoto, M, Paukkunen, J., Pykala, J., Raatikainen, K. & Kuussaari, M. (2006) Different responses of plants and herbivore insects to a gradient of vegetation height: an indicator of the vertebrate grazing intensity and successional age. *Oikos* 115, 401–412.

Rastall, K., Kondo, V., Strazanac, J. & Butler, L. (2003) Lethal effects of biological insecticide applications on nontarget lepidopterans in two Appalachian forests. *Environmental Entomology* 32, 1364–1369.

Relf, M. & New T.R. (2009) Conservation needs of the Altona skipper butterfly, *Hesperilla flavescens flavescens* Waterhouse (Lepidoptera: Hesperiidae) near Melbourne, Victoria. *Journal of Insect Conservation* 13, 143–149.

Rickert, C., Fichtner, A., van Klink, R. & Bakker, J.P. (2012) α- and β- diversity in moth communities in salt marshes is driven by grazing management. *Biological Conservation* 146, 24–31.

Ricouart, F., Cereghino, R., Gers, C., Winterton, P. & Legal, L. (2013) Influences of fire prevention management strategies on the diversity of butterfly fauna in the eastern Pyrenees. *Journal of Insect Conservation* **17**, 95–111.

Rudolph, D.C., Ely, C.A., Schaefer, R.R., Williamson, J.H. & Thill, R.E. (2006a) Monarch (*Danaus plexippus* L., Nymphalidae) migration, nectar resources and fire regime in the Ouachita Mountains of Arkansas. *Journal of the Lepidopterists' Society* **60**, 165–170.

Rudolph, D.C., Ely, C.A., Schaefer, R.R., Williamson, J.H. & Thill, R.E. (2006b) The Diana fritillary (*Speyeria diana*) and Great spangled fritillary (*S. cybele*): dependence on fire in the Ouachita Mountains of Arkansas. *Journal of the Lepidopterists' Society* **60**, 218–226.

Runquist, E.B. (2011) Butterflies, cattle grazing, and environmental heterogeneity in a complex landscape. *Journal of Research on the Lepidoptera* **44**, 61–76.

Russell, C. & Schultz, C.B. (2010) Effects of grass-specific herbicides on butterflies: an experimental investigation to advance conservation efforts. *Journal of Insect Conservation* **14**, 53–63.

Saarinen, K. & Jantunen, J. (2005) Grassland butterfly faunas under traditional animal husbandry: contrasts in diversity in mown meadows and grazed pastures. *Biodiversity and Conservation* **14**, 3201–3213.

Schellhorn, N.A., Lane, C.P. & Olson, D.M. (2005) The co-occurrence of an introduced biological control agent (Coleoptera: *Coccinella septempunctata*) and an endangered butterfly (Lepidoptera: *Lycaeides melissa samuelis*). *Journal of Insect Conservation* **9**, 41–47.

Schtickzelle, N., Choutt, J., Goffort, P., Fichofel, V. & Baguette, M. (2005). Metapopulation dynamics and conservation of the marsh fritillary butterfly: population viability analysis and management options for a critically endangered species in Western Europe. *Biological Conservation* **126**, 569–581.

Schultz, C.B. (2001) Restoring resources for an endangered butterfly. *Journal of Applied Ecology* **38**, 1007–1019.

Schultz, C.B. & Crone, E.E. (1998) Burning prairie to restore butterfly habitat: a modeling approach to management tradeoffs for the Fender's blue. *Restoration Ecology* **6**, 244–252.

Schultz, C.B. & Dlugosch, K.M. (1999) nectar and host plant scarcity limit populations of an endangered Oregon butterfly. *Oecologia* **129**, 231–238.

Schultz, C.B., Russell, C. & Wynn, L. (2008) Restoration, reintroduction and captive propagation for at-risk butterflies: a review of British and American conservation efforts. *Israel Journal of Ecology and Evolution* **54**, 41–61.

Sears, M.K., Hellmich, R.L., Stanley-Horn, D.E. et al. (2001) Impact of Bt corn pollen on monarch butterfly populations: a risk assessment. *Proceedings of the National Academy of Sciences* **98**, 11937–11942.

Severns, P.M. (2003) The effects of a fall prescribed burn on *Hemileuca eglanterina* Boisduval (Saturniidae). *Journal of the Lepidopterists' Society* **57**, 137–143.

Shirai, Y. & Takahashi, M. (2005) Effects of transgenic Bt corn pollen on a non-target lycaenid butterfly, *Pseudozizeeria maha*. *Applied Entomology and Zoology* **40**, 151–159.

Shuey, J.A. (1997) Dancing with fire: ecosystem dynamics, management, and the Karner blue (*Lycaeides melissa samuelis* Nabokov) (Lycaenidae). *Journal of the Lepidopterists' Society* **51**, 263–269.

Singer, M.C., Thomas, C.D. & Parmesan, C. (1993) Rapid human-induced evolution of insect-host associations. *Nature* **366**, 681–683.

Skorka, P., Lenda, M., Moron, D., Kalarus, K. & Tryjanowski, P. (2013) Factors affecting road mortality and the suitability of road verges for butterflies. *Biological Conservation* **159**, 148–157.

Sparks, T.H., Dennis, R.L.H., Croxton, P.J. & Cade, M. (2007) Increased migration of Lepidoptera linked to climate change. *European Journal of Entomology* **104**, 139–143.

Stark, J.D., Chen, X.D. & Johnson, C.S. (2012) Effects of herbicides on Behr's metalmark butterfly, a surrogate species for the endangered butterfly, Lange's metalmark. *Environmental Pollution* **164**, 24–27.

Summerville, K.S. & Crist, T.O. (2008) Structure and conservation of lepidopteran communities in managed forests of northeastern North America; a review. *Canadian Entomologist* **140**, 475–494.

Swengel, A.B. (1996) Effects of fire and hay management on abundance of prairie butterflies. *Biological Conservation* **76**, 73–85.

Swengel, A.B. & Swengel, S.R. (1996) Factors affecting abundance of adult Karner blue (*Lycaeides melissa samuelis*) (Lepidoptera: Lycaenidae) in Wisconsin surveys 1987–95. *Great Lakes Entomologist* **29**, 93–105.

Swengel, A.B. & Swengel, S.R. (2007) Benefit of permanent non-fire refugia for Lepidoptera conservation in fire-managed areas. *Journal of Insect Conservation* **11**, 263–279.

Thomas, C.D., Bodsworth, E.J., Wilson, R.J. et al. (2001) Ecological and evolutionary processes at expanding range margins. *Nature* **411**, 577–581.

USFWS (United States Fish and Wildlife Service) (1999) Schaus swallowtail Butterfly *Heraclides aristodemus ponceanus*. pp. 4.743–4.766 in Multi-Species Recovery Plan for South Florida. Atlanta, Georgia.

USFWS (United States Fish and Wildlife Service) (2007) Kern Primrose sphinx moth (*Euproserpinus euterpe*). 5-year review: Summary and evaluation. Sacramento, California.

Van Driesche, R.G., Nunn, C. & Pasquale, A. (2004) Life history pattern, host plants, and habitat as determinants of population survival of *Pieris napi oleracea* interacting with an introduced parasitoid. *Biological Control* **29**, 278–287.

van Langevelde, F., Ettema, J.A., Donners, M. & WallisDeVries, M. (2011) Effect of spectral composition of artificial light on the attraction of moths. *Biological Conservation* **144**, 2274–2281.

van Swaay, C.A.M. & Warren, M.S. (2012) Developing butterflies as indicators in Europe: current situation and future options. De Vlinderstichting/Dutch Butterfly Conservation, Butterfly Conservation UK, Butterfly Conservation Europe, Wageningen. Report no. VS2012.012.

Veldtman, R., McGeoch, M.A. & Scholtz, C.H. (2002) Variability in cocoon size in southern African wild silk moths. *African Entomology* **10**, 127–136.

Vogel, J.A., Debinski, D.M., Koford, R.R. & Miller, J.R. (2007) Butterfly responses to prairie restoration through fire and grazing. *Biological Conservation* **140**, 78–90.

Vogel, J.A., Koford, R.R. & Debinski, D.M. (2010) Direct and indirect responses of tallgrass prairie butterflies to prescribed burning. *Journal of Insect Conservation* **14**, 663–678.

Wang, X.-L. & Yao, Y.-L. (2011) Host insect species of *Ophiocordyceps sinensis*: a review. *ZooKeys* **127**, 43–59.

Warren, M.S. (1994) The UK status and suspected metapopulation structure of a threatened European butterfly, the marsh fritillary, *Eurodryas aurinia*. *Biological Conservation* **67**, 239–249.

Williams, E.H. (1995) Fire-burned habitat and reintroduction of the butterfly *Euphydryas gillettii* (Nymphalidae). *Journal of the Lepidopterists' Society* **49**, 183–191.

Woiwod, I.P. (1997) Detecting the effects of climate change on Lepidoptera. *Journal of Insect Conservation* **1**, 149–158.

Woodcock, B.A., Bullock, J.M., Mortimer, S.R. et al. (2012) Identifying time lags in the restoration of grassland butterfly communities: a multi-site assessment. *Biological Conservation* **155**, 50–58.

Zhou, Y., Cao, C., Chen, H., Yan, F., Xu, C. & Wang, R. (2012) Habitat utilization of the Glanville fritillary in the Tianshan Mountains, China, and its implications for conservation. *Journal of Insect Conservation* **16**, 207–214.

Zimmerman, E.C. (1958) Insects of Hawaii. Volume 7. Macrolepidoptera. University of Hawai'i Press, Honolulu.

13

Assessing Conservation Progress, Outcomes and Prospects

Introduction

The discipline of insect conservation continues to be advanced, and to be enriched, by lessons learned from studies of Lepidoptera. Essentially, Lepidoptera conservation is far more informed than for any other terrestrial invertebrate group, and the lessons from a considerable variety of field studies have furnished the template for adoption and modification in other contexts. In particular, the intensively studied northern temperate region taxa, with some spectacular conservation successes, invite increased conservation emulation in other parts of the world in which perspective, priority, capability and knowledge are far different. There seems little reason to doubt that the levels of losses and declines documented so extensively in western Europe are paralleled (or exceeded) elsewhere, and that recent expansions of human populations and influence in many countries, that until recently harboured much endemic and ecologically restricted native biodiversity, are linked to numerous declines of Lepidoptera and other, less heralded, invertebrates; for the most part they have not been documented, but some examples noted in earlier chapters give cause for immense concern.

Collectively, for both species and biotopes, they indicate some of the 'rules', sometimes considered 'paradigms', developed in wider insect conservation largely from studies on Lepidoptera. The influence of the metapopulation concept, for example, has been pivotal in changing understanding and some practical approaches for species management, with increased attention to landscape ecology issues as well as site-based resource needs, both emphasised in numerous species-level conservation cases. Increasing awareness that both habitat quality and extent, and isolation of metapopulation units operate together and at a variety of scales has emphasised that site-based habitat management should

Lepidoptera and Conservation, First Edition. T.R. New.
© 2014 John Wiley & Sons, Ltd. Published 2014 by John Wiley & Sons, Ltd.

not be sacrificed to substitution by simply maintaining numerous interconnected populations (Thomas et al. 2001). Both approaches are important, perhaps especially in altered landscapes such as cultivated areas. Those cases also demonstrate the difficulties of generalisation, with each species (and even population) needing consideration of individual or idiosyncratic detail within the broader framework of action. Widespread awareness of those differences has itself been a major outcome from Lepidoptera conservation, and contributed to the wider templates of understanding. It has also emphasised the difficulties of extrapolating from one case to another, even though they may appear similar and address closely related taxa.

However, the juxtaposition of detail and generality is itself important to realise, as demonstrated so effectively for the *Maculinea* butterflies of Europe, for which extensive international comparative studies have transformed initially rather parochial conservation exercises into high-profile programmes promoting these five butterfly species as model systems with very wide importance. Kuhn et al. (2005) referred to these large blues as 'superindicators', with the large cooperative programmes for their conservation having the major benefit of 'training across Europe of a new generation of excellent young scientists, highly skilled in butterfly ecology and conservation in numbers that dwarf the previous workers in this field'. The question arises as to how this new enthusiasm and expertise might be deployed most effectively on related topics, with continuing debate over merits of (1) species-level conservation versus wider approaches, and (2) for species whether it is 'better' to capitalise effectively on existing knowledge and public and political sympathies by promoting still more Lepidoptera (especially butterflies), or extending deliberately to encompass wider taxonomic and ecological variety, but for which experience and knowledge are more fragmentary. Whichever path is followed, resources for practical insect conservation will remain inadequate to cover all needs, and some form of species triage selection is inevitable at that level of focus.

Monitoring conservation progress

Whatever approaches or manipulations are undertaken to manage Lepidoptera and their requirements, understanding of the outcomes, whether success or failure, is important in informing future work both on the same species or others in similar biotopes or for which similar approaches may be contemplated. Whether single taxa or assemblages are the main focus, some form of objective monitoring, perhaps over a substantial period, may be needed to achieve this. Historically, this aspect of a conservation programme has often received insufficient attention, but its core relevance implies that provision for monitoring must be part of the initial conservation plan rather than an appeasement afterthought, as currently sometimes seems to be the case. Monitoring depends on regular interval inspections of one or more stages, following replicable methods (Chapter 5), but may need very careful appraisal to detect valid trends in abundance, distribution or assemblage composition. Serial observations or counts are based most commonly on some stage that can be used as an intergenerational marker or 'interseason marker', and that is accessible predictably and can be sampled or inspected without undue difficulty. The immense values of long-term monitoring schemes such as those described earlier for parts of North America and Europe are notable examples through which sound trends have been detected. Such reliability conveys powerful political messages of need, and for many taxa the trajectories of decline are progressively being reversed. In addition to detecting change, understanding success or failure of any such attempt depends on monitoring the outcome.

With restoration monitoring, progress of colonisation from source areas may be a key feature. Comparison with pre-treatment conditions – those initially found wanting – provides some baseline for assessment.

Habitat restoration for the Iolas blue butterfly (*Iolana iolas*) became necessary as the butterfly declined markedly in its sole Swiss locality, due to expansion of vineyard and human settlement around Valais (Heer et al. 2012). The restoration process involved planting batches of 1–12 seedlings of the sole larval food plant (Bladder senna, *Colutea arborescens*) on 38 sites over 2000–2006, with those sites selected as suitable by being sunny, south-facing and relatively free of competing plants. Inspections over 4–10 years from planting showed that butterflies had colonised 19 of those sites, sometimes only in small numbers. Modelling indicated that butterfly abundance was explained best by the senna's blooming intensity, with its flowers being the almost exclusive nectar source for *Iolana*, and the position of sites within the metapopulation area, on a landscape level. These outcomes allowed refinement of future conservation operations, such as plans for further senna plantings within about 550 m of existing populations, and increasing blooming intensity of existing plants by regular pruning, and limiting herbicide applications in the vineyard-dominated landscape. In this example, as a lesson for many others, monitoring, research and future planning go hand-in-hand and also consider a landscape context rather than the focal sites alone. With adequate appraisal, monitoring can hone conservation for the future.

Simple species richness trends alone, as the most frequent measure of restoration trajectories, may not always be a good reflection of restoration success in specialised biotopes or of the 'health' of the restored systems. With this approach, an abundant species is given the same value as a scarce one. Many of the species recorded may not be particularly characteristic of that biotope but be more generalised colonists, in contrast to more specific or ecologically restricted taxa whose presence and abundance may be significant, or core taxa (Chapter 8, p. 144) that essentially define an assemblage in relation to a site or biotope. Likewise, lack of abundance data in many surveys may lead to equating individual occurrence or tiny marginal populations with those that are more secure. Woodcock et al. (2012) measured 'restoration success' by quantifying changes in similarity of species composition and relative abundances between restoration sites and baseline 'target' grasslands. Two main treatments were considered: (1) 'arable reversion sites' (grassland established on bare ground, from seed mixtures), and (2) 'grassland enhancement sites' (degraded grassland sites subjected to scrub removal and establishing cutting and grazing regimes). Butterfly colonisation times were fastest for taxa with widespread host plants or with host plants established directly during restoration. Restoration success increased rapidly over the first 5–10 years on arable reversion sites, whilst enhancement sites showed no such changes – perhaps reflecting that the latter commenced from grasslands and so already contained some grassland butterflies; these, however, were mainly generalists that would be expected to colonise reversion sites within the first year or so. A further observation made, that butterfly colonisation rates were lower when host plants reproduce clonally (so are perhaps able to persist better than those depending on seeds) may also affect management. Initial seeding on sites might be usefully augmented by direct selective plantings ('plug plants') of such desirable larval hosts, for example (Woodcock et al. 2012). Such baseline 'reference sites' to represent target conditions may not incorporate variations between sites, and it becomes important wherever possible to pursue restoration at multiple sites to assess more general progress and refine techniques (Schultz 2001).

Any such monitoring aims to detect long-term trends as reliably as possible, so that the methods used must themselves inspire confidence, to achieve the levels of interpretation demonstrated by the British Butterfly Monitoring Scheme (Chapter 5, p. 50) and on which much conservation activity has devolved. However, the statistical validity of methods employed has only rarely been appraised carefully. For both this United Kingdom scheme and the analogue undertaken in the Netherlands since 1990 – so with about two-thirds of the species (other than extremely rare ones) in common – the power to detect trends (van Strien et al. 1997) depends on the number of sampling sites and the estimates of variance (year-to-year, year-by-site, and incorporating voltinism) was sometimes very limited. Whilst having little effect for common species, this analysis raised questions of reliability for rarer species, those of primary individual conservation interest. Overall results (Table 13.1) indicated that only two of 51 species would reveal a change of less than 25% over 10 years of monitoring, for example. For some species, a decline of more than 50% is needed to give 80% probability of detection over a 20-year period – either because of observations at only few sites, or high year-to-year variance. Whilst one way to

Table 13.1 Butterfly monitoring schemes in the United Kingdom and The Netherlands: characteristics and detection of status change (van Strien et al. 1997. Reproduced with permission of John Wiley & Sons.) (* from British data).

	United Kingdom	The Netherlands
Number of species	51	47
Numbers of species where change of <50% are detected with power = 80%		
In 10 years	23	22
In 20 years	37	29
Median year-to-year variance	0.0015	0.0019*
Median year-site variance	0.1200	0.1089*
Median number of species per site	43	50
Median number of individuals	44	27

increase reliability is simply to increase the number of sampling sites, this expansion will either increase costs, or be simply unavailable as such sites are not available. Much monitoring thus remains somewhat subjective in how trends are interpreted – but those trends, however simplistic, may have far-reaching implications in inducing conservation management and affecting priority of which species (or other entities) are managed.

Since the development of the British scheme, similar monitoring programmes have been initiated in more than 10 European countries, with the common purpose of assessing regional and national trends in butterfly abundance for each species as a signal of its wellbeing, as well as distributional changes. All these schemes have been based on transect walks, with various attempts to reduce sampling intensity from the weekly intervals adopted as a standard in the United Kingdom – for example, by targeting the periods when most, or particularly desirable, species are active (Table 13.2; (van Swaay et al. 2008)). Valid transect counts provide much information for correlations with numerous environmental factors, in addition to monitoring responses to management. Notwithstanding shortcomings noted above, the outcomes of 'Pollard walks' have resulted in butterflies becoming 'the only invertebrate taxon for which it is currently possible to estimate rates of decline in many parts of the world' (van Swaay et al.

Table 13.2 Major characteristics of the 'Traditional' (T) and 'Reduced effort' (RE) Butterfly Monitoring Schemes, and conditions for their use (founded in transect walks) in Europe (largely from van Swaay et al. 2008).

Characteristics. T: based on weekly counts, mostly with free choice of sites. RE: based on higher number of transects, counted only a few times a year, on random or preselected sites

Objectives. T: National, regional and local indices and trends: possibility to compare local indices and trends with regional or national trends, can be used to evaluate conservation measures. RE: National indices and trends for widespread species or targeted at individual rare taxa

Common features. Transects should be as far as possible representative of the sampling unit, be in one 'rough' habitat (such a grassland, woodland, heathland, etc.) to enable trends by habitat to be assessed; standardised for time (15–60 minutes suggested) and length (mostly to a maximum of 2 km dictated by time), preferably 5 m width; with sections homogeneous according to habitat type to enable weighting, and habitat with standard widely recognised classification; weather conditions suitable; adequate reliable and consistent identification of species; not vital, although preferable, to record each transect every year, and if help/volunteers not available, better to count more transects at longer intervals than fewer transects every year

Differences. T: counts preferably each week during flight periods; national trends may be feasible with lower effort but never less than twice a month. RE: 3–5 counts annually, with visits targeted to times that will provide most information; suggestions given are for UK (three visits within 9 weeks with a 1week gap) and France (four visits in 4 months with 15 days inbetween)

Status. T: fully tested, success proven. RE: ongoing assessment

2008), with potential to combine trends for individual species and countries into a more unified measure of biodiversity. One such modelling approach ('TRIM', Chapter 5, p. 70) extends from national level to country combinations to provide information on, in this context, European trends, and subsequently combining indices for selections of species – so that a preliminary grassland indicator discussed by van Swaay et al. was based on trends for 10 grassland specialist butterfly species and seven more widespread grassland species, and offers likelihood of detecting large-scale impacts of agricultural intensification or abandonment.

Much monitoring of Lepidoptera has focused on indicator potential of species, so that the trends observed may reflect both the individual species' wellbeing but also have wider surrogate values for conservation. Thus, information gained from the monitoring of European butterflies contributes to the wider context of monitoring progress toward the European Union target of halting biodiversity loss and halting degradation of ecosystem services by 2020. The major advantages (noted by van Swaay & Warren 2012) include (1) good documentation of the species and, generally, easy recognition to species level; (2) wholly standardised and well tested survey methods, with widespread and increasing capability to enter data directly into assembled and easily disseminated databases; (3) large numbers of informed volunteers, with effective coordination to assure quality control; and (4) overall leadership, education and publicity through Butterfly Conservation Europe, confirming and extending favourable perception of butterflies in the wider community. In the same report, van Swaay and Warren noted also the 'huge potential' to increase the numbers of schemes and improve coverage further, and foresaw also the wide adoption of on-line data entry increasing efficiency considerably.

Indicators

'Indicator values' basically reflect changes in species incidence, abundance and assemblage richness and composition in response to environmental changes, with the predictability of those detectable responses available for monitoring the impacts. Whilst these changes are cited often, the physical appearance of the insects themselves may also reflect environmental influences. The widely publicised example of industrial melanism in the Peppered moth (*Biston betularia*, Geometridae) in Britain, wherein the increased frequency of black individuals was associated with increased industrialisation, is perhaps the best known example. In discussing this case, Majerus (2002) also listed the many other categories of melanism that are found in moths. Whilst some Lepidoptera are notoriously variable in individual appearance, the causes of much such variation are unclear. Some recent examples, however, link changes in clinal variation with temperature changes, as possible indicators of climate changes (Norgate et al. 2009).

Butterflies have been claimed widely to be a useful 'indicator group', whereby their assemblage composition and abundance is an easily assessable and valid surrogate of wider biodiversity in biotopes. In many cases, such claims are difficult to verify objectively, and have flowed commonly simply from butterflies being relatively conspicuous, and easy to identify and count, so that their presence, assemblage composition and relative abundance can be assessed. All, of course, are highly advantageous in seeking to clarify patterns of change and, at times, highly informative in helping to interpret those changes. Knowledge, familiarity and charisma do not necessarily convey indicator value. As Fleishman and Murphy (2009) emphasised, it is very difficult to evaluate the true indicator worth of butterflies in most contexts. Reviewing many such claims, Fleishman and Murphy concluded (2009: 1114) 'There are circumstances in which butterflies can serve effectively as indicators, especially at the assemblage level, but we are not aware of an example in which a single species has provided environmental information that cannot be measured more directly'. Many cases of single-species responses can mirror wider changes, but extensions of claims to wider groups are not always convincing. Often, this simply reflects imprecision, as the term 'indicator' is used in many different contexts with differing objectives (McGeoch 1998, for summary), leading to several published listings of features of 'good indicator groups' (Samways et al. 2010). Decline of any ecologically specialised species in conjunction with decline of habitat extent or quality has sometimes been claimed as their having indicator value, without any direct evidence of wider surrogacy values, and this context is exemplified by designation of particular groups as especially susceptible to habitat change (Chapter 5) and so are valuable tools in assessment of habitat condition and the trajectories of restoration or other management. Groups of forest moths nominated by Kitching et al. (2000, Australia) (Chapter 5, p. 60), Holloway (1984, Sarawak) and others demonstrate the principles of such selection for study, often expedited by relative tractability of the groups available for investigation. As Brown (1991) demonstrated, both the ecological characteristics of the groups and the practical ease of sampling them are relevant. His contrast of two families of larger moths and of colourful butterflies (which he considered an 'ideal group' for monitoring in Brazilian forest ecosystems) in the Neotropics demonstrates the contrasts well, and has much wider relevance in selecting possible 'indicator groups' (Table 13.3).

The advantages of diurnal Lepidoptera for environmental assessments of many kinds reflect the relative ease with which they can be assessed rapidly, identified and differentiated, often by non-destructive approaches and by non-specialists using well illustrated field identification guides. These logistic advantages link with the rapid reproduction of these insects facilitating their rapid responses to changes. Later exploration of this theme (Brown & Freitas 2000) in Brazil's Atlantic Forests was based on >2100 butterfly species assessed over 35 years, with the 21 most studied sites supporting 218–914 species – the most comprehensive survey yet undertaken on tropical butterflies. On any site, half the species could be assessed in a week or less, so that the logistically desirable 'rapid assessments' were indeed possible. Bait-attracted Nymphalidae (comprising 25–29% of species present on sites) correlated well with the entire butterfly fauna as apparently valid surrogates for the assemblages. Various factors influenced the butterflies present, with the importance of climate and disturbance affected by topography, vegetation and soil characteristics – and the

Table 13.3 Comparison of 'less suitable (Saturniidae.Sphingidae) and 'more suitable' (Heliconiini, Ithomiinae) groups of Lepidoptera as indicators in Neotropical ecosystems (data from Brown 1991, as assembled by New 1997b. Reproduced with permission from Springer Science+Business Media.).

Characteristic	Saturniidae/ Sphingidae	Heliconiini/ Ithomiinae
High ecological fidelity	+	++
Relatively sedentary	+	++
Narrowly endemic or well differentiated	+	++
Abundant, easy to find	+	++
Indicators of other species and resources	+	++
Always present	0	+
Easy to obtain large random samples	0	++
Total score*	5/14	13/14

*'0', criterion not met; '+' poor to moderate; '++' good. Maximum score possible is 14 (i.e. '2' × 7 characteristics); the higher the score, the more suitable the group
Note that several other features were also investigated but did not differ between these groups: taxonomically and ecologically diversified (++), taxonomically well known (++), functionally important in ecosystem (+), predictable response to disturbance (+)

different responses of different butterfly groups collectively informative in demonstrating indicator potential for a range of environmental changes. The two 'best' predictors of faunal richness, the families Hesperiidae and Lycaenidae, were the largest components of the fauna, and also the most difficult to assess accurately and comprehensively, so that Nymphalidae, much easier to appraise, were also much more amenable to practical study.

Wider surrogacy values using Lepidoptera are more difficult to investigate. Some years ago, Hammond and McCorkle (1984) suggested that *Speyeria* fritillaries and their *Viola* larval food plants were 'among the best indicator organisms of native, undisturbed ecological communites in North America', because they were among the first species to disappear from anthropogenic disturbance. As examples, they noted the Diana fritillary (*Speyeria diana*) declining due to destruction of old-growth hardwood forest, and its resurgence with planned regrowth; the Regal fritillary (*S. idalia*), formerly widespread on tall-grass prairies in the central United States with its *Viola petadifida* food plant, but largely extirpated for agriculture, to leave only a few patches of prairie suitable for habitation; and *S. nokomis*, reduced extensively through destruction of its boggy marsh habitats. With several other *Speyeria* taxa (some of them western forms amongst the substantial number of California butterflies of concern), collectively encompassing a variety of different biotopes, the commonality of need becomes a clear and viable taxonomic focus for conservation attention.

Whilst it is intrinsically satisfying to suggest that lowered richness of butterflies or moths is likely to mirror parallels in other taxa, such correlation has only rarely been investigated carefully, and sometimes found misleading. The basic, and widely appealing, assumption commonly adopted for butterflies is that diversity patterns of the purported indicator group are correlated with those of

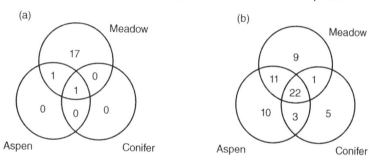

Fig. 13.1 Species distribution of (a) butterflies and (b) moths among three major habitats. Numbers of species, excluding singletons, of species recorded from each habitat (Meadow; Aspen; Conifer) shown exclusively from each, in each combination of two habitats, and in all three (Source: Ricketts et al. 2002. Reproduced with permission from Elsevier.).

other taxonomic groups or with environmental conditions or changes. But, as Ricketts et al. (2002) demonstrated in attempting to correlate butterflies with moths, no clear correlation was evident across the 19 sites (representing the major subalpine Colorado habitats of meadows, conifer forest and aspen forests) they surveyed. This lack of correspondence reflected that butterflies (sampled by timed searches, pooled $n = 29$ species) were predominantly found in the meadows, whilst moths (sampled by light traps, pooled $n = 91$ species) were distributed more generally across all three biotopes, with several species restricted to each, and many species transcending two or three categories (Fig. 13.1). No butterfly host plants occurred within the conifer forests, and meadows were by far the most hospitable regime for these insects, both for consumable resources and thermal suitability. Patterns of diversity of even these two related taxonomic groups thereby differed fundamentally in these landscapes, and reliance on butterflies as indicators (surrogates) of wider diversity may thus be very misleading. Ricketts et al. advocated a more habitat-based approach to conservation, rather than incorporating the imprecise presumption of effective surrogacy even between such related taxa.

In contrast, burnet moths (Zygaenidae) may have greater indicator value of butterfly richness in semi-natural pastures in Europe (Franzen & Ranius 2004) where, in this context, they were categorised as 'an outstanding group of indicators'. Although only five species of these brightly coloured and conspicuous diurnal moths occurred in the Swedish study area, all had low mobility and narrow ecological requirements. They also showed a 'nested pattern', with nested assemblages being those in which the species comprising impoverished assemblages are subsets of the species occurring in successively richer faunas.

More widely, Orsak et al. (2001) found no close correspondence between moth and plant species richness in New Ireland (Papua New Guinea). Even when different taxon groups are each responsive to environmental changes, their changes along disturbance or habitat gradients may differ substantially. In the Ecuadorian Andes, comparisons between mature forest, recovering forest (around 50 years old) and open vegetation, as a disturbance gradient, showed richness of epiphytic lichens largely unchanged, whilst epiphytic bryophytes and vascular plants declined; geometrid moths were significantly richer in recovering forest compared with other habitats, and Arctiidae were richer in both recovering forest and open vegetation compared with mature forest (Noske et al. 2008). Both these moth families had many species represented at only one habitat type (Geometridae, 31% of 829 species; Arctiidae 32% of 282 species), with actual species numbers (mature forest – open vegetation) as (Geometridae) 196, 383, 255, and (Arctiidae) 65, 125, 94. Noske et al. acknowledged that Arctiidae might have been undersampled in forests, because light traps used were confined to understorey so possibly not enticing a richer canopy fauna – notwithstanding that their mobility might be higher than the generally weakly flying Geometridae. The biases introduced by any such possible sampling restrictions are important considerations in interpreting surveys of this kind.

Other examples help to display the variety of contexts and claims for indicator value and use. Wider applications of butterflies in monitoring the effects of land use changes include associations of wider surrogacy, so that 'indication' is primarily of perceived environmental state reflected in internal changes within butterfly assemblages – without direct knowledge of any parallels within other taxa. In Finland, a high proportion of butterfly species (74, or 70% of the fauna) live in agricultural landscapes, and each can be allocated to one of three ecological groups according to habitat preference, as (1) field margins or farmyards: arable; (2) semi-natural grasslands: grassland; or (3) forest clearings or verges: 'forests' (Kuussaari et al. 2005, 2007b). Long-term population trends across four time periods (<1960, 1961–90, 1991–98, 1999–2003), although very uneven in quantity of information accrued, implied the trends of rarity summarised in Fig. 13.2. Declining levels of grassland species (n = 35) contrasted with most arable species (n = 7) increasing, and forest species (n = 32) showing more varied responses (Fig. 13.3). However, a quarter of the forest species showed fluctuating trends in occupancy of the 10 × 10 km recording squares, with clear recent increases following earlier declines. Large-scale changes in distribution pattern were found in all three groups, reflecting changing land use, with severe declines in cattle grazing and areas of semi-natural grasslands, and modern forestry practices producing increased numbers of suitable forest edge habitats.

Despite widespread advocacy for the values of Lepidoptera as indicators, rigid trials to validate this are rather sparse and most rely on correlation rather than direct evidence of causal changes from modified land use or disturbance. Determinations of indicator values (such as by by 'IndVal': for background, see Samways et al. 2010) for individual taxa are also relatively few and, as this approach involves comparing representation of taxa across different habitats or treatments, rare species generally have little such formal value. Nevertheless, butterflies are claimed to 'make excellent bioindicators for forest disturbances and provide rapid indication of habitat quality' (Cleary 2004). Cleary used the butterflies of Borneo to investigate, at the three taxonomic levels of subfamilies, genera and species, their indicator values of logging impacts in forests. The practical consideration that genera may be relatively easy to recognise – for many butterflies, on the wing without need for capture – compared with the often laborious examinations (such as genitalia preparations) needed for species diagnoses in many groups is logistically attractive, not least in allowing greater

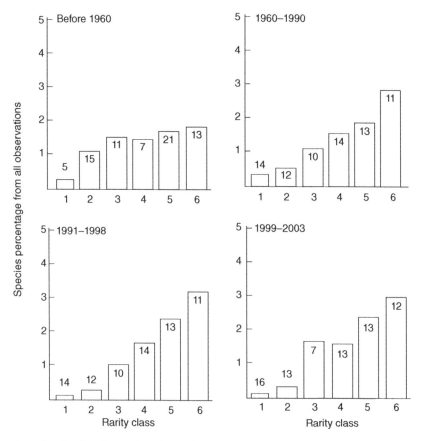

Fig. 13.2 The distributions of 74 butterfly species in Finland across four time periods, divided amongst six rarity classes: 1 'very rare' to 6 'very common'. Within each class (1–6) the bar shows mean percentage of observations per species from all observations over that period; numbers of species in each class also shown (Adapted from Kuussaari et al. 2007b. Reproduced with permission from Springer Science+Business Media.).

reliable participation of non-specialist workers in surveys. However, it is countered by inevitable loss of ecological subtlety in moving upward from species to larger and more varied groupings, particularly amongst faunas where the extent of that variety is unknown. Following other workers, Cleary differentiated 'indicator taxa' (with an IndVal of >70.00) from 'detector taxa' (IndVal 50.00–70.00). The latter have lower levels of habitat 'preference' and are more likely to move to adjacent habitats in response to changed conditions. They may, thereby, be more rapid indicators of change, in Cleary's study of changes in logged forest becoming more similar to conditions in an unlogged forest. Of a pool of 388 butterfly species in 161 genera and 16 subfamilies, a few individual species and genera were each significant indicators of either category. At the broader subfamily level, Miletinae was a significant indicator taxon of logged forest, and Morphinae, Theclinae and Riodininae were significant detector taxa of unlogged

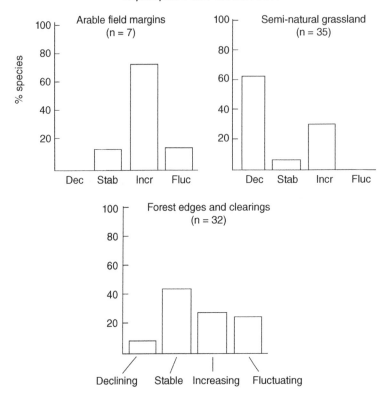

Fig. 13.3 Long-term occupancy trends of butterflies amongst the species of (a) arable field margins, n = 7; (b) semi-natural grasslands, n = 35; and (c) forest edges, n = 32 in Finland. Based on surveys of selected 10 km squares (Adapted from Kuussaari et al. 2007b. Reproduced with permission from Springer Science+Business Media.).

forests. Both individual taxa and wider assemblages can thus have values as indicators of major disturbances.

The very detailed knowledge of changes in status and distribution within the United Kingdom butterflies enabled Brereton et al. (2011) to venture that 'In spite of their imperfections, we conclude that the butterfly indicators have played a valuable role in documenting changes in biodiversity in the UK'. Their interpretation followed allocation of the species to one of four broad categories that facilitated some generalisations about correlation with environmental factors, and as formulated for European birds (Gregory et al. 2005, Table 13.4). 'Type 2' encompassed most of the butterflies, with these separated into 'specialist' and 'generalist' species. Specialist indicators are restricted largely to semi-natural habitats, which are intrinsically rich in species and have major importance for rare species. Generalist indicators represent smaller amounts of 'biodiversity', but extend across wider countryside.

Investigation of single species as surrogates is exemplified by a study of the threatened Alcon blue butterfly (*Maculinea alcon*) on wet heathlands in Belgium (Maes & van Dyck 2005), to examine whether this species was truly an indicator of (1) typical species richness of the biotope and (2) wet heathland

Table 13.4 The four broad categories of United Kingdom butterflies that facilitate correlations with environmental factors, and their roles (Brereton et al. 2011).

All species (up to 53 of the resident UK species)
Measures how specific species or groups of species are doing – e.g. species of conservation concern
Generalist species (wider countryside: relatively mobile species that occur in a wide range of habitats)
Reflect how 'biodiversity' is doing more generally: no direct aim to link indicators to specific environmental factors
Habitat specialists (low-mobility species, mainly restricted to semi-natural habitats)
Use to show how specific species or species groups are responding to environmental variables
Regular migrants
Show how biodiversity is responding to environmental factors in general

quality. Recognising that such usage of single species has been criticised widely in favour of a carefully selected portfolio of taxa for such roles, a suite of other wetland species representing several major taxonomic groups was also investigated. Nine wet heathland areas with the larval food plant, *Erica tetralix*, were surveyed, each by paired sites (with and without *M. alcon*) to assess richness of 20 taxonomic groups, giving a collective total of 624 taxa, including 98 Red List species (RL), and 64 'Typical Wet Habitat' species (TWH, of 90 noted in this category from Belgium). Selecting the latter involved five steps, from which the principles have much wider interest in selecting suites of indicators (Table 13.5). Sites with *M. alcon* were significantly richer in both RL and TWH species, but not in overall diversity. Wet heathland with *M. alcon* present can be considered fairly high quality, as a simple general indicator for both threatened and typical species. However, a drawback is the butterfly's rarity, so that its geographical application is restricted. The other taxa were selected in part on their 'intermediate rarity', rendering the group applicable more widely – but neither entity was a good indicator of total species richness. Presence or absence of particular species within the group may provide some 'signalling function' and hence the benefit of enabling managers to take different scales into account during assessment of biotope quality, impact of management, setting priorities among sites and other related applications.

One of the largest and most foresighted 'indicator investigations' was based on a wider survey of large blues (*Maculinea* spp.) in Europe, with one major objective of 'MacMan' ('Maculinea butterflies of the Habitats Directive and European Red List as Indicators and Tools for Habitat Conservation and Management') essentially expanding from the Belgian work on *M. alcon* (under the banner of '*Maculinea* as indicators') to compare species richness of spiders, Orthoptera, other butterflies, Zygaenidae, ants and vegetation across *Maculinea* sites in Hungary and Poland, complemented by wider studies elsewhere in Europe (summaries in Settele et al. 2005). The impressive outcomes of the programme reflect 4 years of cooperative research by eight collaborating

Table 13.5 Procedures used in compiling a 'multispecies indicator' list spanning five taxonomic groups on wet heathlands in Belgium (Maes & Van Dyck 2005), in the context of endorsing *Maculinea alcon* as a useful indicator for wetland areas and quality.

STEP 1. Decide what ecosystem attributes the indicator taxa should reflect – multispecies groups should contain species that need relatively large areas of wet heathland, are sensitive to fragmentation, desiccation and eutrophication, and are dependent on one or more characteristic biotope attributes of : (1) seasonal humidity (permanently wet, or dry in summer); (2) bare ground (present, absent); (3) scattered trees (territories for butterflies, habitats for insectivorous birds; present, absent); (4) moorland pools (present, absent); (5) microtopography (including vegetation structure and inferring nutrient status; present, absent); (6) seepage (nutrient deposition; present, absent); and (7) typical Sphagnum moss (indicative of low disturbance; present, absent). For each of these, 'presence' is positive, and the group of taxa should encompass all attributes more than once

STEP 2. List all species or taxonomic groups until meet baseline information criteria. Considered significant when taxonomy and species' biology reasonably well known and able to be considered with ecosystem change

STEP 3. Use only intermediately rare and easily detectable species that are evenly distributed in the focal area. Criteria here include being easily observable (diurnal, no traps needed) and identifiable by non-specialists; 'intermediately rare' is defined (in Europe) in terms of range assessed by number of mapping grid cells in which present

STEP 4. List available information on niche and life history, and on sensitivity to environmental stressors – especially information on ecology, habitat relationships and sensitivities

STEP 5. Complete set of complementary species for different groups to satisfy every criterion for STEP 1 by more than one taxon

---

institutions, representing six European nations and involving about 60 ecologists – constituting what Settele et al. hoped would continue as 'a self-sustaining European Network of insect conservation biologists'. The (approximately) 90 projects summarised in their book revealed the immense umbrella values of coordinated appraisal of the five species. The project was initiated with two major premises, flowing from the recognition that the iconic status of these butterflies throughout Europe, and recognition by the World Conservation Union of their global conservation priority, gave them almost unprecedented acknowledged importance. These aims were (1) to generate the basic knowledge needed to conserve the species, by increasing understanding of variations in their functional ecology across the continent's climatic, elevational and phylogeographic gradients, coupled with (2) that their intricate ecology rendered them both sensitive indicators of condition of a wide range of grassland habitats they collectively frequent and also umbrella species whose conservation would benefit numerous other taxa. Their host *Myrmica* ants, for example, are regarded widely as keystone species in those environments. The second theme involved evaluating *Maculinea* as indicators of biodiversity along a European transect.

Future priorities and needs

The serious declines of Lepidoptera described for Europe, linked strongly wih land use changes such as agricultural and forestry intensification, urbanisation, losses of traditional low-intensity land management practices and mosaics, and abandonment of marginal habitats, and coupled with the reality of impacts of alien species and climate change influences increasing in the future, are paralleled – to largely unknown extents – in many other parts of the world. However, nowhere beyond the northern temperate regions have such well informed perspectives been coupled with such extensive conservation efforts to sustain the species and the biotopes they frequent. The templates for butterflies, drawing from the distributional information assembled by Kudrna (1986, see also Kudrna 2002) and many local efforts and collaborators, compiled by van Swaay and Warren as (1) the Red Data Book (1999) and (2) the Prime Butterfly Areas for Europe (2003) have set a globally envied foundation for conservation of invertebrates, with perspective for butterflies now extending to many groups of the larger moths in the region. Despite the massive and continuing conservation efforts, many species continue to decline. In leading Lepidoptera conservation, and setting examples for Lepidoptera conservation elsewhere, two major trends occur: emulation of European approaches in the direct transfer of philosophy and practice to other faunas, and regional innovation as the needs for different approaches and strategy become evident in vastly different environments. The regional differences reflect two broad realities: (1) lack of biological knowledge of most individual species and of capability to increase and apply this to inadequately documented taxa in extensive, remote and ecologically complex areas, coupled with (2) lack of dedicated personnel, resources for conservation and local political priority. These inevitably lead to greater emphasis on biotope conservation, as places and habitats in which ecological information can progressively be sought. However, notwithstanding these impediments, the advances made in Lepidoptera conservation science and practice can only bode well for future progress.

Looking forward for Europe, Warren and Bourn (2011) assembled a list of 10 'challenges' for Lepidoptera ecologists and conservationists, as likely to be of key importance for European Lepidoptera in the next few decades. These (summarised in Table 13.6), some posed in the form of questions for debate, illustrate some of the broad principles involved, and how these might apply to the 'rest of the world' merit serious and urgent consideration. Warren and Bourn noted that their challenges are ambitious (and they or their modifications will be even more so beyond Europe!), but essential aspects of attempting to halt wider biodiversity loss, through using Lepidoptera as a focal group. They combine policy and practicality, but a major regional bias arises from the fate of many European Lepidoptera being tied closely to survival of long-term traditional agricultural or forestry land-use systems and to the restoration of these combined with deintensification of other practices that have largely replaced these.

Emphasis in Europe is necessarily largely on altered biotopes, rather than incorporating large areas of relatively pristine biotopes (such as tropical forest and savannas) that have not yet undergone equivalent levels of anthropogenic changes – however rapidly that may be proceeding at present. It is thus (just!) possible to think in terms of protecting some such key areas from major interference, but in many places this will demand major political goodwill and economic restitution. The butterfly ranching exercises noted in Chapter 10 are one such effort, and may open opportunities for wider aspects of ecotourism to develop, associated with increasing awareness of conservation need and, in some instances, raising funds for this to occur. Again from Europe, the 'best sites' (point 6 in Table 13.6) are already largely documented, and many are already on conservation agendas, often in conjunction with other conservation foci as well as butterflies. Equivalent priority, other than designation of national parks and other reserves, over much of the tropics has not heeded Lepidoptera specifically, but many species are fortuitously included in such areas – their protection from major changes and interferences is important, but may be very difficult to enforce in relation to local needs. Many declared reserves are in essence 'paper parks' not exempt from human intrusions and exploitation, and with external impacts not monitored. Larsen (1995), for example, emphasised the importance of 'totem forests' as butterfly sanctuaries in West Africa as constituting major forest remnant patches in highly affected landscapes. However, deliberate planning for **any** insect conservation in many tropical areas is rare, and largely utopian when compared with European or North American measures.

This situation is unlikely to improve markedly, despite some welcome expressions of concern, and most vocal concerns are likely to continue from people elsewhere in the world and, in some cases, lacking practical experience of the complex social and practical arenas occupied by the Lepidoptera for which they are advocates. The primary needs in many of the places of greatest concern for Lepidoptera conservation, such as tropical south-east Asia and the Neotropics are to cater for burgeoning human population food and living. As over much of the Pacific region, accessible lowland areas, in particular, have been changed extensively for cropping and urban development, and little natural vegetation characteristic of formerly complex and characteristic systems remains – the

Table 13.6 Summary of the 'ten challenges' formulated to aid conservation of Lepidoptera in Europe in future decades (after Warren & Bourn 2011).

1. How do we reform agricultural policy to foster sustainable production and reverse the loss of biodiversity?
2. How do we ensure sustainable management of woodlands?
3. How do we manage the matrix between semi-natural areas?
4. How do we ensure habitat management is coordinated on landscape scale to ensure the long-term survival of metapopulations?
5. How (can) we mitigate for climate change?
6. A robust planning system throughout Europe that protects our best sites and fosters wildlife conservation within urban areas
7. A comprehensive system for recording and monitoring European Lepidoptera
8. Long-term funding for nature conservation
9. How do we make wildlife matter to politicians?
10. How do we make wildlife matter to people?

plight of the Hawai'ian moths noted in Chapter 9 exemplifies much wider reality, but also that lack of detailed figures of numbers of taxa involved may weaken conservation cases made to government agencies. Whether near-complete or indicative, the impressive numbers of taxa gained from inventory surveys carry political weight. Measures to harmonise these needs with conservation are urgent to redress these impacts, to some extent.

The trade in ranched Lepidoptera has in some places contributed to this, with numerous taxa unlikely to become marketable benefiting from the reduced destruction of natural vegetation. To some extent linking with this, the diversity of global ecotourism interests is also proliferating. Thus, recent tourist surveys for South Africa (Huntly et al. 2005) have indicated that perpectives may expand from simply seeing the 'big five' mammals to other taxa, including invertebrates. Spectacles of massed insects such as the Jersey tiger moths on Rhodes (Chapter 4, p. 45) or overwintering Monarch butterflies in Mexico (Chapter 4, p. 45) have both 'scenic' and educational components for conservation. Expansions of interest may be stimulated substantially by well illustrated field guides enabling visitors to identify the animals they see – that by Woodhall (2005) for South Africa, in conjunction with the recently published conservation summary and atlas for this notable fauna (SABCA 2013) has considerable potential to increase conservation interest. Somewhat unusually, the former book has been displayed at airport bookstores in South Africa, a far more conspicuous outlet than usual for books of that kind!

As an outcome of the geographically widespread interest partially summarised in this book, sympathy for Lepidoptera conservation is widespread. They (at least some of the butterflies!) are acknowledged widely, and at high political levels, as 'worthy', and the number of signalled threatened species – even though highly unrepresentative of the real scale of needs – is sufficient to demonstrate the reality of human impacts and of need for conservation. Sufficient information is available on the order's biology, and taxonomic and ecological variety, to display that need across a considerable variety of biotopes, as well as the current vulnerability both of individual taxa and diverse assemblages. More optimistically, the conservation successes achieved demonstrate that, with determination, persistence and support, remedial and proactive measures are indeed possible, and that public and community support for those measures can be garnered, with people engaged in many ways and for many reasons. Lepidoptera will continue to lead by examples flowing from this interest in the developing discipline of invertebrate conservation (New 2012c). The ecological and political contexts in which they may be involved continue to diversify and to generate complexity and dispute. As examples, the progressive pressures for 'habitat offsets' to counter development of sites occupied by threatened species have involved debate over grassland fragments occupied by the Golden sun-moth in Australia (Chapter 9, p. 174, New 2012a), with polarised opinions for or against development and need for very careful appraisal of techniques, quality assessment and monitoring. A related issue involves the values of threatened species on habitat patches within the principle of 'biodiversity credits', whereby notable threatened species may confer formal value in providing 'reward' (such as tax relief) to land owners to protect the site under covenent. The principles for butterfies were noted by New (2006), largely as a possible means by which 'political weight' might be added to cases to protect rare or restricted biotopes through acknowledging the values of different life forms.

There is no room for complacency in the continuing endeavour to conserve Lepidoptera and to use them as tools in wider ecological assessments and political arguments, despite the notable advances in understanding achieved, and the publicity and spectacular success attending some past efforts. The ramifications of future climate change and other human impacts remain highly uncertain. Whilst strategies such as assisted migration and patch restoration are attractive and currently rewarding, they may be doing little more than 'buying time', in the face of environmentally forced range changes and novel interactions. That time, however, might be critical in enabling adaptation to new circumstances. Considerable urgency exists to consider long-term and large-scale conservation measures that might help to counter the impacts of these widespread and intangible changes. Lepidoptera may help us to do so.

References

Brereton, T., Roy, D.B., Middlebrook, I., Botham, M. & Warren, M. (2011) The development of butterfly indicators in the United Kingdom and assessments in 2010. *Journal of Insect Conservation* 15, 139–151.

Brown, K.S. Jr (1991) Conservation of Neotropical environments: insects as indicators. pp. 349–404 in Collins, N.M. & Thomas, J.A. (eds) The Conservation of Insects and Their Habitats. Academic Press, London.

Brown, K.S. Jr & Freitas, A.V.L. (2000) Atlantic Forest butterflies: indicators for landscape conservation. *Biotropica* 32, 934-956.

Cleary, D.F.R. (2004) Assessing the use of butterflies as indicators of logging in Borneo at three taxonomic levels. *Journal of Economic Entomology* 97, 429–435.

Fleishman, E. & Murphy, D.D. (2009) A realistic assessment of the indicator potential of butterflies and other charismatic taxonomic groups. *Conservation Biology* 23, 1109–1116.

Franzen, M. & Ranius, T. (2004) Habitat associations and occupancy patterns of burnet moths (Zygaenidae) in semi-natural pastures in Sweden, *Entomologica Fennica* 15, 91–101.

Gregory, R.D., Van Strien, A.J., Vorisek, P. et al. (2005) Developing indicators for European birds. *Philosophical Transactions of the Royal Society, series B* 360, 269–288.

Hammond, P.C. & McCorkle, D.V. (1984) The decline and extinction of *Speyeria* populations resulting from human environmental disturbances (Nymphalidae: Argynninae). *Journal of Research on the Lepidoptera* 22, 217–224.

Heer, P., Pellet, J., Sierro, A. & Arlettaz, R. (2012) Evidence-based assessment of butterfly habitat restoration to enhance management practices. *Biodiversity and Conservation* DOI 10.1007/s 10531-012-0417-9.

Holloway, J.D. (1984) Moths as indicator organisms for categorizing rain-forest and monitoring changes and regeneration processes. pp. 235–242 in Chadwick, A.C. & Sutton, S.L. (eds) Tropical Rain-Forest. The Leeds Symposium. Leeds Philosophical and Literary Society, Leeds.

Huntly, P.M., Van Noort, S. & Hamer, M. (2005) Giving increased value to invertebrates through ecotourism. *South African Journal of Wildlife Research* 35, 53–62.

Kitching, R.L., Orr, A.G., Thalib, I., Mitchell, H., Hopkins, M.S. & Graham, A.W. (2000) Moth assemblages as indicators of environmental quality in remnants of upland Australian rain forest. *Journal of Applied Ecology* 37, 284–297.

Kudrna, O. (1986) Butterflies of Europe. Volume 8. Aspects of the conservation of butterflies in Europe. AULA-Verlag, Weisbaden.

Kudrna, O. (2002) The Distribution Atlas of European Butterflies. *Oedippus* 20, 1–342.

Kuhn, E., Feldmann, R., Thomas, J. & Settele, J. (eds) (2005) Studies on the Ecology and Conservation of Butterflies in Europe. Volume 1. General concepts and case studies. Pensoft, Sofia.

Kuussaari, M., Heliola, J., Poyry, J., Saarinen, K. & Hulden, L. (2005) Developing indicators for monitoring biodiversity in agricultural landscapes: differing status of butterflies associated with semi-natural grasslands, field margins and forest edges. pp. 89–92 in Kuhn, E., Feldmann, R., Thomas, J. & Settele, J. (eds) Studies on the Ecology and Conservation of Butterflies in Europe. Vol. 1: General concepts and case studies. Pensoft, Sofia and Moscow.

Kuussaari, M., Heliola, J., Poyry, J. & Saarinen, K. (2007b) Contrasting trends of butterfly species preferring semi-natural grasslands, field margins and forest edges in northern Europe. *Journal of Insect Conservation* **11**, 351–366.

Larsen, T.B. (1995) Butterfly biodiversity and conservation in the Afrotropical region. pp. 290–303 in Pullin, A.S. (ed.) Ecology and Conservation of Butterflies. Chapman & Hall, London.

Maes, D. & Van Dyck, H. (2005) Habitat quality and biodiversity indicator performance of a threatened butterfly versus a multispecies group for wet heathlands in Belgium. *Biological Conservation* **123**, 177–187.

Majerus, M.E.N. (2002) Moths. Harper-Collins, London.

McGeoch, M.A. (1998) The selection, testing and application of terrestrial insects as bioindicators. *Biological Reviews* **73**, 181–201.

New, T.R. (2006) Is there any future for butterfly credits in conservation planning? *Journal of Insect Conservation* **10**, 1–3.

New, T.R. (2012a) The Golden sun-moth, *Synemon plana* Walker (Castniidae): continuing conservation ambiguity in Victoria. *Victorian Naturalist* **129**, 109–113.

New, T.R. (ed.) (2012c) Insect Conservation: Past, Present and Prospects. Springer, Dordrecht.

Norgate, M., Chamings, J., Pavlova, A., Bell, J.K., Murray, N.D. & Sunnucks, P. (2009) Mitochondrial DNA indicates late Pleistocene divergence of populations of *Heteronympha merope*, an emerging model in environmental change biology. *PLoS ONE* **4**, e7950. DOI 10.1371/journal.pone.0007950.

Noske, N.M., Hilt, N., Werner, F.A. et al. (2008) Disturbance effects on diversity of epiphytes and moths in a montane forest in Ecuador. *Basic and Applied Ecology* **9**, 4–12.

Orsak, L.J., Eason, N. & Kosi, T. (2001) A preliminary assessment of the insects of southern New Ireland, with special focus on moths and butterflies. pp. 40–49 in Beehler, B.L. & Alonso, L.E. (eds) Southern New Ireland, Papua New Guinea – A Biodiversity Assessment. Bulletin of Biological Assessment 21. Conservation International, Washington DC.

Ricketts, T.H., Daily, G.R. & Ehrlich, P.R. (2002) Does butterfly diversity predict moth diversity? Testing a popular indicator group at local scales. *Biological Conservation* **103**, 361–370.

SABCA (Southern African Butterfly Conservation Assessment project) (2013) Conservation Assessment of Butterflies of South Africa, Lesotho and Swaziland: Red List and Atlas. University of Cape Town, Cape Town.

Samways, M.J., McGeoch, M.A. & New, T.R. (2010) Insect Conservation. A Handbook of Approaches and Methods. Oxford University Press, Oxford.

Schultz, C.B. (2001) Restoring resources for an endangered butterfly. *Journal of Applied Ecology* **38**, 1007–1019.

van Strien, A.J., van de Pavert, R., Moss, D., Yates, T.J., van Swaay, C.A.M. & Vas, P. (1997) The statistical power of two butterfly monitoring schemes to detect trends. *Journal of Applied Ecology* **34**, 817–828.

van Swaay, C.A.M, Nowicki, P., Settelle, J. & van Strien, A.J. (2008) Butterfly monitoring in Europe: methods, applications and perspective. *Biodiversity and Conservation* **17**, 3455–3469.

van Swaay, C.A.M. & Warren, M.S. (1999) Red Data Book of European Butterflies (Rhopalocera). Nature and Environment Series, no. 199, Council of Europe, Strasbourg.

van Swaay, C.A.M. & Warren, M.S. (2003) Prime Butterfly Areas of Europe: Priority Sites for Conservation. National Reference Centre for Agriculture, Nature and Fisheries; Ministry of Agriculture, Nature Management and Fisheries, The Netherlands.

van Swaay, C.A.M. & Warren, M.S. (2012) Developing butterflies as indicators in Europe: current situation and future options. De Vlinderstichting/Dutch Butterfly Conservation, Butterfly Conservation UK, Butterfly Conservation Europe, Wageningen. Report no. VS2012.012.

Warren, M.S. & Bourn, N.A.D. (2011) Ten challenges for 2010 and beyond to conserve Lepidoptera in Europe. *Journal of Insect Conservation* **15**, 321–326.

Woodcock, B.A., Bullock, J.M., Mortimer, S.R. et al. (2012) Identifying time lags in the restoration of grassland butterfly communities: a multi-site assessment. *Biological Conservation* **155**, 50–58.

Index

Lepidoptera and Conservation, First Edition. T.R. New.
© 2014 John Wiley & Sons, Ltd. Published 2014 by John Wiley & Sons, Ltd.

Printed and bound by CPI Group (UK) Ltd, Croydon, CR0 4YY

27/10/2024

14580359-0002